T0320788

Symmetry Studies

Experimental data can often be associated with or indexed by certain symmetrically interesting structures or sets of labels that appear, for example, in the study of short symbolic sequences in molecular biology, in preference or voting data, in visual field and corneal topography arrays, or in experimental refractive optics. The symmetry studies introduced in this book describe the interplay among symmetry transformations that are characteristic of these sets of labels, the resulting algebraic decomposition of the data that are indexed by them, and the research questions that are induced by those transformations. The overall purpose is to facilitate and guide the statistical study of the structured data from both a descriptive and inferential perspective. The text combines notions of algebra and statistics and develops a systematic methodology to better explore the many different data-analytic applications of symmetry.

CAMBRIDGE SERIES IN STATISTICAL AND PROBABILISTIC MATHEMATICS

This series of high-quality upper-division textbooks and expository monographs covers all aspects of stochastic applicable mathematics. The topics range from pure and applied statistics to probability theory, operations research, optimization, and mathematical programming. The books contain clear presentations of new developments in the field and also of the state of the art in classical methods. While emphasizing rigorous treatment of theoretical methods, the books also contain applications and discussions of new techniques made possible by advances in computational practice.

Already Published

1. *Bootstrap Methods and Their Application*, by A. C. Davison and D. V. Hinkley
2. *Markov Chains*, by J. Norris
3. *Asymptotic Statistics*, by A. W. van der Vaart
4. *Wavelet Methods for Time Series Analysis*, by Donald B. Percival and Andrew T. Walden
5. *Bayesian Methods*, by Thomas Leonard and John S. J. Hsu
6. *Empirical Processes in M-Estimation*, by Sara van de Geer
7. *Numerical Methods of Statistics*, by John F. Monahan
8. *A User's Guide to Measure Theoretic Probability*, by David Pollard
9. *The Estimation and Tracking of Frequency*, by B. G. Quinn and E. J. Hannan
10. *Data Analysis and Graphics Using R*, by John Maindonald and John Braun
11. *Statistical Models*, by A. C. Davison
12. *Semiparametric Regression*, by D. Ruppert, M. P. Wand, and R. J. Carroll
13. *Exercise in Probability*, by Loic Chaumont and Marc Yor
14. *Statistical Analysis of Stochastic Processes in Time*, by J. K. Lindsey
15. *Measure Theory and Filtering*, by Lakhdar Aggoun and Robert Elliott
16. *Essentials of Statistical Inference*, by G. A. Young and R. L. Smith
17. *Elements of Distribution Theory*, by Thomas A. Severini
18. *Statistical Mechanics of Disordered Systems*, by Anton Bovier
19. *The Coordinate-Free Approach to Linear Models*, by Michael J. Wichura

Symmetry Studies

An Introduction to the Analysis of Structured Data in Applications

MARLOS A. G. VIANA
University of Illinois at Chicago

CAMBRIDGE
UNIVERSITY PRESS

CAMBRIDGE
UNIVERSITY PRESS

University Printing House, Cambridge CB2 8BS, United Kingdom

One Liberty Plaza, 20th Floor, New York, NY 10006, USA

477 Williamstown Road, Port Melbourne, VIC 3207, Australia

314-321, 3rd Floor, Plot 3, Splendor Forum, Jasola District Centre, New Delhi - 110025, India

103 Penang Road, #05-06/07, Visioncrest Commercial, Singapore 238467

Cambridge University Press is part of the University of Cambridge.

It furthers the University's mission by disseminating knowledge in the pursuit of education, learning and research at the highest international levels of excellence.

www.cambridge.org
Information on this title: www.cambridge.org/9780521841030

First published 2008

A catalogue record for this publication is available from the British Library

Library of Congress Cataloging in Publication data
Viana, Marlos A. G.
Symmetry studies : an introduction to the analysis of structured data in applications / Marlos Viana.
p. cm. – (Cambridge series in statistical and probabilistic mathematics)
Includes bibliographical references and index.
ISBN-13: 978-0-521-84103-0 (hardback)
ISBN-10: 0-521-84103-8 (hardback)
1. Symmetry (Mathematics) 2. Statistics – Data processing.
3. Probabilities – Data processing. 4. Data structures (Computer science)
I. Title. II. Series.
QA174.7.S96V53 2008
519.5–dc22 2008002983

ISBN 978-0-521-84103-0 Hardback

Contents

Preface

This text is an introduction to data-analytic applications of symmetry principles and arguments or symmetry studies. Its motivation comes from a variety of disciplines in which these principles continue to play a significant role in describing natural phenomena, and from the goal of methodologically applying them to classification, description, and analysis of data. The product of the methodology presented here is a broader class of data-analytic tools derived from well-established and theoretically related areas in algebra and statistics, such as group representations and analysis of variance.

The principles discussed in the text reflect many defining aspects of symmetry, a Greek conception dating from the Hellenic Era and part of a class of terms and forms of expression that designated harmony, rhythm, balance, stability, good proportions, and evenness of structure. Early Greek art and architecture often capture the outstanding dualism intrinsic to the original notion of symmetry – that of retaining the static uniqueness of one's being and at the same time promoting its dynamical multipresent realizations. This dualism is only apparently hidden in the 12th-century Athenian detail shown in the front cover. I invite you to recognize the presence of these pleasant concepts in the methodology to be introduced in the coming pages.

The text is divided as follows. Chapter 1 gives a complete overview of the methodology, including an introduction to the concepts of data indexed by symmetries, finite groups, group actions, orbits and classification, and representations in the data space. At the same time it outlines the step-by-step connection between the algebra and statistical inference, in the context of analysis of variance. It also emphasizes the fact that the same symmetry arguments useful here for classification and data reduction are part of the common language of the chemist, the geneticist, and the physicist.

Chapters 2, 3, and 4 cover the algebraic background, presented along with characteristic data-analytic applications. Readers with a basic course in linear or abstract algebra will have the required preparedness to follow these chapters, with focus on the applications and motivation to complete the proposed exercises. Their familiarity with the principles of analysis of variance will make more evident, broader, and attractive the applications introduced in these chapters.

Chapter 5 describes a number of prototypic applications, ranging from classical examples in experimental designs to specific symmetry studies of data indexed by objects with dihedral and cyclic symmetries. The more detailed applications, described in Chapters 6, 7, and 8, are symmetry studies for data indexed by short symbolic sequences, corneal curvature, and the study of handedness in simple planar images.

The chapters include briefly annotated suggestions for further reading and complementary exercises, followed by a chapter with basic computing algorithms in ©Maple and a glossary of main symbols and notations. The text also includes a selected number of short biographical citations, abstracted from very valuable sources, including The MacTutor History of Mathematics Archives at the University of St. Andrews and The Nobel Lectures (Elsevier, Amsterdam). The reader is invited to send notice of errors of any form to the author at viana@uic.edu.

This text evolved from lecture notes and talks prepared for short courses and special sessions held at several institutions since 2003, including the Instituto Nacional de Matemática Pura e Aplicada (IMPA) in my hometown (Rio de Janeiro), the EURANDOM, the Greek Statistical Institute, the International Commission for Optics, the Sociedad Chilena de Estadística, the Universidad de Antofagasta, University of Connecticut, Universidad de Costa Rica, University of Cyprus, Eindhoven University of Technology, Indiana University, University of Piraeus, Universidade Federal do Rio de Janeiro, Universidade de São Paulo, Universidad Simón Bolívar, and St. Petersburg State (ICMO).

Throughout this project, I benefited from the dedicated suggestions, conversations, enthusiasm, and inspiration of many colleagues, to whom I owe my gratitude. In particular, to Steen Andersson, Henrik Aratyn, Arjeh Cohen, Persi Diaconis, Alessandro Di Bucchianico, Joe Glaz, Markos Koutras, Vasudevan Lakshminarayanan, Gérard Letac, Peter McCullagh, Ingram Olkin, Takis Papaioannou, Carlos de B. Pereira, Michael Perlman, Donald Richards, Stephen Smith, Peter van de Ven, and Henry Wynn. This text also reflects the constant enthusiasm of my students at the Honors College who attended our weekly "Symmetry in Science and Applications" seminar.

I am thankful to D. Azar, J. Bauman, and L. Kaufman for their timely facilitation of the institutional support for both research and teaching, essential to the completion of this text.

My special thanks go to my editor, Lauren Cowles with Cambridge University Press, for her professional guidance, encouragement, and constant attention to the project.

This work is dedicated to my wife, Grace, and our children, Alice, Andrew, and Alex, who together gave me the strength to pursue it over the years.

Marlos A. G. Viana
Chicago, April 23, 2008.

1

Symmetry, Classification, and the Analysis of Structured Data

1.1 Introduction

George Pólya, in his introduction to mathematics and plausible reasoning, observes that

> A great part of the naturalist's work is aimed at describing and classifying the objects that he observes. A good classification is important because it reduces the observable variety to relatively few clearly characterized and well-ordered types.

Pólya's (1954, p. 88) remark introduces us directly to the practical aspect of partitioning a large number of objects by exploring certain rules of equivalence among them. This is how symmetry will be understood in the present text: as a set of rules with which we may describe certain regularities among experimental objects or concepts. The classification of crystals, for example, is based on the presence of certain symmetries in their molecular framework, which in turn becomes observable by their optical activity and other measurable quantities.

The delicate notion of measuring something on these objects and recording their data is included in the naturalist's methods of description, so that the classification of the objects may imply the classification or partitioning of their corresponding data. Pólya's picture also includes the notion of interpreting, or characterizing, the resulting types of varieties. That is, the naturalist has a better result when he can explain why certain varieties fall into the same type or category.

This chapter is an introduction to the interplay among symmetry, classification, and experimental data, which is the driving motive underlying any symmetry study and is often present in the basic sciences. The purpose here is to demonstrate that principles derived from such interplay often lead to novel ways of looking at data, particularly of planning experiments and, potentially, of facilitating contextual explanation. We will observe the intertwined presence of symmetry, classification, and experimental data in a number of examples from chemistry, biology, and physics. Many principles and techniques will repeat across different disciplines, and it is exactly this cross-section of knowledge that constitutes the higher motivation and basis for these symmetry studies.

1.2 Symmetry and Classification

In grade school we were amused (for a little while at least!) by drawings and games with colorful patterns repeated periodically along straight lines and contours. These bands can be classified according to their distinct generating rules, such as horizontal translations, line and point reflections and rotations. These rules for symmetry in two dimensions are explored in wallpaper, textile, and tapestry designs, with the technical constraint of artistically and graphically designing these repeating motifs within the finite boundaries of the work.

The common understanding and perception of symmetry developed from our collective sensory and cultural experience with repetition or constancy can guide us in classifying, for example, the uppercase roman font printing of the English alphabet, imagined as subsets of the Euclidean plane. For example, the letters N, S, and Z are characterized by having a center of reflection symmetry whereas the letters H, I, O, and X have line (horizontal and vertical) and point reflection symmetry.

When a letter and its transformed image under a vertical line reflection $v : (y_1, y_2) \mapsto (-y_1, y_2)$ are indistinguishable, we say that the letter has the symmetry of v. If, in addition, the letter has the symmetry of a horizontal line reflection $h : (y_1, y_2) \mapsto (y_1, -y_2)$, then, consequently, it must have the symmetry of the iterated transformation (vh) of these two symmetries. Because the iterated transformation of h and v is a point reflection $o : y \mapsto -y$, we then learn that the letter has the symmetries of v, h, and o. Trivially, all letters have the symmetry of the identity transformation $1 : y \mapsto y$, often indicated simply as 1.

The resulting symmetries in $G = \{1, v, h, o\}$ *multiply* according to Table (1.1) and share the algebraic properties of a finite group: the product $(*)$ of two symmetries is a symmetry; the product is associative; 1 is the identity element and all symmetries have an inverse symmetry also in G.

$*$	1	v	h	o
1	1	v	h	o
v	v	1	o	h
h	h	o	1	v
o	o	h	v	1

(1.1)

We observe, in addition, that any $f \in G$ is a bijective transformation of the Euclidean plane preserving its algebraic properties, in the sense that $f(x + y) = f(x) + f(y)$ for all vectors x, y in the plane. These are called automorphisms of the plane.

Any two letters are then classified together when they share the same set of symmetries or automorphisms. For example, the letters $\ell \in \{H,I,O,X\}$ are classified together by sharing the symmetries of G. We then say that G is their automorphism group and write $\text{Aut}\{\ell\} = G$ for all $\ell \in \{H,I,O,X\}$. In summary, after classifying the letters of the English alphabet, we have the following:

ℓ	$\text{Aut}(\ell)$
F,G,J,K,L,P,Q,R	1
A,M,T,U,V,W,Y	$1, v$
B,C,D,E	$1, h$
N,S,Z	$1, o$
H,I,O,X	$1, h, v, o$

1.3 Data Indexed by Symmetries

The lines in the left-hand side of Table (1.2) were abstracted from a visual acuity testing chart developed for the Early Treatment Diabetic Retinopathy Study, or ETDRS (Ferris III et al., 1993, Table 5). The 10 different letters {Z,N,H,V,R,K,D,S,O,C} that appear in the actual chart differ only in that they are printed with specially created Sloan fonts (Sloan, 1959) and are presented according to an experimental protocol.

C	O	H	Z	V
S	Z	N	D	C
V	K	C	N	R
K	C	R	H	N
Z	K	D	V	C
H	V	O	R	K
R	H	S	O	N
K	S	V	R	H

ℓ	$\text{Aut}(\ell)$	$p(\ell)$	entropy(ℓ)	$-\log \text{CS}(\ell)$
Z	$1, o$	0.844	0.433	0.63
N	$1, o$	0.774	0.535	0.53
H	$1, o, v, h$	0.688	0.619	0.44
V	$1, v$	0.636	0.656	0.56
R	1	0.622	0.663	0.46
K	1	0.609	0.669	0.57
D	$1, h$	0.556	0.687	0.43
S	$1, o$	0.516	0.693	0.44
O	$1, o, v, h$	0.470	0.692	0.34
C	$1, h$	0.393	0.673	0.36

(1.2)

The individual letters are shown in the adjacent table, along with their automorphisms, estimated probability (p) of correct identification, corresponding entropy $-[p \log p + (1-p)\log(1-p)]$, and estimated ($-\log$) contrast sensitivity. The entropy of a letter is a measure of the relative uncertainty in its correct identification. Its value is zero in the absence of uncertainty, and it is positive otherwise and attains its maximum value ($\log 2 = 0.693$) when the events are equally like, that is, $p = 1/2$. The probabilities of correct identification were estimated from a large sample of test subjects reported by Ferris III et al. (1993). The letter contrast sensitivity is a direct measure of the subject's visual performance. It is estimated from psychophysical experiments to determine the threshold of perception under varying levels of background contrast (Alexander et al., 1997). The smaller is the contrast needed to see the letter, the larger is the sensitivity.

We are interested in describing the connection among font symmetry, letter entropy, and contrast sensitivity from samples of Sloan lines similar to those shown in (1.2).

To each symmetry t in $G = \{1, v, h, o\}$, indicate by fix_t the subset of letters in a selected line with the symmetry of t and by $x_t = |\text{fix}_t|$ the number of elements in fix_t. For example, the first line C O H Z V in the chart gives

$$(1, v, h, o) \xrightarrow{x} (5, 3, 3, 3), \tag{1.3}$$

which is an example of data indexed by the elements in G, and a point in the vector space $\mathcal{V} = \mathbb{R}^4$. If $|\text{fix}_t| \neq 0$ then the mean line entropy

$$\frac{1}{|\text{fix}_t|} \sum_{\ell \in \text{fix}_t} \text{entropy}(\ell)$$

based on those letters with the symmetry of t leads to a different indexing of data by the elements of G. In this case, for the same line, the new indexing is

$$(1, v, h, o) \xrightarrow{x} (0.512, 0.655, 0.661, 0.575). \tag{1.4}$$

Similarly, when averaging the ($-$ log) contrast sensitivity over the letters with same symmetry, the indexing is

$$(1, v, h, o) \xrightarrow{x} (0.466, 0.446, 0.380, 0.476). \tag{1.5}$$

Note that the first components in (1.3), (1.4), and (1.5) are, respectively, the total number (5) of letters in each line, the line mean entropy and mean contrast sensitivity. These are examples of data indexed by a particular structure (a finite group in this case) or, simply, examples of structured data.

If similar lines are sampled from a larger set of charts, then x is a random vector and statistical summaries of the resulting sample are of interest. For example, Figure 1.1 summarizes the distributions of the four entropy components in (1.4) based on a sample of 42 lines similar to those in (1.2). The distributions should be interpreted along with the symmetry content of the underlying set of Sloan letters and the likely distribution of these symmetries over the 42 lines. Table (1.6) summarizes the underlying joint distribution of the 10 reference letters and symmetries. The marginal column and row sums are, respectively, the number

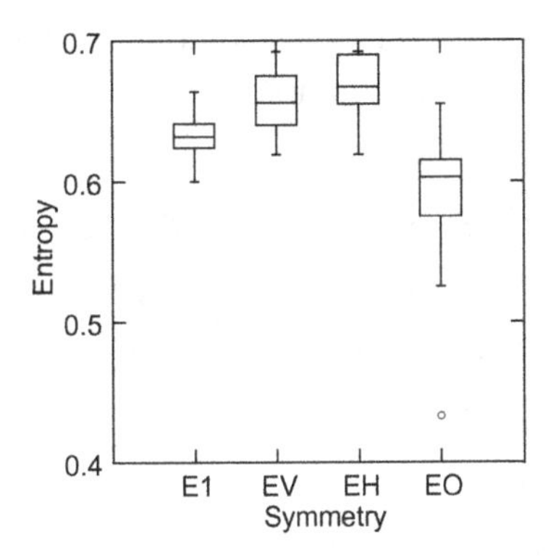

Figure 1.1: Distribution of line mean letter entropy by symmetry type.

$|\text{Aut}(\ell)|$ of automorphisms of ℓ and the number $|\text{fix}_t|$ of letters with the symmetry of t.

$t\backslash\ell$	Z	N	H	V	R	K	D	S	O	C	$\|\text{fix}_t\|$
1	1	1	1	1	1	1	1	1	1	1	10
v	0	0	1	1	0	0	0	0	1	0	3
h	0	0	1	0	0	0	1	0	1	1	4
o	1	1	1	0	0	0	0	1	1	0	5
$\|\text{Aut}(\ell)\|$	2	2	4	2	1	1	2	2	4	2	22

(1.6)

It is observed that point symmetry is present in the largest number ($|\text{fix}_t| = 5$) of reference letters and that at the same time the two letters with the smallest entropy (Z and N) have $|\text{Aut}(\ell)| = 2$ characterized precisely by the same symmetry.

1.4 Symmetry and Data Reduction

Classical physical measurements are understood, mathematically, as real vectors x in the usual Euclidean vector space. Consequently, it is of natural interest to represent the symmetries described by $G = \{1, h, v, o\}$ into the vector space $\mathcal{V} = \mathbb{R}^4$ for the data, shown in (1.3), (1.4), or (1.5), indexed by G. These representations are accomplished by associating to each element t in G a linear transformations ρ_t in $\mathcal{V}$.

Specifically, using the multiplication table of G shown in (1.1), to each element t in G associate the permutation matrix

$$\{e_1, e_v, e_h, e_o\} \xrightarrow{\rho_t} \{e_{t*1}, e_{t*v}, e_{t*h}, e_{t*o}\}, \tag{1.7}$$

in which the entry $(\rho_t)_{sf}$ of ρ_t at row s and column f is equal to 1 if and only if $f = t * s$, for $f, t, s \in G$. For example, $(\rho_v)_{ho} = 1$ indicates that $v * h = o$. Therefore,

$$\rho_1 = \begin{bmatrix} 1 & 0 & 0 & 0 \\ 0 & 1 & 0 & 0 \\ 0 & 0 & 1 & 0 \\ 0 & 0 & 0 & 1 \end{bmatrix}, \quad \rho_v = \begin{bmatrix} 0 & 1 & 0 & 0 \\ 1 & 0 & 0 & 0 \\ 0 & 0 & 0 & 1 \\ 0 & 0 & 1 & 0 \end{bmatrix},$$

$$\rho_h = \begin{bmatrix} 0 & 0 & 1 & 0 \\ 0 & 0 & 0 & 1 \\ 1 & 0 & 0 & 0 \\ 0 & 1 & 0 & 0 \end{bmatrix}, \quad \rho_o = \begin{bmatrix} 0 & 0 & 0 & 1 \\ 0 & 0 & 1 & 0 \\ 0 & 1 & 0 & 0 \\ 1 & 0 & 0 & 0 \end{bmatrix}.$$

These resulting linear transformations then connect the symmetries in the group G with the vector space $\mathcal{V}$ for (1.3), (1.4), or (1.5) in a way that the multiplication in G described by (1.1) is now represented as multiplication of nonsingular linear transformations in $\mathcal{V}$, that is,

$$\rho_{t*t'} = \rho_t \rho_{t'} \quad \text{for all } t, t' \in G. \tag{1.8}$$

This is the homomorphic property, characteristic of these linear representations.

The algebraic aspects developed in the next chapters will show that certain linear combinations of $\{\rho_1, \rho_v, \rho_h, \rho_o\}$ then lead to four algebraically orthogonal projection matrices $\mathcal{P}_1, \ldots, \mathcal{P}_4$, given by

$$1/4 \begin{bmatrix} 1 & 1 & 1 & 1 \\ 1 & 1 & 1 & 1 \\ 1 & 1 & 1 & 1 \\ 1 & 1 & 1 & 1 \end{bmatrix}, \quad 1/4 \begin{bmatrix} 1 & 1 & -1 & -1 \\ 1 & 1 & -1 & -1 \\ -1 & -1 & 1 & 1 \\ -1 & -1 & 1 & 1 \end{bmatrix},$$

$$1/4 \begin{bmatrix} 1 & -1 & 1 & -1 \\ -1 & 1 & -1 & 1 \\ 1 & -1 & 1 & -1 \\ -1 & 1 & -1 & 1 \end{bmatrix}, \quad 1/4 \begin{bmatrix} 1 & -1 & -1 & 1 \\ -1 & 1 & 1 & -1 \\ -1 & 1 & 1 & -1 \\ 1 & -1 & -1 & 1 \end{bmatrix}, \tag{1.9}$$

respectively, which determine statistical summaries $\mathcal{P}_1 x, \ldots, \mathcal{P}_4 x$ characterized by the particular representation (1.7) of G. We will refer to these summaries, in general, as the canonical invariants in the study – a concept that will be developed throughout the text. In the present case, these projections directly identify four invariants, namely,

$$\mathcal{I}_1 = x_1 + x_o + x_v + x_h, \quad \mathcal{I}_v = x_1 + x_v - x_o - x_h,$$
$$\mathcal{I}_h = x_1 + x_h - x_o - x_v, \quad \mathcal{I}_o = x_1 + x_o - x_v - x_h, \tag{1.10}$$

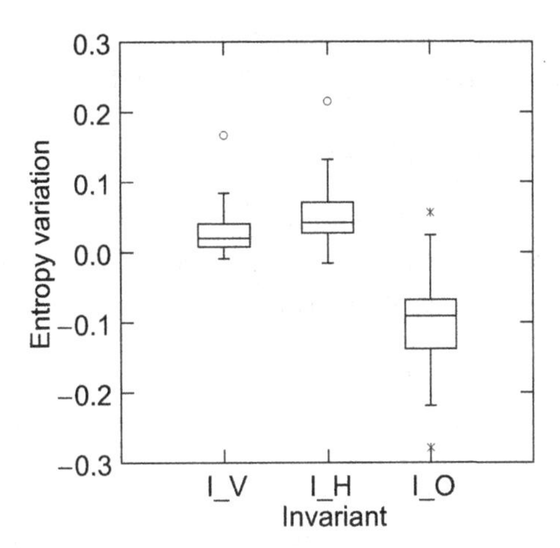

Figure 1.2: Distribution of the canonical invariants $\mathcal{I}_v, \mathcal{I}_h, \mathcal{I}_o$ for the mean line entropy data.

each one taking values on subspaces in the dimension of 1. These summaries depend on the labels (provided by G) only up to companion points determining a linear subspace of the data space (called invariant subspace). For example, the summary $x_1 + x_o - x_v - x_h$ is such that

$$x_{t*1} + x_{t*o} - x_{t*v} - x_{t*h} = \pm(x_1 + x_o - x_v - x_h) \quad \text{for all } t \in G.$$

The summaries of the data induced by G can then be interpreted as of exactly two types:

(1) The overall sum of responses ($\mathcal{I}_1$) and
(2) The three pairwise comparisons ($\pm\mathcal{I}_v$, $\pm\mathcal{I}_h$, $\pm\mathcal{I}_o$).

These pairwise comparisons are the basis for inferences in this particular symmetry study. Figure 1.2 summarizes the distributions of the canonical invariants $\mathcal{I}_v, \mathcal{I}_h, \mathcal{I}_o$ based on 42 lines of Sloan fonts.

The invariants are the data that should be retained when the arbitrariness of where is left (right) and where is up (down), associated with the action (1.7), is resolved. For example, then, $x_1 + x_o - x_v - x_h$ compares point and line symmetries in a way that depends on the chosen planar orientation only up to an invariant subspace. As effectively suggested by Weyl (1952, p. 144),

> Whenever you have to do with a structure-endowed entity try to determine its group of automorphisms, the group of those element-wise transformations which leave all structural relations undisturbed. You can expect to gain a deep insight into its constitution this way.

We observe that the derivation of these data summaries depends only on the set of labels and the symmetries of interest. Any subsequent statistical analysis, of course, would include the assumptions that apply to a particular experimental condition. For example, if the data indexed by G are the frequency distributions

$$x_1 = (0, 42), \quad x_o = (21, 21), \quad x_v = (39, 3), \quad x_h = (32, 10)$$

with which the corresponding symmetries appeared in at most 2 or in 3 or more of the 5 letters in each line, respectively, summed over 42 Sloan lines, then the invariants may be interpreted as three pairwise comparisons

$$\begin{aligned} x_1 + x_o &= (21, 63) \text{ vs. } x_v + x_h = (71, 13), \\ x_1 + x_h &= (32, 52) \text{ vs. } x_v + x_o = (60, 24), \\ x_1 + x_v &= (39, 45) \text{ vs. } x_o + x_h = (53, 31) \end{aligned} \quad (1.11)$$

between these frequency distributions, which, statistically, could be carried out in many different ways.

1.5 Statistical Aspects

We have remarked that the matrices $\mathcal{P}$ in (1.9) lead to the data summaries $\mathcal{P}x$ shown in (1.10). These matrices are algebraically orthogonal ($\mathcal{P}_i\mathcal{P}_j = \mathcal{P}_j\mathcal{P}_i = 0$ for $i \neq j$) projections ($\mathcal{P}_i^2 = \mathcal{P}_i, i = 1, \ldots, 4$) that reduce the identity operator I in the data vector according to the sum

$$I = \mathcal{P}_1 + \mathcal{P}_2 + \mathcal{P}_3 + \mathcal{P}_4,$$

so that, consequently, the theory of statistical inference for (real symmetric) quadratic forms can be applied to study the decomposition

$$x'x = x'\mathcal{P}_1x + \cdots + x'\mathcal{P}_4x$$

of the sum of squares $x'x$ of x.

To illustrate, consider the data shown in (1.12). Each row is a sample of size 5, obtained from 5 different Sloan chart lines, of the corresponding mean line entropy $x_t = \sum_{\ell \in \text{fix}_t} \text{entropy}(\ell)/|\text{fix}_t|$, indexed by the symmetry element t.

$t\backslash$Sample	1	2	3	4	5
1	0.614	0.636	0.632	0.624	0.66
v	0.675	0.619	0.692	0.640	0.619
h	0.655	0.619	0.660	0.690	0.667
o	0.603	0.603	0.553	0.603	0.635

(1.12)

The application of the algebraic arguments outlined above and detailed in the next chapters resulted in the analysis of variance table shown in (1.13), where the degrees of freedom (df) are the traces of the corresponding canonical projections and the F-ratios derived from the ratios of the mean sum of squares $x'\mathcal{P}x/df$ relative to the mean error sum of squares.

Component	$x'\mathcal{P}x$	df	$x'\mathcal{P}x/df$	F-ratio
$\mathcal{I}_1$	8.0633	1	8.0633	
$\mathcal{I}_v$	0.000757	1	0.000757	1.036
$\mathcal{I}_h$	0.002312	1	0.002312	3.165
$\mathcal{I}_o$	0.006956	1	0.006956	9.525
Error	0.011684	16	0.000730	

(1.13)

Here, the decomposition of the sum of squares is the consequence of jointly shuffling the rows and columns of the table in (1.12) using $G = \{1, h, v, o\}$ and the permutations of $\{1, 2, 3, 4, 5\}$, respectively.

Shuffling the rows in (1.12) according to G means relabeling them according to

$$\begin{bmatrix} v \\ 1 \\ o \\ h \end{bmatrix}, \quad \begin{bmatrix} h \\ o \\ 1 \\ v \end{bmatrix}, \quad \text{or} \quad \begin{bmatrix} o \\ h \\ v \\ 1 \end{bmatrix},$$

the result of multiplying the original first column by v, h, o respectively. On the other hand, shuffling the columns, indexed by $\{1, 2, 3, 4, 5\}$, simply means performing all their permutations.

Under the usual normality assumptions and corresponding hypotheses of the form $\mathcal{I} = 0$ (in terms of expected values), the indicated F-ratios have a central F-distribution with degrees of freedom 1 and 16 and can be used to test these parametric hypotheses.

It is now evident that the same canonical invariants $\mathcal{I} = \mathcal{P}x$ can be the object of descriptive summaries (Figure 1.2), nonparametric comparisons (1.11), or parametric hypotheses (1.13) for the structured data.

The analysis of variance (1.13) points to a significant distinction in mean line entropy when the differentiation (among chart lines) is due to point vs. line symmetries ($\mathcal{I}_o \neq 0$). The explanation of this finding, expressed in terms of the invariant $\mathcal{I}_o$, may then be found in the theories of eye movement, for example.

1.6 Algebraic Aspects

The role of algebra in the analysis of structured data is that of ascertaining its methodological aspects, of providing a well-defined sequence of steps leading to predictable data-analytic tools. We illustrate this with the following preliminary summary.

The mean line entropy data $x' = (x_1, x_v, x_h, x_o)$ shown in Table (1.12) were introduced as an example of data indexed by the elements of a finite group $G = \{1, v, h, o\}$. It was then possible to identify

(1) a set (G) of labels with the algebraic properties of a finite group;
(2) a set of data (x) indexed by those labels (the structured data);
(3) a group action, defined in (1.7), with which the symmetries in G were applied to itself;
(4) a linear representation (ρ) of that action connecting the labels and the data vector space ($\mathcal{V}$);
(5) the projection matrices $\mathcal{P}_1, \ldots, \mathcal{P}_4$ shown in (1.9);
(6) the canonical invariants $\mathcal{P}_1 x, \ldots, \mathcal{P}_4 x$ in the data, described in (1.10), and their interpretations, and
(7) the resulting analysis of variance $x'x = x'\mathcal{P}_1 x + \cdots + x'\mathcal{P}_4 x$ based on the decomposition $I = \mathcal{P}_1 + \cdots + \mathcal{P}_4$, shown in (1.13).

Note that the effect of reordering the basis used in the construction (1.7) of the representation ρ is such that the new decomposition is now

$$I = \eta\mathcal{P}_1\eta' + \cdots + \eta\mathcal{P}_4\eta'$$

where η is the corresponding permutation matrix. The new decomposition is in fact the same as (1.9), but relabeled. For example, if the entries had been written in the order of $1, o, v, h$ instead of the original order $1, v, h, o$, then $\eta\mathcal{P}_4\eta' = \mathcal{P}_3, \eta\mathcal{P}_3\eta' = \mathcal{P}_2$, and $\eta\mathcal{P}_2\eta' = \mathcal{P}_4$. Consequently, the invariants (1.10), their interpretation, and the resulting analysis of variance (1.13) would remain exactly the same.

However, the algebra has more to say here. A quick review of the projection matrices in (1.9) reveals that they can be written in terms of the matrices

$$\mathcal{A} = \frac{1}{2}\begin{bmatrix} 1 & 1 \\ 1 & 1 \end{bmatrix}, \quad \mathcal{Q} = \frac{1}{2}\begin{bmatrix} 1 & -1 \\ -1 & 1 \end{bmatrix} \tag{1.14}$$

which combine and compare the two components of a point in $\mathbb{R}^2$ and orthogonally reduce, or decompose, the identity matrix in that space into the sum $\mathcal{A} + \mathcal{Q}$. This reduction in $\mathbb{R}^2$ is an example of a *standard reduction* and will be used many times in these studies.

We have, using the symbol $\otimes$ to indicate the Kronecker product of two matrices, that

$$\mathcal{P}_1 = \mathcal{A} \otimes \mathcal{A}, \quad \mathcal{P}_2 = \mathcal{Q} \otimes \mathcal{A}, \quad \mathcal{P}_3 = \mathcal{A} \otimes \mathcal{Q}, \quad \mathcal{P}_4 = \mathcal{Q} \otimes \mathcal{Q}.$$

If, in addition, the data x can justifiably be indexed by a product $f \otimes g$ of two two-level labels f and g, then the data (briefly identified here with the labels) decompose as $\mathcal{A}f \otimes \mathcal{A}g$, $\mathcal{Q}f \otimes \mathcal{A}g$, $\mathcal{A}f \otimes \mathcal{Q}g$, and $\mathcal{Q}f \otimes \mathcal{Q}g$. This, more elementary, construction of the projections $\mathcal{P}_1, \ldots, \mathcal{P}_4$ is explained in terms of smaller component symmetry groups acting (by simple transpositions) on the bivariate component labels f, g. It leads, precisely, to the well-known concepts of factors and factor levels in simple factorial experiments. It is only when these component groups are introduced that a distinction between the projections $\{\mathcal{P}_2, \mathcal{P}_3\}$ and $\mathcal{P}_4$ can be envisioned.

The projections $\{\mathcal{P}_2, \mathcal{P}_3\}$ average the data over the levels of one factor and compare over the levels of the other factor, whereas $\mathcal{P}_4$ describes what is known as an interaction between the two factors. For example, it is now clear that if the two components of f or the two components of g are equal then $(\mathcal{Q} \otimes \mathcal{Q})(f \otimes g) = \mathcal{Q}f \otimes \mathcal{Q}g = 0$, or conversely, if $(\mathcal{Q} \otimes \mathcal{Q})(f \otimes g) \neq 0$ then the components of either factors must be unequal. Only an eventual interpretation of $G = \{1, v, h, o\}$ as a product of smaller symmetries may then justify the distinctions just described.

The tables in (1.15) show the analysis of variance for the original data of Table (1.12) now indexed by the labels $F_1 \times F_2$.

F_1	F_2	1	2	3	4	5
1	1	0.614	0.636	0.632	0.624	0.660
−1	1	0.675	0.619	0.692	0.64	0.619
1	−1	0.655	0.619	0.660	0.69	0.667
−1	−1	0.603	0.603	0.553	0.603	0.635

Source	SS	df	MS	F-ratio
F_1	0.000757	1	0.000757	1.036
F_2	0.002312	1	0.002312	3.169
$F_1 * F_2$	0.006956	1	0.006956	9.525
Error	0.011648	16	0.000730	

(1.15)

With this design, the original variance component assigned (by actual experimental construction) to point symmetry would be now interpreted as an interaction component.

1.7 Structured Data

The previous sections illustrated the reduction of data indexed by a group of symmetries, consequence of shuffling the labels according to the multiplication table of the group. This type of group action, called *regular action*, has many data-analytic applications and deserves a special introduction. Data from studies of voting preferences are naturally indexed by the group of permutations of the rankings under consideration. Table (1.16) shows the frequencies (x_τ), abstracted and adapted from Diaconis (1989), with which voters chose each one of the 24 different rankings (τ) of four candidates $\{a, g, c, t\}$.

For example, 29 voters ranked candidates in the order of (a, g, c, t) whereas 37 voters chose the order (t, g, c, a). Consequently, the frequency $x_\tau = 37$ can be associated to the permutation transposing candidates a and t, relative to the word $agct$.

The choice of $agct$ as a reference (or of any one of the 24 rankings) is certainly arbitrary and should then be resolved by the resulting invariants. This is in analogy with the symmetry study of the Sloan fonts, which can be interpreted as an example of data indexed by the permutations {*agct, gatc, atcg, ctag*}, each one of which can be arbitrarily assigned to be the reference one.

τ	x_τ	τ	x_τ	τ	x_τ	τ	x_τ
agct	29	*tgca*	37	*ctga*	22	*gcta*	26
agtc	11	*actg*	25	*tacg*	43	*gtac*	44
acgt	19	*atgc*	46	*tgac*	24	*ctga*	57
atcg	50	*gcat*	24	*gatc*	35	*catg*	34
gact	50	*gtca*	54	*ctag*	49	*tagc*	26
cgat	22	*cagt*	26	*tcga*	29	*tcag*	67

. (1.16)

These two studies are then examples of data indexed by certain permutations of $\{1, 2, 3, 4\}$, which are particular functions $\{1, 2, 3, 4\} \mapsto \{a, g, c, t\}$. This leads us to the consideration of the case in which the set of labels is any finite set (V) endowed with certain symmetries of experimental interest. In the previous examples, we had $V = G$. This extension requires, however, a broader interpretation of some of the steps in the summary presented in Section 1.6.

To illustrate, suppose that the labels (V) of interest are the 16 purine-pyrimidine sequences in length of 4 shown in Table (1.21), where the symbols $\{u, y\}$ represent the classes of purines $\{a, g\}$ and pyrimidines $\{c, t\}$, respectively, translated from the original sequence written with the alphabet $\{a, g, c, t\}$ of adenines (a), guanines (g), thymines (t), and cytosines (c). The structured data are the frequencies with which these sequences appear in 10 subsequent 200-bp-long regions of the BRU isolate of the Human Immunodeficiency Virus Type I (HIV-1). The entire 9229-bp-long DNA sequence is available from the National Center for Biotechnology Information[1] database using the accession number K02013.

The symmetries of potential interest, among others, are all the permutations of the four positions in $L = \{1, 2, 3, 4\}$ or the permutations of the two symbols in $C = \{u, y\}$. These sets of permutations, together with the operation of composition of functions, are examples of *full symmetric groups*, indicated by S_L, S_C, or simply by S_ℓ when only the number (ℓ) of elements in the set is of interest. An important distinction, however, is the fact that now the group operation in general is not commutative, as the reader may verify by composing different pairs of permutations in S_3. The consequences of this fact for the analysis of data indexed by S_ℓ and other noncommutative groups will become evident later on during the algebraic considerations introduced in Chapter 2.

[1] http://www.ncbi.nlm.nih.gov/

Given a sequence $s \in V$, a permutation τ in S_4, and a permutation σ in S_2, we observe that the composites

$$s\tau^{-1} : L \xrightarrow{\tau^{-1}} L \xrightarrow{s} C \quad \text{and} \quad \sigma s : L \xrightarrow{s} C \xrightarrow{\sigma} C$$

are also binary sequences in length of 3 and consequently a point in V. The resulting subsets $\{s\tau^{-1}; \tau \in S_4\}$ and $\{\sigma s; \sigma \in S_2\}$ of V are examples of symmetry *orbits*. The collection of all orbits in V obtained under a particular group action is generally indicated by V/G.

The action of the permutations in S_4 on the positions of the symbols of each word in V generates a partition $V = \mathcal{O}_0 \cup \mathcal{O}_1 \cup \mathcal{O}_2 \cup \mathcal{O}_3 \cup \mathcal{O}_4$ of V in which each orbit $\mathcal{O}_k$ has exactly $\binom{4}{k}$ elements, specifically,

$$\mathcal{O}_0 = \{yyyy\}, \quad \mathcal{O}_4 = \{uuuu\}, \tag{1.17}$$

$$\mathcal{O}_1 = \{yyyu, yyuy, yuyy, uyyy\}, \quad \mathcal{O}_3 = \{yuuu, uyuu, uuyu, uuuy\}, \tag{1.18}$$

and

$$\mathcal{O}_2 = \{yyuu, yuyu, yuuy, uyyu, uyuy, uuyy\}, \tag{1.19}$$

and can be characterized by the distribution

$$(k, \ell - k) \tag{1.20}$$

of purines (u) and pyrimidines (y) in the sequences.

Word\region	1	2	3	4	5	6	7	8	9	10
yyyy	5	8	3	5	7	8	5	25	16	6
uuuu	52	29	36	35	30	34	44	35	37	17
yuuu	18	16	20	16	20	20	16	18	17	17
uyuu	12	16	19	14	20	14	15	11	16	14
uuyu	15	14	21	17	21	12	13	10	16	12
uuuy	17	16	20	16	20	19	16	18	17	17
yyuu	16	11	11	10	10	14	12	15	11	15
yuyu	6	12	9	11	6	8	8	2	4	10
yuuy	10	11	9	8	10	8	11	8	10	11
uyyu	11	14	10	11	8	12	14	10	11	15
uyuy	9	10	11	14	7	6	6	1	4	9
uuyy	12	14	8	8	8	15	14	16	11	16
yyyu	5	6	5	7	8	8	6	10	7	10
yyuy	1	9	4	8	7	7	7	5	7	10
yuyu	4	6	7	11	8	5	5	4	7	9
uyyy	5	6	5	7	8	8	6	10	7	10

(1.21)

The reader may want to identify, in this example, all the steps described in Pólya's reasoning, introduced earlier on in the chapter, namely: description, classification, and interpretation of the objects of interest.

> A good classification is important because it reduces the observable variety to relative few clearly characterized and well-ordered types.

The practical aspect of this simple example, and its consequences for the planning and analysis of experimental data, is the varied structural classifications that can be obtained from the same initial set of labels by introducing different symmetries, and different actions on the sets of positions $L = \{1, 2, 3, 4\}$ and symbols $C = \{u, y\}$. In each case a new partition of V can be obtained, with a corresponding new partition in the data space.

In each one of the elementary orbits $\mathcal{O}_0, \mathcal{O}_1, \ldots,$ a representation of G is then determined, in analogy with the original case in which $V = G$ so that steps (4)–(7) in the summary of Section 1.6 would then lead to the canonical projections. Consequently, the identification of elementary orbits, in a way that will be made more specific later on, constitutes a methodological step of interest in our studies. These orbits will be referred to as sets in which the group acts *transitively*. To illustrate further the notion of an elementary orbits, transitive actions, and the role of the underlying group of symmetries, consider the words in the orbit

$$\mathcal{O}_2 = \{yyuu, uuyy, yuyu, uyuy, yuuy, uyyu\} \equiv \{a, \alpha, b, \beta, c, \gamma\} \subset V$$

introduced above. Indicate by $v = (12)(34)$ the transposing of positions $1, 2$ and $3, 4$, and similarly $o = (13)(24)$, $h = (14)(23)$. These permutations multiply according to (1.1) and the resulting action

	a	α	b	β	c	γ
1	a	α	b	β	c	γ
h	α	a	β	b	c	γ
v	a	α	β	b	γ	c
o	α	a	b	β	γ	c

on $\mathcal{O}_2$ identifies three elementary orbits

$$\mathcal{O}_{21} = \{a, \alpha\}, \quad \mathcal{O}_{22} = \{b, \beta\}, \quad \mathcal{O}_{23} = \{c, \gamma\}$$

decomposing $\mathcal{O}_2$, each one of which giving a linear representation of G in the dimension of 2. With the algebraic tools of Chapters 2 and 3, it will be seen that the decomposition of the identity matrix in each of these three subspaces of the original space $\mathcal{V}$ (in the dimension of 6) indexed by orbit $\mathcal{O}_2$ is simply the standard reduction

$$I = \mathcal{A} + \mathcal{Q},$$

introduced earlier in (1.14) on page 10. The identity matrix in $\mathcal{V}$ then reduces according to the sum $\mathbb{P}_{\mathcal{A}} + \mathbb{P}_{\mathcal{Q}}$ of two canonical projections

$$\mathbb{P}_{\mathcal{A}} = \text{Diag}(\mathcal{A}, \mathcal{A}, \mathcal{A}), \quad \mathbb{P}_{\mathcal{Q}} = \text{Diag}(\mathcal{Q}, \mathcal{Q}, \mathcal{Q}) \tag{1.22}$$

in $\mathcal{V}$. In the above expression, the two projections are block diagonal matrices with blocks $\mathcal{A}$ and $\mathcal{Q}$, respectively.

The nontrivial canonical invariants of interest are, clearly, the within-orbit comparisons. For (log) frequency data $\{x_a, x_\alpha, x_b, x_\beta, x_c, x_\gamma\}$, say, the broader interpretation of the canonical invariants would suggest the study of the ratios

$$\log \frac{x_a}{x_\alpha}, \quad \log \frac{x_b}{x_\beta}, \quad \log \frac{x_c}{x_\gamma}.$$

This example suggests that the orbits $\mathcal{O}_{21}$, $\mathcal{O}_{22}$, and $\mathcal{O}_{23}$ are essentially the same and could be classified as of the same type. Similarly, the orbits in (1.17), (1.18) and (1.19) define three classes or types of orbits. In the following section, we give an account of these important facts from two complementary perspectives.

1.8 Partitions

Consider first the set V of all binary sequences or words in length of ℓ and let P indicate a probability model in V, where a group G of symmetries is identified. We say that P has the symmetry of G if P is constant (uniform) over each one of the orbits of V. For example, if $P(s) = P(s\tau^{-1})$ for all sequences s in V and permutations τ in S_ℓ, then the probability law P should be constant in the position-symmetry orbits, characterized in (1.20). If the sequences are random, the probability laws

$$\mathcal{L}_i = \left(\frac{i}{\ell}, \frac{\ell - i}{\ell}\right), \quad i = 0, 1, \ldots, \ell, \tag{1.23}$$

associated with the position symmetry orbits are also random. The likelihood of each one of the possible probability laws

$$\mathcal{L}_0 = (0, 1),\ \mathcal{L}_1 = \left(\frac{1}{4}, \frac{3}{4}\right),\ \mathcal{L}_2 = \left(\frac{2}{4}, \frac{2}{4}\right),\ \mathcal{L}_3 = \left(\frac{3}{4}, \frac{1}{4}\right),\ \mathcal{L}_4 = (1, 0)$$

derived from V is therefore determined by the probability of seeing a sequence that is associated with that law. Because all sequences in the orbit $\mathcal{O}_i$ lead to the law $\mathcal{L}_i$ and conversely, we see that $\mathcal{L}_i$ occurs with probability $P(\mathcal{O}_i)$. Clearly, if the law P is such that all sequences are equally likely (P is said to be uniform), then $P(s) = P(s\tau^{-1})$ for all sequences $s \in V$ and permutations $\tau \in S_\ell$ and

$$\text{Probability of law } \mathcal{L}_i = P(\mathcal{O}_i) = \frac{|\mathcal{O}_i|}{|V|} = \frac{\binom{\ell}{i}}{|V|}, \tag{1.24}$$

so that the most likely distribution under uniformly distributed sequences in V is $\mathcal{L}_2 = (1/2, 1/2)$.

Consider now the set V of all sequences in length of 3 in the symbols $\{a, c, g, t\}$ so that V has 64 sequences. If these sequences are random, then the probability laws

$$\mathcal{L}_\lambda = \left(\frac{f_a}{3}, \frac{f_c}{3}, \frac{f_g}{3}, \frac{f_t}{3}\right),$$

where (f_a, f_c, f_g, f_t) are frequency distributions with $f_a + f_c + f_g + f_t = 3$ are also random. The index λ in $\mathcal{L}_\lambda$ indicates the corresponding orbit type, in analogy with expression (1.23), in which $\mathcal{O}_0$ and $\mathcal{O}_4$ belong to class $\mathcal{O}_{40}$, $\mathcal{O}_1$ and $\mathcal{O}_3$ belong to the class $\mathcal{O}_{31}$, and $\mathcal{O}_2$ coincides with $\mathcal{O}_{22}$.

Integer partitions

The indices λ in $\mathcal{L}_\lambda$ are the possible integer partitions of 3 in length of 4, namely the nonnegative integers $\{n_1, \ldots, n_4\}$ with $n_1 \geq n_2 \geq n_3 \geq n_4 \geq 0$ satisfying $n_1 + \cdots + n_4 = 3$. Consequently, there are three types of orbits, namely $\mathcal{O}_{3000}$, $\mathcal{O}_{2100}$ and $\mathcal{O}_{1110}$, and corresponding laws

$$\mathcal{L}_{3000} = (1, 0, 0, 0), \quad \mathcal{L}_{2100} = \left(\frac{2}{3}, \frac{1}{3}, 0, 0\right), \quad \mathcal{L}_{1110} = \left(\frac{1}{3}, \frac{1}{3}, \frac{1}{3}, 0\right).$$

Similarly to expression (1.24), we now obtain

$$\text{Probability of a law type } \mathcal{L}_\lambda = P(\mathcal{O}_\lambda) = \begin{cases} \binom{3}{3,0,0,0}/|V| = 1/64, & \text{if } \lambda = 3000 \\ \binom{3}{2,1,0,0}/|V| = 3/64, & \text{if } \lambda = 2100 \\ \binom{3}{1,1,1,0}/|V| = 6/64, & \text{if } \lambda = 1110, \end{cases}$$

so that, under the assumption that all 64 sequences are equally likely (uniform probability), the most probable distribution comes from the *class* of distribution given by $\mathcal{L}_{1110}$, each of which has the highest probability, 6/64. Simple combinatorics show that there are $4!/3!1! = 4$ orbits of type $\lambda = 1110$, namely

$$\left(\frac{1}{3}, \frac{1}{3}, \frac{1}{3}, 0\right), \quad \left(\frac{1}{3}, \frac{1}{3}, 0, \frac{1}{3}\right), \quad \left(\frac{1}{3}, 0, \frac{1}{3}, \frac{1}{3}\right), \quad \left(0, \frac{1}{3}, \frac{1}{3}, \frac{1}{3}\right). \tag{1.25}$$

These are the most probable probability laws describing the 64 sequences after their position symmetry classification.

A view from mechanics

In physics as in chemistry, we find that certain physical properties of a system remain unchanged under certain transformations of such system. Riley et al. (2002) observe that

> If a physical system is such that after application of a particular symmetry transformation the final system is indistinguishable from the original system then its behavior, and hence the functions that describe its behavior,

must have the corresponding property of invariance when subject to the same transformations.

The study of these transformations is a study of the symmetries of the system. Bacry (1963) remarks that the study of the symmetries of a physical system often suggests the study of the symmetries of certain physical laws and theories, and not infrequently, leads to symmetry-related principles, such as Kepler's Law of planetary motion, the principle of time-reversal invariance, or the Relativity Principle.

The following quote is from von Mises (1957, p. 200), with the notation partially adapted. The theory studies the distributions of a certain number ℓ of molecules over c regions in the velocity space under the assumption that all possible c^ℓ distributions have the same probability. Given two molecules with labels in $\{1, 2\}$, and three different locations x,y,z then the number of different distributions is nine, since each of the three locations of molecule 1 can be combined with each of molecule 2. According to the classical theory, all these distributions, as random events, have the same probability, $1/9$.

A new theory, first suggested by the Indian physicist Bose,[2] and developed by Einstein, chooses another assumption regarding the equal probabilities. Instead of considering single molecules and assuming that each molecule can occupy all location in the velocity space with equal probability, the new theory starts with the concept of repartition. This is given by the number of molecules at each location of the velocity space, without paying attention to the individual molecules. From this point of view, only six "partitions" are possible for two molecules on three locations, namely, both molecules may be together at locations x, y, or z, or they may be separated, one at location x and one at y, one at x and one at z, or one at y and one at z. According to the Bose-Einstein theory, each of these six cases has the same probability, $1/6$. In the classical theory, each of these three possibilities would have the probability of $1/9$, each of the other three, however, $2/9$, because, in assuming individual molecules, each of the last three possibilities can be realized in two different ways: molecule 1 can be at location x, and 2 at y, or vice versa.

The Italian physicist Fermi[3] advanced still another hypothesis. He postulated that only such distributions are possible – and possess equal probabilities – in which all molecules occupy different places. In our example of two molecules and three locations, there would only be three

[2] Satyendranath Bose, born: January 1, 1894, in Calcutta, India, died: February 4, 1974, in Calcutta, India.

[3] Enrico Fermi was born in Rome on September 29, 1901. The Nobel Prize for Physics was awarded to Fermi for his work on the artificial radioactivity produced by neutrons, and for nuclear reactions brought about by slow neutrons. He died in Chicago on November 29, 1954.

possibilities, each having the probability $1/3$; i.e., one molecule at x and one at y; one at x and one at z; one at y and one at location z.

To identify the symmetry argument in von Mises' narrative, let $L = \{1, 2\}$, $C = \{x, y, z\}$, and

$$V = \{xx, yy, zz, xy, yx, xz, zx, yz, zy\}$$

the set of all ternary sequences $s : L \to C$ in length of 2.

Under the Maxwell-Boltzmann (MB) model, it is assumed that all points or configurations in V are equally likely, or uniformly distributed, that is:

$$P(s) = \frac{1}{|V|} = \frac{1}{9}, \quad \text{for all } s \in V.$$

The number $|V| = c^\ell$ of points in V is called the MB statistic.

Under the Bose-Einstein (BE) model, it is assumed that the sets of points, obtained from V by shuffling the molecules' labels in L, are equally likely. Thus, in the BE model, the uniform probability applies to the resulting orbits

$$\mathcal{O}_x = \{xx\}, \quad \mathcal{O}_y = \{yy\}, \quad \mathcal{O}_z = \{zz\},$$
$$\mathcal{O}_{xy} = \{xy, yx\}, \quad \mathcal{O}_{xz} = \{xz, zx\}, \quad \mathcal{O}_{yz} = \{yz, zy\},$$

each one having probability of $1/6$. A probability law in V/S_2 such as

$$P(s) = \begin{cases} 1/6 & \text{when } s \in \mathcal{O}_x \cup \mathcal{O}_y \cup \mathcal{O}_z, \\ 1/12 & \text{when } s \in \mathcal{O}_{xy} \cup \mathcal{O}_{xz} \cup \mathcal{O}_{yz}, \end{cases}$$

would be consistent with the assumptions of the BE model. The BE statistic is the number

$$\binom{c + \ell - 1}{\ell}$$

of distinct orbits. In the example, there are $\binom{4}{2} = 6$ distinct orbits.

The Fermi-Dirac (FD) model assumes that only the injective mappings $V_I = \{xy, yx, xz, zx, yz, zy\}$ are admissible representations of the physical system, and that a uniform probability law is assigned to the resulting orbits $\mathcal{O}_{xy}$, $\mathcal{O}_{xz}$ and $\mathcal{O}_{yz}$ in V_I/S_2, each of these assigned with a probability of $1/3$. In the present example, a probability law in V_I/S_2 given by $P(s) = 1/6$ when s is injective and $P(s) = 0$ otherwise, would be consistent with the assumptions of the FD model. The FD statistic is the number $\binom{c}{\ell}$ of distinct orbits. In the example, we observed $\binom{3}{2} = 3$ distinct orbits.

Table 1.1 summarizes the domains of the uniform law in each of the models discussed above.

Table 1.1. *Maxwell-Boltzmann, Bose-Einstein and Fermi-Dirac Probabilities.*

Model	Domain of the Uniform Law	Probability
Maxwell-Boltzmann	V	$1/c^{\ell}$
Bose-Einstein	V/G	$1/\binom{c+\ell-1}{\ell}$
Fermi-Dirac	V_I/G	$1/\binom{c}{\ell}$

Macrostates and microstates in thermodynamics. Consider, to illustrate, six distinguishable molecules and four energy levels, leading to $4^6 = 4{,}096$ accessible microstates described by the mappings

$$s : \{1, 2, 3, 4, 5, 6\} \to \{\mathcal{E}_1, \mathcal{E}_2, \mathcal{E}_3, \mathcal{E}_4\}.$$

Microstates become measurable macrostates by the effect of similarities among the molecules when their identifying labels are erased. That is, when their labels are shuffled by the action $s\tau^{-1}$ of the permutations τ in the symmetric group S_6. The resulting classes $\mathcal{O}_\lambda$ of orbits, indexed by the possible integer partitions (n_1, n_2, n_3, n_4) of 6, are then the energy macrostates realized by the system. The resulting classes, their volume $|\mathcal{O}_\lambda|$, usually indicated by Ω_λ in the thermodynamics context, and their number Q_λ of quantal states are described in Table (1.26).

λ	Ω_λ	Q_λ	$\Omega_\lambda \times Q_\lambda$
6000	1	4	4
5100	6	12	72
4200	15	12	180
4110	30	12	360
3300	20	6	120
3210	60	24	1,440
3111	120	4	480
2220	90	4	360
2211	180	6	1,080
Total	522	84	4,096

. (1.26)

There are $Q = 6$ quantal states associated with the most probable ($\Omega = 180$) orbit type, $\lambda = 2211$. Also, note that

$$\sum_\lambda Q_\lambda = \binom{c + \ell - 1}{\ell} = \binom{9}{6} = 84,$$

where c in the number of energy levels and ℓ the number of molecules, is the Bose-Einstein statistic.

Boltzmann's Entropy Theorem. In Boltzmann's model, the mean energy level

$$\overline{\mathcal{E}} = \frac{1}{\ell}\sum_i \mathcal{E}_i f_i,$$

where f_i indicates the number of molecules at the energy level $\mathcal{E}_i$, of any configuration in V, is an invariant under the composition rule $s\tau^{-1}$ and, therefore, depends only on the orbit (macrostate) realized by the configuration. Boltzmann reasoned that the molecule-energy configurations in V evolved from least probable configurations to most probable configurations, so that the quest for describing the equilibrium energy distribution in the ensemble requires the determination of the most likely configurations in V. This, in turn, requires the determination of the macrostate (orbit) with the largest volume Ω, conditioned on the fact that mean energy of the isolated ensemble must remain constant. Given a configuration s with f_1 particles at the energy level $\mathcal{E}_1$, f_2 particles at the level $\mathcal{E}_2$, f_3 particles at the level $\mathcal{E}_3$, etc., its orbit $\mathcal{O}_s$ has volume

$$|\mathcal{O}_s| = \frac{\ell!}{f_1!f_2!f_3!\ldots}.$$

We have then a well-defined mathematical problem: find the macrostate identified by $f_1, f_2, \ldots$, which maximizes $|\mathcal{O}_s|$ for a given mean energy level $\overline{\mathcal{E}}$. The solution is $f_i = \ell\ P(\mathcal{E}_i)$, where

$$P(\mathcal{E}_i) = \frac{e^{-\beta\mathcal{E}_i}}{\sum_j e^{-\beta\mathcal{E}_j}}$$

is the Maxwell-Boltzmann canonical distribution or partition function. It describes the most likely energy distribution of the ensemble. Similar calculations, e.g., Huang (1987), and Reif (1965, pp. 343-350), can be obtained for the models of Fermi-Dirac and Bose-Einstein.

In addition, the entropy

$$H = -\sum_i \frac{f_i}{\ell}\ln\left(\frac{f_i}{\ell}\right)$$

of the probability law associated with the orbit of $f_1, f_2, f_3, \ldots$, is a physical characteristic (such as temperature, mass) of the gas and, at the same time, a measure of uniformity in its thermodynamical probability law. The canonical distribution corresponds to an ensemble configured to its maximum entropy. Boltzmann's statistical expression

$$S = k\ln\Omega$$

for the equilibrium entropy (usually indicated by S in thermodynamics) relates the equilibrium or limit number of accessible microstates, Ω, and k, the (known now as) Boltzmann constant 1.3807×10^{-23} KJ/molecule. A volume of gas, left to itself, will almost always be found in the state of the most probable distribution.

Maxwell-Boltzmann law for velocities in a perfect gas. Maxwell's assumptions, e.g., Ruhla (1989, Ch. 4), lead to the searching of a probability law (F) for the random velocity vector v such that the component-velocities are statistically independent and identically distributed, i.e., $F(v) = f(v_x)f(v_y)f(v_z)$, where f indicates the common probability law for the component-velocities. In addition, the isotropic condition $F(Uv) = F(v)$ should hold for all central space rotations U. This is in analogy with the invariance condition $P(s) = P(s\tau^{-1})$ for all $s \in C^L$ and permutations $\tau \in S_L$ discussed earlier in this section. These two conditions lead to a probability law that has the form

$$F(v) = Ae^{-\mu||v||^2},$$

where $||v|| = \sqrt{v_x^2 + v_y^2 + v_z^2}$ is the *speed* in the velocity vector v and the constants are determined from additional physical considerations, e.g., Reif (1965, p. 267). The speed orbits, then, are simply those velocity vectors with common speed.

Canonical projections

The symmetry among the two molecules imposed by the Bose-Einstein argument introduced on page 18 led to the classification of the ternary sequences in length of 2 into six elementary orbits

$$\mathcal{O}_x = \{xx\}, \quad \mathcal{O}_y = \{yy\}, \quad \mathcal{O}_z = \{zz\},$$
$$\mathcal{O}_{xy} = \{xy, yx\}, \quad \mathcal{O}_{xz} = \{xz, zx\}, \quad \mathcal{O}_{yz} = \{yz, zy\}.$$

In each of the three two-element orbits the associated canonical reduction is $I = \mathcal{A} + \mathcal{Q}$, where $\mathcal{A}$ and $\mathcal{Q}$ are given by (1.14) on page 10, whereas each single-element orbit reduces only trivially as the identity. As a consequence, the original space $\mathcal{V}$ in dimension of 9 reduces into the direct sum of 3 similar subspaces each in dimension of 1 associated with $\mathcal{Q}$ and 6 similar subspaces in dimension of 1 associated with the identity. This collecting together of similar subspaces that appear in the canonical reduction of the data space will be an important aspect of the main theorem on canonical decomposition later on in Chapter 3.

A core concept, discussed later on in the subsequent chapters, refers to the property of certain vector subspaces that are determined by particular linear representations of the groups of interest, called *irreducible* representations. In the present example, this means that $\mathcal{A}x$ is in a subspace in dimension of 1 that transforms according to identity (or symmetric) representation of S_2, whereas the subspace of

$\mathcal{Q}x$ transforms according to another one-dimensional representation of S_2, called the *sign* or alternating representation.

In quantum chemistry, these projections, indexed by the irreducible representations, play a central role in determining whether a chemical bonding can take place in a molecule. The bonding, point Riley et al. (2002, p. 948), is strongly dependent on whether the wavefunction of the two atoms forming a bond transforms according to the same (irreducible) representation. Typically, these reductions take place in a infinite dimension Hilbert space $\mathcal{H}$, in such a way that the invariant subspaces in the reduction of $\mathcal{H}$ define the properties of the quantum system. Consequently, properties are identified with the corresponding projections – one for each irreducible representation – which follow from the physical basis of the system.

For example, the projections I and 0 correspond to the sure property and the impossible property, whereas the projection $I - \mathcal{P}$ corresponds to the negation of the property associated with $\mathcal{P}$. The properties associated with two commuting projections $\mathcal{P}$ and $\mathcal{F}$ are said to be compatible, in which case the projection $\mathcal{P}\mathcal{F} = \mathcal{F}\mathcal{P}$ represents the conjunction of $\mathcal{P}$ and $\mathcal{F}$, whereas the projection $\mathcal{P}\mathcal{F} + \mathcal{P}(I - \mathcal{F}) + (I - \mathcal{P})\mathcal{F} = \mathcal{P} + \mathcal{F} - \mathcal{P}\mathcal{F}$ is associated with the disjunction of the two properties. If, in addition, $\mathcal{P}\mathcal{F} = 0$, the properties are compatible, mutually exclusive, and the disjunction is given by the sum $\mathcal{P} + \mathcal{F}$. See, for example, Omnès (1994) or Faris (1996) for a review of Omnès' work.

Unit vectors y in the Hilbert space are associated with the *states* of the system, and determine a mathematical specification of probabilities for all properties. These probabilities are obtained from the fact that associated to each set of mutually exclusive properties $\mathcal{P}_1, \ldots, \mathcal{P}_h$ whose disjunction is sure, i.e., $I = \mathcal{P}_1 + \cdots \mathcal{P}_h$, there is a decomposition

$$1 = ||y||^2 = y'y = y'\mathcal{P}_1 y + \cdots y'\mathcal{P}_h y,$$

which is interpreted as a probability distribution among the corresponding properties $\mathcal{P}_1, \ldots, \mathcal{P}_h$. Each state $y \in \mathcal{H}$ then provides a probabilistic description,

$$y \to (y'\mathcal{P}_1 y, \ldots, y'\mathcal{P}_h y) = (||\mathcal{P}_1 y||^2, \ldots, ||\mathcal{P}_h y||^2),$$

of the system[4]. In the present finite analogy, in which $I = \mathcal{A} + \mathcal{Q}$, we have $y \mapsto (||\mathcal{A}y||^2, ||\mathcal{Q}y||^2)$.

The projections on one-dimensional subspaces are *pure states*, characteristic of exactly two types of one-dimensional representations: those associated with $\mathcal{A}$ are symmetric, i.e., $\rho_\tau \mathcal{A}y = \mathcal{A}y$ for all $\tau \in S_2$; and those associated with $\mathcal{Q}$ are anti-symmetric, i.e., $\rho_\tau \mathcal{Q}y = \mathcal{Q}y$ if $\tau = 1$ and $\rho_\tau \mathcal{Q}y = -\mathcal{Q}y$ if τ is a transposition in S_2.

[4] The notation $y'\mathcal{P}y$ has the interpretation of the $< y, \mathcal{P}y >$ under the appropriate inner product in $\mathcal{H}$.

1.9 Summary

In this chapter we illustrated the key steps and most of the concepts present in the analysis of structured data. These steps were first introduced in Section 1.6 and are now reviewed to include the additional structures introduced in Section 1.7.

A symmetry study includes the identification of

(1) a set V of labels;
(2) a set of data (x) indexed by those labels (the structured data);
(3) a rule or group action, with which the symmetry transformations in G are applied to V;
(4) the classes and multiplicities of the resulting elementary orbits, subsets of V, where G acts transitively;
(5) the resulting linear representations of these actions in the corresponding data vector spaces;
(6) the canonical (algebraically orthogonal) projection matrices $\mathcal{P}_1, \mathcal{P}_2, \ldots$;
(7) the canonical invariants $\mathcal{P}_1 x, \mathcal{P}_2 x, \ldots$ on the data, and their interpretations; and
(8) a statistical analysis of the canonical invariants and, if applicable, their analysis of variance $x'x = x'\mathcal{P}_1 x + x'\mathcal{P}_2 x + \ldots$ based on the canonical decomposition $I = \mathcal{P}_1 + \mathcal{P}_2 + \ldots$ of the identity operator in the data spaces.

When necessary, these canonical projections may be grouped together to describe the original data space $\mathcal{V}$, as outlined in (1.22) on page 15 and in analogy with the partitioning calculations shown in (1.26) on page 19. Specifically, if there are q_λ orbits of type λ, each with o_λ elements, classifying the $v = \sum_\lambda q_\lambda o_\lambda$ points in $\mathcal{V}$ then, for each class of orbits with $v_\lambda = q_\lambda o_\lambda$ points, the identity matrix in the corresponding vector space reduces according to

$$I_{v_\lambda} = \mathbb{P}_1^\lambda + \mathbb{P}_2^\lambda + \cdots$$

where $\mathbb{P}_j^\lambda$ is the canonical projection constructed with q_λ copies of the canonical projection $\mathcal{P}_j^\lambda$ in the vector space associated with the elementary orbits of type λ, and indexed by the irreducible representations $j = 1, 2, \ldots,$ of the underlying group G. The reduction in the original space $\mathcal{V}$ is then

$$I_v = \text{Diag}(\ldots, I_{v_\lambda}, \ldots)$$

with as many components as the different types of orbits identified by the action of G on V.

Further Reading

The classical introductory work of Weyl (1952) on symmetry includes the notions of bilateral, translatory, rotational, ornamental, and crystal symmetry.

> A thing that is symmetrical . . . if there is something that you can do to it, so that after you have finished doing it, it still looks the same as it did before you did it.

Hermann Weyl was born on November 9, 1885, in Elmshorn, Germany, and died on December 8, 1955, in Zürich, Switzerland. He was a student of Hilbert at Göttingen, and from 1933 until he retired in 1952 he worked at the Institute for Advanced Study at Princeton. From 1923 to 1938 he evolved the concept of continuous groups, using matrix representations and its applications to quantum mechanics. Weyl (1950) is the English translation of the original German text first published in 1931. Weyl (1953) is a revised and supplemented edition of his 1939 publication on the invariants and representations of the classical groups. The work of Weyl, Wigner[5] and van der Waerden[6], pioneered the methods of group representation to quantum mechanics – *the three W's* of quantum mechanics. The English translation of Wigner's work by J. J. Griffin appeared in 1959 under the title *Group Theory and Its Application to Quantum Mechanics and Atomic Spectra*, Academic Press, New York.

A connection between the notions of symmetry and prior (to experiment) predictions or statements is described in Weyl (1952, p. 126) where he argues that all a priori statements in physics have their origin in symmetry. If conditions that uniquely determine their effect possess certain symmetries, then the effect will exhibit the same symmetry. For example, equal weights balance in scales of equal arms, concluded Archimedes a priori; in casting dice which are perfect cubes, each side is perceived as equally likely. In contrast, the law of equilibrium for scales with arms of different lengths can be settled only by experience or by physical principles based on experience.

The propagation of symmetry from cause to effect appears in Rosen's symmetry principle (Rosen, 1995) and earlier on in the work of Jaeger (1919). See also Rosen (1975), Bryan (1920), and Sarton (1921).

George Pólya, an American mathematician of Hungarian origin, was born in Budapest, Hungary, on December 13, 1887, and died in Palo Alto, the USA on September 7, 1985. He worked on a variety of mathematical topics, including series, number theory, combinatorics, and probability. Geometric symmetry and the enumeration of symmetry classes of objects was a major area of interest for Pólya over many years. He added to the understanding of the 17 plane crystallographic groups in 1924 by illustrating each with tilings of the plane. Pólya's work using generating functions and permutation groups to enumerate isomers in organic chemistry was of fundamental importance. Pólya's remark on the consequences of a

[5] Gruppentheorie und ihre Anwendungen auf die Quantenmechanik der Atomspektren, Braunschweig: Vieweg, 1931.

[6] Die gruppentheorietische Methode in der Quantenmechanik, Berlin: Springer, 1932.

good classification, introduced at the opening of this chapter, is further illustrated in the words of the Nobel Laureate Sir Anthony James Leggett:

> In trying to make sense of the bewildering variety of observed particles, and the ways in which they interact and decay into one another, it is impossible to overemphasize the role played by symmetry (or invariance) principles and the related conservation laws. (Leggett, 1987, p. 54)

In particular, invariance under time translation implies that the total energy of an isolated system of particles is preserved; invariance under space translation implies that its total momentum is conserved. Similarly, for an isolated system all directions in space appear to be indistinguishable – this isotropy, or invariance under rotation in space leads to the conservation of the system's angular momentum.

The text by Aigner (1979) on combinatorial theory has a comprehensive discussion on symmetry operations on the set of functions on finite sets.

Vibrational spectroscopy is a perfect example illustrating the objective connection between symmetry and observable measurements. The reader may refer to Harris and Bertolucci (1978), where the authors review the classical symmetry operations applied to molecules and their resulting classification according to the symmetries of point groups. The algebraic methods of symmetry studies are an integral part of the contemporary language with which the theory of vibrational spectroscopy can be explained. See, for example, Sternberg (1994).

The notion of points as labels identifying potential events appears in modern-day physics, in contrast to Newton's views in which points are essentially indistinguishable. A comment in that direction is found in Cartier (2001).

The structure of short nucleotide sequences introduced in this chapter appears, implicitly, in the work of Doi (1991) on evolutionary molecular biology. Related examples with applications of algebraic arguments appear in the works of Evans and Speed (1993) on phylogenetic trees and Dudoit and Speed (1999) on linkage analysis, among others.

Exercises

Exercise 1.1. The group $G = \{1, v, h, o\}$ of the symmetries of the rectangle, introduced in this chapter, is know as the Klein[7] four-group and is often indicated by K_4. Show that its elements are bijective transformations of the Euclidean plane preserving its additive structure, distances, and angles.

Exercise 1.2. Classify the equations $\{y = x^2,\ \ y^2 = x,\ \ y = x^3,\ \ x^2 + 2y^2 = 1,\ y = x + x^4\}$, from Pólya (1954, p. 89), according to the transformations in K_4.

[7] Christian Felix Klein, a German mathematician 1849–1925.

Exercise 1.3. Determine the group of symmetries of a rigid parallelepiped in each one of the following cases: (i) all three dimensions are unequal; (ii) two dimensions are equal, and (iii) all dimensions are equal.

Exercise 1.4. Evaluate Table (1.6) for the five vowels $\{A, E, I, O, U\}$ and then for the five equations in Exercise 1.2 and show that $\sum_t |\text{fix}_t| = \sum_\ell |\text{Aut}(\ell)|$. This argument, of summing over rows and over columns will appear later on again in Section 2.9 on page 58 of Chapter 2.

Exercise 1.5. Show, using Table (1.1), that $x_{t*1} + x_{t*o} - x_{t*v} - x_{t*h} = \pm(x_1 + x_o - x_v - x_h)$, for all $t \in G$.

Exercise 1.6. Show that the c^ℓ mappings $s : \{1, \ldots, \ell\} \to \{1, \ldots, c\}$ can be enumerated by $i(s) = 1 + \sum_{j=1}^{\ell}(s(j) - 1)c^{j-1}$ and that, conversely, $(s(1), \ldots, s(\ell))$ is the base c expansion of $i(s) - 1$. To obtain the unumeration shown in Appendix A, to this chapter make $y = 1$ and $u = 2$, $\ell = 4$ and $c = 2$, so that, e.g., the mapping with label $i(s) = 13$ corresponds to the base 2 expansion $[0, 0, 1, 1]$ of 12.

Exercise 1.7. If $r(t) = (ut\cos(t + \alpha), vt\sin(t + \alpha), 0)$ describes the moving in the xy plane of an object of mass m, evaluate its momentum $p = mdr/dt$ and show that the angular momentum $r \times p$ remains invariant under rotations in the xy plane around the z axis. Also, show that the trajectories of constant angular momentum are elliptic orbits in the xy plane.

Exercise 1.8. The diagram[8] of a basic Wheatstone bridge circuit, shown in Figure 1.3, contains four resistances $\{r_1, r_2, r_3, r_4\}$, a constant unit voltage input V_{in}, and an output voltage V_g, related by

$$V_g = \frac{r_1 r_3 - r_2 r_4}{(r_1 + r_2)(r_3 + r_4)}.$$

Shuffle the resistances in the circuit by applying the permutations in $G = \{1, (12)(34), (13)(24), (14)(23)\}$ (see page 14) to their labels, so that $x_\tau = [r_{\tau 1} r_{\tau 3} - r_{\tau 2} r_{\tau 4}]/[(r_{\tau 1} + r_{\tau 2})(r_{\tau 3} + r_{\tau 4})]$ is an example of data indexed by the elements in G and show that x can be written as $x_\tau = \chi(\tau) x_1$, where $\chi(\tau) \in \{1, -1\}$ and satisfies $\chi(\tau\sigma) = \chi(\tau)\chi(\sigma)$ for all τ, σ in G. Following Section 1.4 on page 5, interpret χ as a one-dimensional representation of G. As it turned out, $\chi(\tau)$ is the *direction* of the electric current through the potential V_g determined by the resistance configuration τ.

[8] e.g., http://www.efunda.com/designstandards/sensors/methods/

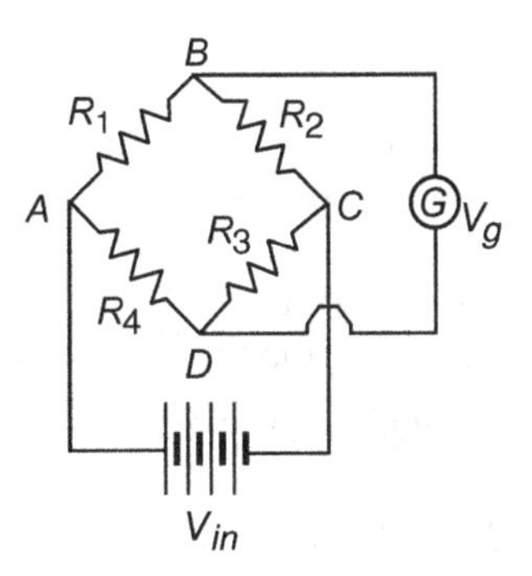

Figure 1.3: Wheatstone bridge circuit.

Exercise 1.9. Appendix A to this chapter shows the set (V) of binary sequences in length of 4, numbered according to the rule described in Exercise 1.6, the permutations of $\{1, 2, 3, 4\}$ defining the group S_4 and the resulting words obtained by shuffling the positions of the symbols in each word in V according to the rule $s \to s\tau^{-1}$. For example, the permutation $\tau = \binom{1234}{3214} = \tau^{-1}$, indicated by 3214, takes $s = uuyu$ (word number 12) into $yuuu$ (word number 15). Following Section 1.7 on page 11, identify the resulting position symmetry orbits of V and, for each sequence s in a given orbit, determine the group (indicate it by G_s) of those permutations in S_4 fixing s. How are these groups related to each other within each position orbit of S_4 in V? In analogy with the calculations shown in (1.6), evaluate the sum of the number $|G_s|$ of elements in G_s when s varies in V and compare it with the sum of the number $|\text{fix}(\tau)|$ of sequences fixed by the permutation τ when τ varies in S_4.

Exercise 1.10. Following Exercise 1.9, derive the symbol symmetry orbits for V and their G_s groups.

Exercise 1.11. Table (1.27) shows the entropy (top half) and the corresponding – log contrast sensitivity (bottom half) obtained from a sample of 5 Sloan chart lines.

$t\backslash$sample	1	2	3	4	5
1	0.614	0.636	0.632	0.624	0.66
v	0.675	0.619	0.692	0.640	0.619
h	0.655	0.619	0.660	0.690	0.667
o	0.603	0.603	0.553	0.603	0.635
1	0.476	0.496	0.482	0.51	0.484
v	0.450	0.490	0.34	0.525	0.490
h	0.415	0.490	0.427	0.385	0.420
o	0.470	0.470	0.50	0.470	0.450

(1.27)

For example,

$$x_1 = (0.614, 0.476), \quad x_v = (0.675, 0.45),$$
$$x_h = (0.655, 0.45), \quad x_o = (0.603, 0.47)$$

is the first bivariate sample indexed by $\{1, v, h, o\}$. Use the canonical projections shown in (1.9) to decompose and describe the association between the entropy and contrast sensitivity with the language symmetry studies.

Exercise 1.12. Indicate by $\mathcal{A}$ the $n \times n$ matrix with all entries equal to $1/m$, and let $\mathcal{Q} = I - \mathcal{A}$, so that $I = \mathcal{A} + \mathcal{Q}$ is the n-dimensional equivalent of the standard reduction introduced in (1.14). Applying Step 8 in the summary of Section 1.9 to the standard reduction, show that the resulting canonical components depend only on the sample mean and the sample standard deviation.

Appendix A

The action $s\tau^{-1}$ of S_4 on the binary sequences in length of 4

$s(1)$	y	u	y	u	u	u	y	y	y	u	u	u	y	y	y	u
$s(2)$	y	u	u	y	u	u	y	u	u	y	y	u	y	y	u	y
$s(3)$	y	u	u	u	y	u	u	y	u	y	u	y	y	u	y	y
$s(4)$	y	u	u	u	u	y	u	u	y	u	y	y	u	y	y	y
$S_4\backslash s$	1	16	15	14	12	8	13	11	7	10	6	4	9	5	3	2
1234	1	16	15	14	12	8	13	11	7	10	6	4	9	5	3	2
1243	1	16	15	14	8	12	13	7	11	6	10	4	5	9	3	2
1324	1	16	15	12	14	8	11	13	7	10	4	6	9	3	5	2
1432	1	16	15	8	12	14	7	11	13	4	6	10	3	5	9	2
2134	1	16	14	15	12	8	13	10	6	11	7	4	9	5	2	3
3214	1	16	12	14	15	8	10	11	4	13	6	7	9	2	3	5
4231	1	16	8	14	12	15	6	4	7	10	13	11	2	5	3	9
1342	1	16	15	12	8	14	11	7	13	4	10	6	3	9	5	2
1423	1	16	15	8	14	12	7	13	11	6	4	10	5	3	9	2
2314	1	16	14	12	15	8	10	13	6	11	4	7	9	2	5	3
2431	1	16	14	8	12	15	6	10	13	4	7	11	2	5	9	3
3124	1	16	12	15	14	8	11	10	4	13	7	6	9	3	2	5
3241	1	16	12	14	8	15	10	4	11	6	13	7	2	9	3	5
4232	1	16	8	15	12	14	7	4	6	11	13	10	3	5	2	9
4213	1	16	8	14	15	12	6	7	4	13	10	11	5	2	3	9
2143	1	16	14	15	8	12	13	6	10	7	11	4	5	9	2	3
3212	1	16	12	8	15	14	4	11	10	7	6	13	3	2	9	5
4321	1	16	8	12	14	15	4	6	7	10	11	13	2	3	5	9
2341	1	16	14	12	8	15	10	6	13	4	11	7	2	9	5	3
2413	1	16	14	8	15	12	6	13	10	7	4	11	5	2	9	3
3421	1	16	12	8	14	15	4	10	11	6	7	13	2	3	9	5
3142	1	16	12	15	8	14	11	4	10	7	13	6	3	9	2	5
4123	1	16	8	15	14	12	7	6	4	13	11	10	5	3	2	9
4312	1	16	8	12	15	14	4	7	6	11	10	13	3	2	5	9

2

Sorting the Labels: Group Actions and Orbits

2.1 Introduction

The analysis of variance shown in Table (1.13) on page 9 of Chapter 1 was a consequence of the joint action of $G = \{1, h, v, o\}$ and S_5 shuffling, respectively, the rows and columns of Table (1.12). The study and data-analytic applications of these rules for shuffling the experimental labels, called group actions, are among the main objectives of the present chapter. The study of group actions and orbits is an integral part of any symmetry study, and was identified earlier in the summary of Chapter 1, on page 23, steps 3, and 4. The algebraic aspects in this and the next two chapters follow closely from Serre (1977, Part I).

2.2 Permutations

In the classification of the binary sequences in length of 4 introduced in Chapter 1, the symmetries of interest were the permutations of the four positions and the permutations of the two symbols $\{u, y\}$. These sets of permutations, together with the operation of composition of functions, share all the defining algebraic properties identified in the multiplication table of $\{1, v, h, o\}$, characteristics of a finite group. In Section 1.7 of Chapter 1, it was shown that permutations appear in many, if not all, steps of a symmetry study. They appear as labels for the voting preference data on page 12, in matrix form as linear representations, on page 6, and as shuffling machanisms with which sets of labels could be classified and interpreted for the purpose of describing the data indexed by those labels.

Recall, from Chapter 1, that we denote by S_L the set of all bijective mappings defined on a set L and write S_ℓ when only the number (ℓ) of elements in L is of interest. When ℓ is finite, the $\ell!$ mappings in S_L are called the permutations in L.

The cycle notation for permutations

To illustrate the notation, the $3! = 6$ permutations

$$\begin{matrix}1&2&3\\ \downarrow&\downarrow&\downarrow\\ 1&2&3\end{matrix},\quad \begin{matrix}1&2&3\\ \downarrow&\downarrow&\downarrow\\ 2&3&1\end{matrix},\quad \begin{matrix}1&2&3\\ \downarrow&\downarrow&\downarrow\\ 3&1&2\end{matrix},\quad \begin{matrix}1&2&3\\ \downarrow&\downarrow&\downarrow\\ 2&1&3\end{matrix},\quad \begin{matrix}1&2&3\\ \downarrow&\downarrow&\downarrow\\ 3&2&1\end{matrix},\quad \begin{matrix}1&2&3\\ \downarrow&\downarrow&\downarrow\\ 1&3&2\end{matrix}$$

in S_3 are denoted, respectively, by $\{1, (123), (132), (12), (13), (23)\}$, or also by $\{123, 231, 312, 213, 321, 132\}$.

Permutations such as (12) or (132) are called cycles. Every permutation can be expressed as the composition of disjoint cycles. For example,

$$\sigma = \begin{matrix}1&2&3&4&5&6&7&8&9\\ \downarrow&\downarrow&\downarrow&\downarrow&\downarrow&\downarrow&\downarrow&\downarrow&\downarrow\\ 3&8&7&6&1&2&5&4&9\end{matrix} = (1375)(2846), \tag{2.1}$$

observing that cycles with a single element are excluded. Equivalently, $\sigma = (2846)(1375)$, a consequence of the fact that the composition of any two disjoint (without common elements) cycles is commutative. We note that the composition $\tau\sigma$ of two permutations τ and σ is the permutation obtained by first applying σ followed by τ, e.g., if $\tau = (23)$ and $\sigma = (13)$, then $\tau\sigma = (123)$. Because $\sigma\tau = (132)$, the composition of permutations is not commutative in general.

Parity of a permutation

Two-element cycles, such as (13), are called transpositions. Note that

$$(1375) = (15)(17)(13), \quad (2846) = (26)(24)(28). \tag{2.2}$$

In general, every cycle can be expressed as a composition of transpositions. This is trivially true for all two-element cycles. Assuming that it holds for cycles of length $n - 1$, direct evaluation shows that $(12\ldots n) = (1n)(12\ldots n-1)$, thus proving that the stated decomposition holds for cycles of length n. From (2.1) and (2.2) we obtain

$$\sigma = (1375)(2846) = (15)(17)(13)(26)(24)(28),$$

and observe that σ, of length 8, is the composition of 2 disjoint cycles. These two numbers are sufficient to characterize the permutation. The difference $8 - 2 = 6$ is called the *decrement* and corresponds to the number of transpositions expressing σ. To see this in general, note that every cycle of length n is the composition $(12\ldots n) = (1n)(1\ n-1)\ldots(12)$ of $n - 1$ transpositions, so that a permutation σ of length m with h disjoint cycles of length $n_1, \ldots, n_h$ can be written as the composition of $\sum_{i=1}^{h}(n_i - 1) = m - h$ transpositions, equal to its decrement.

The parity of a permutation is defined as the parity of its decrement. Consequently, an even (respectively odd) permutation is the composition of an even (respectively odd) number of transpositions.

Proposition 2.1. *For all permutations σ and transpositions τ, decrement $(\tau\sigma) =$ decrement $(\sigma) \pm 1$.*

Proof. Following Bacry (1963), there are 2 cases to consider: When $\{a, b\}$ belong to a common cycle of σ, the fact that $(ab)(a \ldots xb \ldots y) = (a \ldots x)(b \ldots y)$ shows that the number of cycles of σ is increased by one unit, so that the decrement of σ decreases by one unit; Similarly, when $\{a, b\}$ belong to a different cycle, the fact that $(a \ldots xb \ldots y) = (ab)(a \ldots x)(b \ldots y)$ shows that the number of cycles is decreased by one unit, and consequently the decrement of σ is increased by one unit. In both cases, because disjoint cycles commute, it is always possible to write the transposition to the left of the cycle(s) of interest. ∎

The sign (Sgn), or signature, of a permutation is given by

$$\operatorname{Sgn}(\sigma) = \begin{cases} +1 & \text{if } \sigma \text{ is even,} \\ -1 & \text{if } \sigma \text{ is odd.} \end{cases}$$

As a consequence of Proposition 2.1, the reader may verify that the parity of the composition $\sigma\tau$ of any two permutations σ, τ is given by

$\sigma \backslash \tau$	even	odd
even	even	odd
odd	odd	even

,

and that, consequently, for any two permutations σ, τ,

$$\operatorname{Sgn}(\sigma\tau) = \operatorname{Sgn}(\sigma)\operatorname{Sgn}(\tau). \tag{2.3}$$

Clearly, Sgn $(1) = 1$ and Sgn $(\tau\sigma) = -$Sgn (σ) for all transpositions τ.

Conjugacy classes

Two permutations σ and η are conjugate when $\sigma = \tau\eta\tau^{-1}$ for some permutation τ. Conjugacy defines an equivalence relation among permutations and the resulting classes are called conjugacy classes. Also note that the only effect the operation of conjugacy has on the cycle structure of a permutation is that of eventually renaming the elements within each cycle. For example, in S_3, if $\eta = (12)$ and $\tau = (23)$, then the conjugacy $\tau\eta\tau^{-1}$ transforms the cycle (12) into the cycle (13), that is, $\tau\eta\tau^{-1} = (13)$. The conjucacy classes of S_3 are $\{1\}$, $\{(123), (132)\}$ and $\{(12), (13), (23)\}$. We have:

Proposition 2.2. *Conjugacy classes determine the same cycle structure.*

Integer partitions and Young frames

The conjugacy classes of S_ℓ, and hence by Proposition 2.2, their cycle structure, can be described by the integer partitions $\lambda = (n_1, n_2, \ldots, n_\ell)$ of ℓ, defined by the integers $n_1 \geq \ldots \geq n_\ell \geq 0$ with $n_1 + \cdots + n_\ell = \ell$. For example, the cycle structure of S_3 described above is determined by $\lambda = (1, 1, 1)$ indicating the cycle structure with 3 cycles of length 1, also indicated by $\lambda = 111$. Its single representative element is the identity permutation; $\lambda = (2, 1, 0)$ or 211 indicating the cycle structure with one cycle of length 2 and one of length 1. A representative is the transposition (12); $\lambda = (3, 0, 0)$, or 300, having one cycle of length 3. A representative is the permutation (123).

These classes of cycle structures can be represented by their respective Young[1] frames

$$111 : \begin{array}{|c|} \hline - \\ \hline - \\ \hline - \\ \hline \end{array}\,, \quad 210 : \begin{array}{|c|c|} \hline - & - \\ \hline - & \\ \cline{1-1} \end{array}\,, \quad 300 : \begin{array}{|c|c|c|} \hline - & - & - \\ \hline \end{array}$$

which, in general, have $n_1, \ldots, n_\ell$ boxes in rows $1, \ldots, \ell$ respectively. Indicating by $(m_\ell, \ldots, m_1)$ the multiplicities with which the distinct cycle lengths of each frame occur, each frame can be written, uniquely, as $\lambda = \ell^{m_\ell} \ldots 2^{m_2} 1^{m_1}$.

Data indexed by Young tableaux. The Young tableaux are the distinct $n!$ assignments of permutations to Young frames. They have a number of particularly useful data-analytic interpretations. To illustrate, consider the case $n = 4$, the voting data of page 12 and $\lambda = 31$. A possible assignment of names of candidade to the corresponding Young frame is

$$\begin{array}{|c|c|c|} \hline a & g & c \\ \hline t & \\ \cline{1-1} \end{array}\,,$$

which may be utilized to sort out the ranking preferences for two types of positions, say, committee president, first and second vice presidents from committee secretary, to single out the least favored candidate from the others, or to denote a partial or incomplete rankings. This is obtained by defining two tableaux as equivalent when their rows differ only by a permutation of their entries. These "order free" tableaux can be identified with the sets

$$\{a, g, c\}, \quad \{a, g, t\}, \quad \{a, c, t\}, \quad \{g, c, t\},$$

[1] Alfred Young, a mathematician at Cambridge University, 1873–1940.

to each one of which a summary of the data indexed by their elements may be obtained. If the summary is the total number of votes, then we would have the following indexing:

$$x_{\{a,g,c\}} = 170, \quad x_{\{a,g,t\}} = 186, \quad x_{\{a,c,t\}} = 268, \quad x_{\{g,c,t\}} = 225.$$

These structured data can then be further reduced with the methods to be developed in the coming chapters.

Young frames and tableaux are also in one-to-one correspondence with the patterns of ties in ordered observations and give a classification of the space of all mappings of a finite set into itself. We will return to these useful constructs later on in Section 3.9 on page 82.

2.3 Groups and Homomorphisms

In Chapter 1 we remarked that the composition table

$*$	1	v	h	o
1	1	v	h	o
v	v	1	o	h
h	h	o	1	v
o	o	h	v	1

of $G = \{1, v, h, o\}$ summarizes the defining properties of a finite group, namely, the composition of two symmetries is also a symmetry, the composition is associative, the identity element 1 is an element in G and each element in G has an inverse element also in G. More specifically,

Definition 2.1. A group is a nonempty set G equipped with an associative operation $(\sigma, \tau) \in G \times G \to \sigma * \tau \in G$, an (identity) element $1 \in G$, satisfying $1 * \tau = \tau * 1 = \tau$, for all $\tau \in G$ and such that for every $\tau \in G$, there is an (inverse) element $\tau^{-1} \in G$ such that $\tau * \tau^{-1} = \tau^{-1} * \tau = 1$.

The group operation is generally referred to as multiplication so that the n-th power τ^n of a group element τ means its n-fold $\tau * \ldots * \tau$ product.

A word about notation. We adopt the convention that when a group multiplication $\tau * \sigma$ is defined by the composition of functions that then $\tau * \sigma = \sigma\tau$, for $\tau, \sigma \in G$. For example, $(123) * (23) = (13)$, which is the result of (123) followed by (23). Otherwise, the group operation ($*$) is omitted and we write, for simplicity, $\tau\sigma$ instead. The complete multiplication table for S_3 is shown in (2.4), constructed using Algorithm 2 on page 221.

*	1	(123)	(132)	(12)	(13)	(23)
1	1	(123)	(132)	(12)	(13)	(23)
(123)	(123)	(132)	1	(23)	(12)	(13)
(132)	(132)	1	(123)	(13)	(23)	(12)
(12)	(12)	(13)	(23)	1	(123)	(132)
(13)	(13)	(23)	(12)	(132)	1	(123)
(23)	(23)	(12)	(13)	(123)	(132)	1

(2.4)

Indicating by $|S|$ the cardinality of a finite set S, the group order is the number $|G|$ of elements in the group. A commutative, or Abelian, group G is one in which the operation is commutative, that is, $\tau\sigma = \sigma\tau$ for all $\sigma, \tau \in G$. The group $\{1, v, h, o\}$ is Abelian, whereas S_3 is not.

A subgroup of G is any subset of G that is a group under the operation of G restricted to that subset. For example, $\{1, v\}$, $\{1, h\}$, and $\{1, o\}$ are subgroups of $\{1, v, h, o\}$ of order 2. $C_3 = \{1, (123), (132)\}$ is a commutative subgroup of S_3 of order 3.

The order of a group element τ is the smallest positive integer n such that $\tau^n = 1$. The elements v, h, and o have order 2. The permutation (123) has order 3.

Group homomorphisms

Given two groups G, H, a homomorphism from G to H is a function $\rho : G \to H$ preserving their group structure, that is, $\rho_{\tau\sigma} = \rho_\tau \rho_\sigma$ for all τ, σ in G. Homomorphisms map the identity of G into the identity of H and the inverse of $\tau \in G$ into the inverse in H of its image, for all τ in G. An injective homomorphism is called a monomorphism. An isomorphism is an invertible homomorphism. We write $G \simeq H$ to indicate that G is isomorphic with H.

A simple, however important, fact is that the symmetric group S_ℓ is isomorphic with the group of $\ell \times \ell$ permutation matrices. It follows from associating to each permutation $\tau \in S_\ell$ the permutation matrix r_τ defined by

$$(r_\tau)_{ij} = 1 \iff j = \tau i. \tag{2.5}$$

We then have $r(\tau * \sigma) = r(\tau)r(\sigma)$, for all $\tau, \sigma \in S_\ell$. It is often more convenient, if not computationally more attractive, to think of permutations in terms of their corresponding permutation matrices.

The 3 x 3 permutation matrices. Applying Algorithm 1 on page 221 gives the permutation matrices

$$r_1 = \begin{bmatrix} 1 & 0 & 0 \\ 0 & 1 & 0 \\ 0 & 0 & 1 \end{bmatrix}, \quad r_{(123)} = \begin{bmatrix} 0 & 1 & 0 \\ 0 & 0 & 1 \\ 1 & 0 & 0 \end{bmatrix}, \quad r_{(132)} = \begin{bmatrix} 0 & 0 & 1 \\ 1 & 0 & 0 \\ 0 & 1 & 0 \end{bmatrix},$$

$$r_{(12)} = \begin{bmatrix} 0 & 1 & 0 \\ 1 & 0 & 0 \\ 0 & 0 & 1 \end{bmatrix}, \quad r_{(13)} = \begin{bmatrix} 0 & 0 & 1 \\ 0 & 1 & 0 \\ 1 & 0 & 0 \end{bmatrix}, \quad r_{(23)} = \begin{bmatrix} 1 & 0 & 0 \\ 0 & 0 & 1 \\ 0 & 1 & 0 \end{bmatrix}$$

for S_3. Isomorphic groups can be considered as algebraic copies of each other, or different realizations of a common abstract group defining their algebraic characteristics. Consequently, groups become distinct only up to isomorphisms.

When $G = H$, an isomorphism of G is called an automorphism of G. The kernel, indicated by $\ker_\rho$ of a homomorphism $\rho : G \to H$ is the set of those elements in G mapped into the identity element of H, whereas its range or image is the set $\text{im}_\rho = \{\rho_\tau; \tau \in G\}$. Clearly $\ker_\rho$ and im_ρ are subgroups of G and H, respectively. Moreover, if $\ker_\rho = \{1\}$ and $\rho_\tau = \rho_\sigma$ then $\rho_{\tau\sigma^{-1}} = \rho_\tau \rho_{\sigma^{-1}} = \rho_\sigma \rho_{\sigma^{-1}} = 1$, so that $\tau\sigma^{-1} \in \{1\}$ and $\sigma = \tau$. That is, G is isomorphic with im_ρ.

The complex numbers

The set $\mathbb{C}$ of the complex numbers is an additive Abelian group and the sets $\mathbb{R}$ and $\mathbb{Z}$ of the real and integer numbers, respectively, are subgroups of $\mathbb{C}$. The unit-norm complex numbers with the operation of multiplication form a group called the circle group.

Cyclic groups

The sets $C_n = \{1, \omega, \omega^2, \ldots, \omega^{n-1}\}$, for $n = 1, 2, \ldots,$ and $\omega = e^{2\pi i/n}$ are multiplicative groups of order n. We refer to C_n and any of its equivalent realizations as cyclic groups. They are characterized here as any finite group with some element of order equal to the order of the group. Any such element is called a group generator. For example, C_3 may be realized, isomorphically, as

$$\{1, (123), (132)\} \simeq \{1, \omega, \omega^2\} \simeq \{3\mathbb{Z}, 3\mathbb{Z}+1, 3\mathbb{Z}+2\} \simeq \{1, r, r^2\} \simeq \{1, \beta, \beta^2\},$$

where $3\mathbb{Z} + k = \{3m + k; m \in \mathbb{Z}\}, k = 0, 1, 2,$

$$r = \begin{bmatrix} \cos 2\pi/3 & -\sin 2\pi/3 \\ \sin 2\pi/3 & \cos 2\pi/3 \end{bmatrix}, \quad \beta = \begin{bmatrix} \omega & 0 \\ 0 & \overline{\omega} \end{bmatrix},$$

with group generators, (123), $3\mathbb{Z}+1$, r and β, respectively. Similarly, (132), $3\mathbb{Z}+2$, r^2 and β^2 are also generators. The groups $\{1, r, r^2\}$ and $\{1, \beta, \beta^2\}$ are realizations of C_3 as groups of rotations in the plane.

Note that (1234) generates $C_4 = \{1, (1234), (13)(24), (1432)\}$, whereas (13)(24) does not. When $n > 1$ is prime, however, C_p can be generated, without distinction, by any one of its members distinct from the identity. Equivalently (Weyl, 1952, p. 140), because the points in C_n may be realized as the roots in $\mathbb{C}$ of the equation $z^n - 1 = 0$ then C_n can be generated by any one of its nontrivial roots. The roots of the equation $z^n - 1 = 0$ form the vertices of a regular n-sided polygon, and since $z^n - 1 = (z-1)(z^{n-1} + z^{n-2} + \cdots + z + 1)$, the roots of $z^{n-1} + z^{n-2} + \cdots + z + 1$ are said to be algebraically indiscernible.

Data indexed by cyclic labels. The joint frequencies of mining disasters during a certain period, shown in Matrix (2.6) and discussed in Wit and McCullagh (2001), constitute an example of data indexed by a set product (yearly seasons $\times$ weekly days) the components of which are chronological cycles. This fact suggests the joint action of the cyclic groups C_4 and C_7 to analyze the joint frequencies. The study is presented in Chapter 5 on page 153.

	Mon	Tue	Wed	Thu	Fri	Sat	Sun	total
Autumn	7	10	5	5	6	7	1	41
Winter	5	9	10	10	11	7	0	52
Spring	3	7	10	12	13	9	2	56
Summer	4	8	8	9	5	6	2	42
total	19	34	33	36	35	29	5	191

(2.6)

Cosets

The sets $\mathcal{O}_k = n\mathbb{Z} + k, k = 0, 1, \ldots, n-1$ are examples of cosets of a subgroup, in this case $n\mathbb{Z} \subset \mathbb{Z}$. They decompose the original group $\mathbb{Z}$ into their disjoint union. Direct calculation shows that the sets $\mathcal{O}_k$, together with the operation $\mathcal{O}_\ell + \mathcal{O}_m = \mathcal{O}_{\ell+m \mod n}$, form a group, called the quotient group and denoted by $\mathbb{Z}/n\mathbb{Z}$. More generally, the (left) cosets of a subgroup H of G are the subsets $\tau H = \{\tau\sigma : \sigma \in H\} \subseteq G$, defined for all $\tau \in G$. Because an element $\tau \in G$ belongs to exactly one coset, namely τH, it follows that any two cosets are disjoint sets and the union of the collection of all cosets decomposes the set G. Moreover, any two cosets have the same (eventually infinite) number of elements, called the index of H in G and indicated by $|G : H|$. This follows from the fact that $\tau\sigma \to \tau'\sigma$ is a bijective correspodence between τH and $\tau' H$. There is no need to distinguish between left and right cosets, which are defined equivalently

Data indexed by cosets. The bijective correspondence between a subgroup and its cosets has some fundamental consequences for the analysis of data indexed by Abelian groups, as demonstrated in great detail, in Van de Ven and Di Bucchianico (2006) and Van de Ven (2007).

To illustrate a simple noncommutative situation, consider the observed frequencies

	1	2	3	4	5	6	7	8	9
act	8	16	16	7	17	11	12	6	14
cta	15	8	14	9	14	15	8	5	16
tac	7	17	13	15	9	11	18	5	17
cat	14	15	16	14	21	17	15	10	8
tca	11	18	10	17	11	16	14	9	13
atc	7	15	9	13	11	11	11	12	10
total	62	89	78	75	83	81	78	47	78

(2.7)

with which the nucleotide sequences $V = \{act,\ cta,\ tac,\ cat,\ tca,\ atc\}$ appear in nine subsequent 900-bp-long regions along the BRU isolate of the HIV-1 described earlier on in Chapter 1 on page 13. Note that V describes all permutations of the nucleotides $\{a, c, t\}$, so that (2.7) is an example of data indexed by S_3. Observing that $H = \{act, cta, tac\}$ and $F = \{cat, tca, atc\}$ are isomorphic to C_3, it then follows H and F are cosets to each other in S_3. Also note that one set is obtained from the other by transposing, for example, the symbols in positions 2 and 3, or equivalently, the result of the action $s\tau^{-1}$, with $s \in V$ and $\tau = (23)$.

These two cosets can be clearly identified in Table (2.8), which is simply the multiplication table of S_3 shown in (2.4), relabeled with reference to $s = act$.

act	*cta*	*tac*	*cat*	*tca*	*atc*
cta	*tac*	*act*	*atc*	*cat*	*tca*
tac	*act*	*cta*	*tca*	*atc*	*cat*
cat	*tca*	*atc*	*act*	*cta*	*tac*
tca	*atc*	*cat*	*tac*	*act*	*cta*
atc	*cat*	*tca*	*cta*	*tac*	*act*

(2.8)

The two top blocks make evident the fact that each row transforms according to the permutations

$$r_1 = \begin{bmatrix} 1 & 0 & 0 \\ 0 & 1 & 0 \\ 0 & 0 & 1 \end{bmatrix}, \quad r_{(123)} = \begin{bmatrix} 0 & 1 & 0 \\ 0 & 0 & 1 \\ 1 & 0 & 0 \end{bmatrix}, \quad r_{(132)} = \begin{bmatrix} 0 & 0 & 1 \\ 1 & 0 & 0 \\ 0 & 1 & 0 \end{bmatrix},$$

of the elements in the first row in the block. These matrices, indexed by the elements of C_3, are group homomorphisms introduced on page 36.

The reduction of the data indexed by each coset will follow from linear combinations $\mathcal{P}_1$, $\mathcal{P}_2$, and $\mathcal{P}_3$ of these permutations, with respective sets of coefficients

$$\chi_1 = \{1, 1, 1\}, \quad \chi_2 = \{1, \omega, \omega^2\}, \quad \chi_3 = \{1, \omega^2, \omega\},$$

where $\omega = e^{2\pi i/3}$ is a generator of $C_3 \simeq \{1, \omega, \omega^2\}$. This is in analogy to the construction illustrated earlier in Section 1.4 on page 5 and outlines the theory being developed in this and the next chapter. The resulting matrices

$$\mathcal{P}_1 = 1/3 \begin{bmatrix} 1 & 1 & 1 \\ 1 & 1 & 1 \\ 1 & 1 & 1 \end{bmatrix}, \quad \mathcal{P}_2 = 1/3 \begin{bmatrix} 1 & \omega & \omega^2 \\ \omega^2 & 1 & \omega \\ \omega & \omega^2 & 1 \end{bmatrix}, \quad \mathcal{P}_3 = 1/3 \begin{bmatrix} 1 & \omega^2 & \omega \\ \omega & 1 & \omega^2 \\ \omega^2 & \omega & 1 \end{bmatrix}$$

are the canonical projections of interest. Further evaluating $\mathcal{P}_2 + \mathcal{P}_3$ leads to two real symmetric projections, namely $\mathcal{A} = \mathcal{P}_1$ and

$$\mathcal{Q} = 1/3 \begin{bmatrix} 2 & -1 & -1 \\ -1 & 2 & -1 \\ -1 & -1 & 2 \end{bmatrix},$$

which we recognize as the standard reduction $I = \mathcal{A} + \mathcal{Q}$, introduced on page 10, now defined in $\mathbb{R}^3$.

We see that, consequently, C_3 induces two structurally identical analysis in each *half* fraction of the data indexed by those labels, namely their standard reduction, in which $\mathcal{Q}$ defines the within-coset comparisons, each one taking values on a subspace in dimension of 2. As introduced earlier on in Exercise 1.12, the joint summaries derived from $\mathcal{A}$ and $\mathcal{Q}$ may be expressed then as the within-coset means and standard deviations.

In this particular case, in which the (rank) dimension of $\mathcal{Q}$ is 2, we may in addition consider the graphical display of the bases

$$\{H_1, H_2\} = \left\{\log \frac{act^2}{cta \cdot tac}, \quad \log \frac{cta^2}{act \cdot tac}\right\},$$

$$\{F_1, F_2\} = \left\{\log \frac{cat^2}{tca \cdot atc}, \quad \log \frac{tca^2}{cat \cdot atc}\right\},$$

for the image space of $\mathcal{Q}$, in each coset. These subspaces share the invariance properties introduced in Chapter 1, resolving the arbitrariness in the reference word generating the two cosets, under C_3. Figure 2.1 shows the distribution of the nine regions in each subspace, based on the frequency data in Table (2.7), from which the (apparent) similarities in within-cosets variability among regions can be observed.

The two cosets can also be distinguished by the set $\{(a, c), (c, t), (t, a), (c, a), (a, t), (t, c)\}$ of the directed edges of a regular triangle, where (x, y) indicates a directed edge $x \to y$. The permutation action of S_3 on the vertices gives

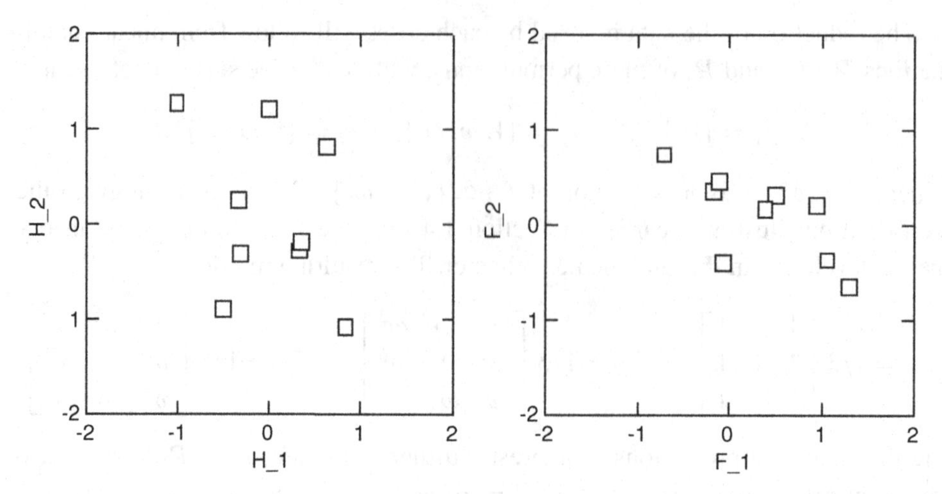

Figure 2.1: Distribution of the nine regions in the $\{H_1, H_2\}$ and $\{F_1, F_2\}$ subspaces.

two orbits in a way that the orientations correspond to cosets, independenly of how the symbols were originally assigned to the vertices.

Semidirect and direct products of groups

Following up with the notation introduced in the previous sections, let $C_3 \simeq \{1, \omega, \omega^2\}$ and $C_2 \simeq \{1, -1\}$, so that if $H = \{act, cta, tac\} \simeq \{1, \omega, \omega^2\}$ then $F = \{cat, tca, atc\} \simeq \{1, \omega^2, \omega\}$. This suggests the definition of

$$\alpha_d : \omega^k \in H \to \omega^{dk} \in F, \quad d = \pm 1,$$

or, isomorphically, $\alpha_d : C_3 \to C_3$ for each $d \in C_2$. Consequently, because

$$\alpha_d(\alpha_{d'}(\omega^k)) = \alpha_d(\omega^{d'k}) = \omega^{dd'k} = \alpha_{dd'}(\omega^k),$$

or $\alpha_d \alpha_{d'} = \alpha_{dd'}$, for all d, d' in C_2, we see that $\alpha : d \to \alpha_d$ is in $\text{Hom}(C_2)$. Moreover,

$$\alpha_d(\omega^k \omega^{k'}) = \alpha_d(\omega^{k+k'}) = \omega^{d(k+k')} = \omega^{dk}\omega^{dk'} = \alpha_d(\omega^k)\alpha_d(\omega^{k'})$$

showing that $\alpha_d \in \text{Hom}(C_3)$. Also, if $\alpha_d(\omega^k) = \alpha_d(\omega^{k'})$ then $dk = dk' \mod 3$ so that $k = k'$ for $k, k' \in \{0, 1, 2\}$. Since α_d is clearly surjective, then $\alpha_d \in \text{Aut}(C_3)$, and, in summary, $\alpha \in \text{Hom}(C_2)$ taking values in $\text{Aut}(C_3)$.

Semidirect products. With these properties in mind, we can define an operation in $C_3 \times C_2$, called the semidirect product of C_3 and C_2, and indicated by $\times_\alpha$. Specifically

$$(\omega^k, d) \times_\alpha (\omega^{k'}, d') = (\omega^k \alpha_d(\omega^{k'}), dd'). \tag{2.9}$$

It simply states that the effect $\omega^{k+dk'}$ of a reversal is that of introducing a *phase* shift of $\omega^{\pm k'}$ to ω^k. The reader may want to verify that the product set $C_3 \times C_2$ endowed with the semidirect product is a group, in which the identity is $(1, 1)$ and $(\alpha_d(\omega^{-k}), d)$ is the inverse of (ω^k, d) in $C_3 \times C_2$. This definition can be extended to any two finite groups.

Direct products. If *alpha* is the trivial ($\equiv 1$) homomorphism, the product set is then endowed with the componentwise product operation

$$(\omega^k, d) \times (\omega^{k'}, d') = (\omega^{k+k'}, dd'),$$

and the resulting group is called the direct product of groups C_3 and C_2. This componentwise multiplication extends naturally to three or more groups.

Dihedral groups of the plane. The semidirect product $C_n \times_\alpha C_2$ for $n \geq 3$ gives a group of order $2n$, called the dihedral group, and indicated by D_n. It will prove useful, in many data-analytic interpretations, to think of the dihedral groups as two-sided groups of (planar) rotations. To illustrate, with the case $n = 4$, consider the rotations

$$\begin{array}{ccc} & agct: & \\ a & \longrightarrow & g \\ \uparrow & & \downarrow \\ t & \longleftarrow & c \end{array} \quad \begin{array}{ccc} & tagc: & \\ t & \longrightarrow & a \\ \uparrow & & \downarrow \\ c & \longleftarrow & g \end{array} \quad \begin{array}{ccc} & ctag: & \\ c & \longrightarrow & t \\ \uparrow & & \downarrow \\ g & \longleftarrow & a \end{array} \quad \begin{array}{ccc} & gcta: & \\ g & \longrightarrow & c \\ \uparrow & & \downarrow \\ a & \longleftarrow & t \end{array}, \tag{2.10}$$

of C_4, generated, respectively, by shuffling the symbols in the word *agct* according to the permutations $\{1, (1234), (13)(24), (1432)\}$ of their positions. Each diagram in (2.10) displays the letters of the corresponding word according to

$$\begin{array}{ccc} a:1 & \longrightarrow & g:2 \\ \uparrow & & \downarrow \\ t:4 & \longleftarrow & c:3 \end{array},$$

respectively, whereas the arrows indicate the orientation of the edges. Composing each rotation with (12)(34) has the effect of transposing or substituting the symbols $\{a, c\}$, thus leading to the coset

$$\begin{array}{ccc} & cgat: & \\ c & \longleftarrow & g \\ \downarrow & & \uparrow \\ t & \longrightarrow & a \end{array} \quad \begin{array}{ccc} & tcga: & \\ t & \longleftarrow & c \\ \downarrow & & \uparrow \\ a & \longrightarrow & g \end{array} \quad \begin{array}{ccc} & atcg: & \\ a & \longleftarrow & t \\ \downarrow & & \uparrow \\ g & \longrightarrow & c \end{array} \quad \begin{array}{ccc} & gatc: & \\ g & \longleftarrow & a \\ \downarrow & & \uparrow \\ c & \longrightarrow & t \end{array}. \tag{2.11}$$

The orientation now appears as if seen "from the other side." Together, the resulting permutations define the group D_4. The semidirect product (2.9) applied to $C_4 \times C_2$, where $C_2 \simeq \{1, (12)(34)\} \simeq \{-1, 1\}$ leads to its multiplication table,

1	*agct*	*tagc*	*ctag*	*gcta*	*cgat*	*tcga*	*atcg*	*gatc*
(1432)	*tagc*	*ctag*	*gcta*	*agct*	*gatc*	*cgat*	*tcga*	*atcg*
(13)(24)	*ctag*	*gcta*	*agct*	*tagc*	*atcg*	*gatc*	*cgat*	*tcga*
(1234)	*gcta*	*agct*	*tagc*	*ctag*	*tcga*	*atcg*	*gatc*	*cgat*
(13)	*cgat*	*tcga*	*atcg*	*gatc*	*agct*	*tagc*	*ctag*	*gcta*
(14)(23)	*tcga*	*atcg*	*gatc*	*cgat*	*gcta*	*agct*	*tagc*	*ctag*
(24)	*atcg*	*gatc*	*cgat*	*tcga*	*ctag*	*gcta*	*agct*	*tagc*
(12)(34)	*gatc*	*cgat*	*tcga*	*atcg*	*tagc*	*ctag*	*gcta*	*agct*

, (2.12)

in analogy to the multiplication table (2.8) for $D_3 \simeq S_3$ on page 38. These permutations, clearly, define the symmetry permutations for the square with vertices labeled according to the diagrams above.

With the data-analytic applications to be developed later on in Chapter 7, however, we will make extensive use of the following realization of the dihedral groups.

Dihedral groups in the plane

Writing $\phi = 2\pi/n$ and $\omega = e^{i\phi}$, let

$$\beta_{k,d} = \frac{1}{2}\begin{bmatrix} (1+d)\omega^{k-1} & (1-d)\omega^{k-1} \\ (1-d)\omega^{-k+1} & (1+d)\omega^{-k+1} \end{bmatrix}, \quad k = 1, \ldots, n, \; d = \pm 1. \tag{2.13}$$

These $2n$ matrices are the canonical or phase space (Bacry, 1967, p. 203) versions of the matrices

$$\begin{aligned} X\beta_{k,1}X^{-1} &= \begin{bmatrix} \cos((k-1)\phi) & -\sin((k-1)\phi) \\ \sin((k-1)\phi) & \cos((k-1)\phi) \end{bmatrix}, \\ X\beta_{k,-1}X^{-1} &= \begin{bmatrix} \cos((k-1)\phi) & \sin((k-1)\phi) \\ \sin((k-1)\phi) & -\cos((k-1)\phi) \end{bmatrix}, \end{aligned} \tag{2.14}$$

in the Euclidean plane, where

$$X = \sqrt{2}\begin{bmatrix} 1 & i \\ 1 & -i \end{bmatrix}/2.$$

In that space, $\beta_{k,1}$ are planar rotations around the origin in the counterclockwise direction by an angle of $k\phi$ radians, whereas $\beta_{k,-1}$ are planar reflections along lines oriented at (dihedral) angles of $k\phi/2$ radians. The notation in (2.13) is consistent with the fact that $d = \det \beta_{k,d} = \pm 1$ (Cartan, 1966, Ch. 1, Sec. II). We will use the same notation $\beta_{k,d}$ to indicate either representation unless there is need to distinguish between them.

The reader may want to verify that these matrices, in either representation, multiply according to the semidirect product (2.9), so that the mapping

$$(k, d) \to \beta_{k,d} \tag{2.15}$$

is a group homomorphism of $D_n = ((C_n, C_2), \times_\alpha)$. That is, $(k, d) \times_\alpha (k', d') \to \beta_{k,d}\beta_{k'd'}$. The association,

$$\beta_{k,d} \to \begin{cases} (1\ldots m \ldots n)^k \underbrace{[(1\ n)(2\ n-1)\cdots(m-1\ m+1)]}_{reversal}{}^{(3-d)/2}, & \text{if } n \text{ is odd,} \\ (1\ldots m\ m' \ldots n)^k \underbrace{[(1\ n)(2\ n-1)\cdots(m\ m')]}_{reversal}{}^{(3-d)/2}, & \text{if } n \text{ is even,} \quad k = 1, \ldots, n,\ d = \pm 1, \end{cases}$$

gives an isomorphism between the permutation group and the planar rotations/reversals realization of D_n.

The dihedral group D_4. Using the representation given by (2.14), and writing $r = r_{1,1}$ the four-fold planar rotations fixing the square are then

$$\beta_{k,1} = \begin{bmatrix} 1 & 0 \\ 0 & 1 \end{bmatrix}, \begin{bmatrix} 0 & -1 \\ 1 & 0 \end{bmatrix}, \begin{bmatrix} -1 & 0 \\ 0 & -1 \end{bmatrix}, \begin{bmatrix} 0 & 1 \\ -1 & 0 \end{bmatrix}, \quad k = 1, 2, 3, 4,$$

whereas the reversals are the line reflections

$$\beta_{k,-1} = \begin{bmatrix} 1 & 0 \\ 0 & -1 \end{bmatrix}, \begin{bmatrix} 0 & 1 \\ 1 & 0 \end{bmatrix}, \begin{bmatrix} -1 & 0 \\ 0 & 1 \end{bmatrix}, \begin{bmatrix} 0 & -1 \\ -1 & 0 \end{bmatrix}, \quad k = 1, 2, 3, 4$$

defined, respectively, by the lines $y = 0$, $y = x$, $x = 0$, and $y = -x$. The notation,

$$D_4 = \{\beta_{k,d}; k = 1, \ldots, 4, d = \pm 1\} \tag{2.16}$$

will be followed from now on. The reader may want to verify that, not surprisingly, $\beta_{k,-1} = \beta_{k,1}\beta_{1,-1}$.

Data indexed by D_n. In optical applications, the representations (2.13) and (2.14) can be used to construct dihedral labels, such as

$$x_{k,d} = \frac{1}{n}\mathrm{Tr}\,[\beta_{-kd,d}\mathcal{M}], \tag{2.17}$$

for experimental 2×2 data matrices $\mathcal{M}$. These labels will appear later on in applications of dihedral Fourier analyses.

Matrix groups

The general linear group is defined by the set $GL(n, \mathbb{F})$ of $n \times n$ nonsingular (i.e., nonzero determinant) matrices with real ($\mathbb{F} = \mathbb{R}$) or complex ($\mathbb{F} = \mathbb{C}$) entries, under the operation of matrix multiplication. Equivalently, we may consider the underlying vector space $\mathcal{V} = \mathbb{F}^n$ and simply write $GL(\mathcal{V})$. Some classical subgroups: the $n \times n$ permutation matrices define a finite subgroup of $GL(\mathcal{V})$, of order $n!$, also called the Weyl subgroup; the special linear group of the matrices in $GL(n, \mathbb{F})$ of determinant 1, denoted by $SL(n, \mathbb{F})$. More generally, the general linear group may be considered as the set $GL(\mathcal{V})$ of all invertible linear transformations of the vector space $\mathcal{V}$.

Isometry groups

The bilinear symmetric form $x'y = \sum x_i y_i$ is an example of an inner product of two vectors x, y in $\mathbb{R}^n$. The bilinear Hermitian form $x'\overline{x}$, where $\overline{x}$ indicates the complex conjugate of x, is an inner product in $\mathbb{C}^n$. The form $< x, y >= x_1'y_2 - x_2'y_1$, for $x = (x_1, x_2)$ and $y = (y_1, y_2)$ in $\mathbb{R}^{2n}$ is antisymmetric bilinear, that is $< x, y >= - < y, x >$. The corresponding isometries defined by the different ways of measuring distances and angles (geometries) define the orthogonal, unitary, and symplectic groups, respectively. The corresponding subgroups of $SL(n, \mathbb{F})$ are the special orthogonal and unitary groups. For example, $Sp(2, \mathbb{R})$ is the subgroup of $GL(2, \mathbb{R})$ leaving the symmetric bilinear form

$$(v, w) = \det \begin{bmatrix} r_1 & r_2 \\ p_1 & p_2 \end{bmatrix} = r_1 p_2 - r_2 p_1, \quad v' = (r_1, p_1), \quad w' = (r_2, p_2)$$

invariant. Equivalently, $(v, w) = v' \begin{bmatrix} 0 & 1 \\ -1 & 0 \end{bmatrix} w$. In physics, r and p have the interpretation of position and momentum, respectively, of a particle constrained to move on a line. The set of vectors $(r, p) \in \mathbb{R}^2$ define the particle's phase space and its geometry, determined by the standard form $(.,.)$, is called symplectic geometry and $Sp(2, \mathbb{R})$ is the (real) symplectic group. It respects the geometry of the phase space and is one of the classic isometry groups.

The Lorentz Group. The geometry of the four-dimensional space-time vector space, indicated here by $\mathcal{M}$, is determined by the (Lorentz[2]) fundamental form

$$g(x, y) = x_1 y_1 + x_2 y_2 + x_3 y_3 - x_4 y_4, \quad x, y \in \mathbb{R}^4,$$

where (x_1, x_2, x_3) and (y_1, y_2, y_3) are the spatial coordinates, and x_4 and y_4 are the time coordinates of the point-events x, y, relative to an orthonormal basis typically identified with a frame of reference for an observer of the events. Equivalently, $g(x, y) = x'\eta y$, where $\eta = \text{Diag}(1, 1, 1, -1)$.

Moving from one frame of reference to another without disturbing the geometry of $\mathcal{M}$ has the effect of transforming the point events according to non-singular

[2] Dutch physicist and mathematician Hendrik Antoon Lorentz, 1853–1928.

linear transformations (L) that are consistent with the geometry of $\mathcal{M}$. That is, we must have $g(Lx, Ly) = g(x, y)$, for all $x, y \in \mathcal{M}$. These symmetrically related frames are the admissible frames of reference in the Relativity Principle, stating that all admissible frames of reference are completely equivalent for the formulation of the laws of physics.

The subgroup

$$\mathcal{L} = \{L \in GL(\mathbb{R}^4) : L'\eta L = \eta\} \tag{2.18}$$

of all transformations $L \in GL(\mathbb{R}^4)$ commuting with η defines the Lorentz group.

Defining the distance between two elements in $\mathcal{L}$ in terms of the (maximum) distance among their corresponding matrix entries (L_{ij}), we observe that $M \mapsto ML$ and $L \mapsto L^{-1}$ are continuous transformations so that a continuous Lorentz curve connecting two points M and L can be obtained. However, because $L'\eta L = \eta$ it follows that $(\det L)^2 = 1$, or $\det L = \pm 1$, and that $L_{44}^2 \geq 1$. Therefore, either $L_{44} \geq 1$ or $L_{44} \leq -1$. The continuity argument then leads to the identification and classification of four connected classes of Lorentz symmetries:

$$L_+^\uparrow\colon \det L = +1 \text{ and } L_{44} \geq 1; \quad L_-^\uparrow\colon \det L = -1 \text{ and } L_{44} \geq 1;$$
$$L_+^\downarrow\colon \det L = +1 \text{ and } L_{44} \leq 1; \quad L_-^\downarrow\colon \det L = -1 \text{ and } L_{44} \leq 1.$$

Any two symmetries within the same class are connected, but no Lorentz transformation in one component can be connected with another in a different component. $L_+^\uparrow$ contains the identity matrix and defines a subgroup of $\mathcal{L}$, called the restricted Lorentz group, whereas $L^\uparrow = L_+^\uparrow \cup L_-^\uparrow$ defines the orthochronous Lorentz group. The orthochronous symmetries are the transformations consistent with the geometry of $\mathcal{M}$, preserving time orientation: $x_4 > y_4$ implies $(Lx)_4 > (Ly)_4$. The proper Lorentz group is defined by $L_+ = L_+^\uparrow \cup L_+^\downarrow$, whereas $L_0 = L_+^\uparrow \cup L_-^\downarrow$ defines the orthochorous Lorentz group. When the symmetries of the Lorentz group are enlarged to include spacetime translations, we obtain the so-called inhomogeneous Lorentz group, or Poincaré group.

The group of the quaternions

Given an observation $x' = (x_1, x_2, x_3) \in \mathbb{R}^3$, the matrix

$$X = \begin{bmatrix} x_3 & x_1 - ix_2 \\ x_1 + ix_2 & -x_3 \end{bmatrix}$$

is called the matrix associated with x. These matrices have several remarkable properties of their own (Cartan, 1966, p. 43), for example,

$$\det X = -||x||^2; \quad X^2 = ||x||^2 \begin{bmatrix} 1 & 0 \\ 0 & 1 \end{bmatrix}; \quad XY + YX = 2x'y,$$

for all $x, y \in \mathbb{R}^3$. In particular, if we consider the matrices

$$H_1 = \begin{bmatrix} 0 & 1 \\ 1 & 0 \end{bmatrix}, \quad H_2 = \begin{bmatrix} 0 & -i \\ i & 0 \end{bmatrix}, \quad H_3 = \begin{bmatrix} 1 & 0 \\ 0 & -1 \end{bmatrix}$$

associated with the basis vectors $e_1 = (1, 0, 0)$, $e_2 = (0, 1, 0)$, and $e_3 = (0, 0, 1)$, we observe that $H_1^2 = H_2^2 = H_3^2$,

$$H_1 H_2 + H_2 H_1 = H_1 H_3 + H_3 H_1 = H_2 H_3 + H_3 H_2 = 0,$$

and $H_1 H_2 H_3 = i H_0$, where $H_0 = \begin{bmatrix} 1 & 0 \\ 0 & 1 \end{bmatrix}$. Moreover, because $\sum_{\ell=0}^{3} a_\ell H_\ell = \begin{bmatrix} a_0 + a_3 & a_1 - i a_2 \\ a_1 + i a_2 & a_0 - a_3 \end{bmatrix}$, we conclude that there is no linear relation of the form $\sum_{\ell=0}^{3} a_\ell H_\ell = 0$ with complex coefficients unless all these coefficients are zero. That is, any complex 2×2 matrix can be expressed uniquely as the sum of a scalar matrix $a_0 H_0$ and the matrix associated with a vector. Consequently, we can identify $H_0, \ldots, H_3$ with a basis for a four-dimensional vector space $\mathcal{H}$ in which a vector multiplication has been defined. For all $f, g, h \in \mathcal{H}$ and all scalars α, we have $fg \in \mathcal{H}$, $f(g + h) = fg + fh$, $(f + g)h = fh + gh$, $a(fg) = f(ag) = (af)g$. That is, $\mathcal{H}$ constitutes an algebra. When expressed in terms of $I_0 = H_0$, $I_1 = -i H_1$, $I_2 = -i H_2$, $I_3 = -i H_3$, the algebra is known as Hamilton's algebra[3] of the quaternions. This algebra is associative but not commutative. The elements $1 \equiv H_0, \mathbf{i} \equiv i H_3, \mathbf{j} \equiv i H_2$ and $\mathbf{k} = i H_1$ satisfy the relations

$$\mathbf{ij} = -\mathbf{ji} = \mathbf{k}, \quad \mathbf{jk} = -\mathbf{kj} = \mathbf{i}, \quad \mathbf{ik} = -\mathbf{ki} = -\mathbf{j}, \quad \mathbf{i}^2 = \mathbf{j}^2 = \mathbf{k}^2 = -1,$$

and $\{\pm 1, \pm \mathbf{i}, \pm \mathbf{j}, \pm \mathbf{k}\}$ form a (noncommutative) group known as the group of the quaternions.

We observe that

$$\det \left[\sum_{\ell=0}^{3} a_\ell H_\ell \right] = \det \begin{bmatrix} a_0 + a_3 & a_1 - i a_2 \\ a_1 + i a_2 & a_0 - a_3 \end{bmatrix} = a_0^2 - a_1^2 - a_2^2 - a_3^2 = -||a||^2,$$

where in the above expression $||.||$ indicates the norm defined by the Lorentz fundamental form. Whether or not $||a||^2 = 0$ has nonzero solutions depends on the field of scalars. In particular, when $a_0 = 0$, there are infinitely many complex (isotropic vectors) solutions to $||a||^2 = 0$, whereas only $(0, 0, 0, 0)$ is a solution over the real field. For each nonzero real a_0 the real solutions to $||a||^2 = 0$ transform as the full group of rotations in three dimensions.

[3] Irish mathematician, physicist, and astronomer William Rowan Hamilton, 1805–1865.

Normal subgroups

Fix any member τ of a group G and define the mapping $i_\tau : G \to G$ by $i_\tau(\sigma) = \tau\sigma\tau^{-1}$. Then, for every $\tau \in G$, i_τ is an automorphism in G, and $\tau \mapsto i_\tau$ is a homomorphism from G to $\text{Aut}(G)$. The mapping i_τ is usually called the conjugation by τ, or the inner automorphism generated by τ. A normal or stable subgroup of G is a subgroup N of G satisfying the property $\tau N \tau^{-1} \subset N$ for all $\tau \in G$. Therefore, a normal subgroup contains the complete conjugacy classes of all its elements. For example, $2\mathbb{Z}, 3\mathbb{Z}, \ldots$ are normal subgroups of $(\mathbb{Z}, +)$. This is true for every subgroup of an Abelian group. Homomorphism kernels are normal subgroups. C_3 is a normal subgroup of the symmetric group S_3.

2.4 Group Actions

We have briefly introduced the notion of group action in a number of examples in Chapter 1, such as on pages 5, 14, and 29. We also calculated a number of group orbits such as $\{s\tau^{-1}; \tau \in S_4\}$ and $\{\sigma s; \sigma \in S_2\}$ when discussing the classification of binary sequences in length of 4, or when studying the energy levels and accessible microstates. We now give a precise definition of that important concept.

Definition 2.2 (Group action). The action of a group G on a set V is a mapping $\varphi : G \times V \to V$ satisfying $\varphi(1, s) = s$, for all s in V and $\varphi(\sigma, \varphi(\tau, s)) = \varphi(\sigma\tau, s)$, for all $s \in V$, τ, σ in G.

In short notation, we may also write $\varphi_\tau(s)$ to indicate $\varphi(\tau, s)$ so that G acts on V according to φ when φ_1 is the identity mapping in V and $\varphi_{\tau\sigma} = \varphi_\tau\varphi_\sigma$, for all τ, σ in G.

When $\tau \mapsto \varphi_\tau$ is a monomorphism we say that the corresponding action is faithful, or that G acts on V faithfully.

Regular actions

Earlier on in Chapter 1, Section 1.4 on page 5, we illustrated the reduction of data indexed by a group acting by multiplication on itself. More precisely, when $V = G$ and G acts on itself according to $\varphi_\tau(\sigma) = \tau\sigma$, we say that φ is a (right) regular action. Similar definition applies to left regular actions.

Orbits, stabilizers, and transitive actions

The orbit $\mathcal{O}_s$ of an element $s \in V$ generated by G acting on V according to φ is the set $\mathcal{O}_s = \{\varphi_\tau(s); \tau \in G\}$. We indicate by fix $(\tau) = \{s \in V; \varphi_\tau(s) = s\}$ the set of elements in V that remain fixed by τ under the action φ. The set $G_s = \{\tau \in G; \varphi_\tau(s) = s\}$ of elements $\tau \in G$ fixing the point $s \in V$ is called the stabilizer of s by G under φ. It is then easy to check that $|G| = |\mathcal{O}_s||G_s|$.

G_s is a subgroup of G, often called the isotropy group of s in G under φ.

If $\theta : V \to V$ and $\theta\varphi_\tau = \varphi_\tau\theta$ for all τ in G, we say that θ and φ commute. In this case, as the reader may want to verify, s and $\theta(s)$ have the same group stabilizers, for all $s \in V$. In particular, then, commuting actions share the same group stabilizers.

Note that σ stabilizes $\varphi_\tau(s)$ if and only if $\varphi_\sigma\varphi_\tau(s) = \varphi_\tau(s)$, or equivalently, $\varphi_{\tau^{-1}\sigma\tau}(s) = s$, or, $\tau^{-1}\sigma\tau \in G_s$, or if and only if $\sigma \in \tau G_s\tau^{-1}$. Therefore, $G_{\varphi_\tau(s)} = \tau G_s \tau^{-1}$.

When the orbit $\mathcal{O}_s$ of an element $s \in V$ generated by G under the action φ coincides with V we say that the action φ is transitive, or that G acts transitively on V.

If $f \in \mathcal{O}_s$ then, for some $\sigma \in G$, $\varphi_\tau(f) = \varphi_\tau\varphi_\sigma(s) = \varphi_{\tau\sigma}(s) \in \mathcal{O}_s$, so that the restriction $\varphi_\tau|\mathcal{O}$ of φ_τ to $\mathcal{O}_s$ is a (transitive) group action on $\mathcal{O}_s$.

If $s \neq f$ then either $\mathcal{O}_s \cap \mathcal{O}_f = \emptyset$ or $\mathcal{O}_s = \mathcal{O}_f$. In fact, given $s' \in \mathcal{O}_s$ and an element h in $\mathcal{O}_s \cap \mathcal{O}_f$ then s' is in $\mathcal{O}_f$. Similarly, starting with a point $f' \in \mathcal{O}_f$. Then $\mathcal{O}_s = \mathcal{O}_f$. Moreover, because $s \in \mathcal{O}_s$ for every $s \in V$, we have the following:

Proposition 2.3. *Every group action φ on V decomposes V as the disjoint union of the resulting orbits, in each one of which the restriction of φ acts transitively.*

Applying Proposition 2.3 to the regular action, we conclude that G acts transitively on its cosets. This fact was illustrated earlier with $G = S_3$ on page 38 and with $G = D_4$ on page 42.

Group actions and permutations

Given an action φ of G on V then $\varphi_\tau \in S_V$ for all τ in V. In fact, because $\varphi_\tau(s) = \varphi_\tau(f)$ implies $s = \varphi(\tau^{-1}, \varphi(\tau, s)) = \varphi(\tau^{-1}, \varphi(\tau, f)) = f$ then φ_τ is one-to-one, and since φ_τ takes $\varphi_{\tau^{-1}}(f)$ to f for all $f \in V$, φ_τ is a bijective mapping. Also, by definition, $\varphi_{\tau\sigma} = \varphi_\tau\varphi_\sigma$. Therefore,

Proposition 2.4. $\tau \mapsto \varphi_\tau$ *is a group homomorphism from G to S_V.*

Conversely, it is clear that given a homomorphism η from G to S_V then $\varphi_\tau(s) = \eta(\tau)(s)$ defines a group action of G on V.

Applying Proposition 2.4 to an orbit $\mathcal{O}$ of V, it then follows that $\tau \mapsto \varphi_\tau|\mathcal{O}$ is a group homomorphism from G to $S_{\mathcal{O}}$. Applying it to the regular action, we see that $\tau \mapsto \varphi_\tau$ is a group homomorphism from G to S_G. In view of Propositions 2.3

and 2.4, we will focus our attention, primarily, on the identification of transitive actions.

Contravariant actions

When in Definition 2.2 we have $\varphi_\sigma\varphi_\tau = \varphi_{\tau\sigma}$, we say that the φ is a contravariant action. To illustrate, suppose that G acts on V according to φ and let $\mathcal{F} = \mathcal{F}(V)$ indicate the vector space of scalar-valued functions, x, defined on V. Then, $\theta_\tau(x) = x\varphi_\tau$ is a contravariant action of G on $\mathcal{F}$. In fact, $(\theta_\tau\theta_\sigma)(x) = \theta_\tau(\theta_\sigma(x)) = \theta_\tau(x\varphi_\sigma) = x\varphi_\sigma\varphi_\tau = x\varphi_{\sigma\tau} = \theta_{\sigma\tau}(x)$.

2.5 Actions on Mappings

Binary sequences in length of 2

Consider the set $V = \{uu, yy, uy, yu\}$ of binary sequences in length of 2, so that each sequence is a mapping s from $\{1, 2\}$ into $\{u, y\}$. The reader may verify that

$$\varphi_1 : \{1, 2\} \xrightarrow{\tau^{-1}} \{1, 2\} \xrightarrow{s} \{u, y\}, \quad s \in V, \ \tau \in S_2 \simeq \{1, (12)\},$$

and

$$\varphi_2 : \{1, 2\} \xrightarrow{s} \{u, y\} \xrightarrow{\sigma} \{u, y\}, \quad s \in V, \ \sigma \in S_2 \simeq \{1, (uy)\},$$

are actions of S_2 on V. Action φ_1 classifies the sequences by shuffling the positions $\{1, 2\}$ whereas φ_2 gives a classification by shuffling the symbols $\{u, y\}$. The evaluation

φ_1 :

τ	uu	yy	uy	yu
1	uu	yy	uy	yu
(12)	uu	yy	yu	uy

, φ_2 :

σ	uu	yy	uy	yu
1	uu	yy	uy	yu
(uy)	yy	uu	yu	uy

of these actions shows that φ_1 decomposes V as the disjoint union of three orbits $\{uu\}$, $\{yy\}$, and $\{uy, yu\}$, in each of which (the restriction of) φ_1 acts transitively, whereas, similarly, φ_2 decomposes V as the disjoint union of the two orbits $\{uu, yy\}$ and $\{uy, yu\}$. Table (2.19) shows the orbits of S_4 acting on the binary sequences in length of 4 according to $s\tau^{-1}$ (position symmetry), the resulting isotropy groups and the sequences fixed by each permutation.

$s(1)$	y	u	y	u	u	u	y	y	y	u	u	u	y	y	y	u	
$s(2)$	y	u	u	y	u	u	y	u	u	y	y	u	y	y	u	y	
$s(3)$	y	u	u	u	y	u	u	y	u	y	u	y	y	u	y	y	
$s(4)$	y	u	u	u	u	y	u	u	y	u	y	y	u	y	y	y	
label $\rightarrow$	1	16	15	14	12	8	13	11	7	10	6	4	9	5	3	2	$\|fix\|$
1	1	16	15	14	12	8	13	11	7	10	6	4	9	5	3	2	16
(34)	1	16	15	14	8	12	13	7	11	6	10	4	5	9	3	2	8
(23)	1	16	15	12	14	8	11	13	7	10	4	6	9	3	5	2	8
(24)	1	16	15	8	12	14	7	11	13	4	6	10	3	5	9	2	8
(12)	1	16	14	15	12	8	13	10	6	11	7	4	9	5	2	3	8
(13)	1	16	12	14	15	8	10	11	4	13	6	7	9	2	3	5	8
(14)	1	16	8	14	12	15	6	4	7	10	13	11	2	5	3	9	8
(234)	1	16	15	12	8	14	11	7	13	4	10	6	3	9	5	2	4
(243)	1	16	15	8	14	12	7	13	11	6	4	10	5	3	9	2	4
(123)	1	16	14	12	15	8	10	13	6	11	4	7	9	2	5	3	4
(124)	1	16	14	8	12	15	6	10	13	4	7	11	2	5	9	3	4
(132)	1	16	12	15	14	8	11	10	4	13	7	6	9	3	2	5	4
(134)	1	16	12	14	8	15	10	4	11	6	13	7	2	9	3	5	4
(142)	1	16	8	15	12	14	7	4	6	11	13	10	3	5	2	9	4
(143)	1	16	8	14	15	12	6	7	4	13	10	11	5	2	3	9	4
(12)(34)	1	16	14	15	8	12	13	6	10	7	11	4	5	9	2	3	4
(13)(24)	1	16	12	8	15	14	4	11	10	7	6	13	3	2	9	5	4
(14)(23)	1	16	8	12	14	15	4	6	7	10	11	13	2	3	5	9	4
(1234)	1	16	14	12	8	15	10	6	13	4	11	7	2	9	5	3	2
(1243)	1	16	14	8	15	12	6	13	10	7	4	11	5	2	9	3	2
(1324)	1	16	12	8	14	15	4	10	11	6	7	13	2	3	9	5	2
(1342)	1	16	12	15	8	14	11	4	10	7	13	6	3	9	2	5	2
(1432)	1	16	8	15	14	12	7	6	4	13	11	10	5	3	2	9	2
(1423)	1	16	8	12	15	14	4	7	6	11	10	13	3	2	5	9	2
$\|G_s\|$	24	24	6	6	6	6	4	4	4	4	4	4	6	6	6	6	140

(2.19)

Cyclic orbits for binary sequences in length of 4

Table (2.20) shows the action $s\tau^{-1}$ of $C_4 = \{1, (1234), (13)(24), (1432)\}$ on the mapping space V, abstracted from Table (2.19).

$s(1)$	y	u	y	u	u	u	y	y	y	u	u	u	y	y	y	u
$s(2)$	y	u	u	y	u	u	y	u	u	y	y	u	y	y	u	y
$s(3)$	y	u	u	u	y	u	u	y	u	y	u	y	y	u	y	y
$s(4)$	y	u	u	u	u	y	u	u	y	u	y	y	u	y	y	y
$C_4\backslash s$	1	16	15	14	12	8	13	11	7	10	6	4	9	5	3	2
1	1	16	15	14	12	8	13	11	7	10	6	4	9	5	3	2
(13)(24)	1	16	12	8	15	14	4	11	10	7	6	13	3	2	9	5
(1234)	1	16	14	12	8	15	10	6	13	4	11	7	2	9	5	3
(1432)	1	16	8	15	14	12	7	6	4	13	11	10	5	3	2	9

(2.20)

The resulting orbits (indicating the sequences by their labels) are

$$\begin{gathered}\mathcal{O}_0 = \{1\}, \quad \mathcal{O}_1 = \{9, 5, 3, 2\}, \quad \mathcal{O}_{21} = \{13, 7, 10, 4\}, \\ \mathcal{O}_{22} = \{11, 6\}, \quad \mathcal{O}_3 = \{15, 14, 12, 8\}, \quad \mathcal{O}_4 = \{16\}.\end{gathered}$$

We note that C_4 splits the original orbit $\mathcal{O}_2$ under S_4 into two new orbits, $\mathcal{O}_{21}$ and $\mathcal{O}_{22}$, so that $\mathcal{O}_{21} \cup \mathcal{O}_{22} = \mathcal{O}_2$. Similarly, the action

$G \backslash s$	1	16	15	14	12	8	13	11	7	10	6	4	9	5	3	2
1	1	16	15	14	12	8	13	11	7	10	6	4	9	5	3	2
(13)(24)	1	16	12	8	15	14	4	11	10	7	6	13	3	2	9	5

, (2.21)

$s\tau^{-1}$ of $G = \{1, (13)(24)\}$ on V gives the orbits

$$\begin{gathered}\mathcal{O}_0 = \{1\}, \quad \mathcal{O}_{11} = \{9, 3\}, \quad \mathcal{O}_{12} = \{5, 2\}, \\ \mathcal{O}_{211} = \{13, 4\}, \quad \mathcal{O}_{212} = \{7, 10\}, \quad \mathcal{O}_{221} = \{11\}, \quad \mathcal{O}_{222} = \{6\}, \\ \mathcal{O}_{31} = \{14, 8\}, \quad \mathcal{O}_{32} = \{15, 12\}, \quad \mathcal{O}_4 = \{16\},\end{gathered}$$

and further splits the original orbits.

Dihedral orbits for binary sequences in length of 4

Table (2.22) shows the action $s\tau^{-1}$ of the dihedral group D_4 on the binary sequences in length of 4, and the resulting orbits.

$D_4 \backslash s$	1	16	15	14	12	8	13	11	7	10	6	4	9	5	3	2
1	1	16	15	14	12	8	13	11	7	10	6	4	9	5	3	2
(24)	1	16	15	8	12	14	7	11	13	4	6	10	3	5	9	2
(13)	1	16	12	14	15	8	10	11	4	13	6	7	9	2	3	5
(12)(34)	1	16	14	15	8	12	13	6	10	7	11	4	5	9	2	3
(13)(24)	1	16	12	8	15	14	4	11	10	7	6	13	3	2	9	5
(14)(23)	1	16	8	12	14	15	4	6	7	10	11	13	2	3	5	9
(1234)	1	16	14	12	8	15	10	6	13	4	11	7	2	9	5	3
(1432)	1	16	8	15	14	12	7	6	4	13	11	10	5	3	2	9

. (2.22)

The action leads to the same set of position symmetry orbits generated by C_4. This structure is then said to be *achiral*. The loss of handedness can be seen by translating the Diagrams (2.10) and (2.11) on page 41 into their purine-pyrimidine equivalents, so that the resulting sequences define the orbit $\mathcal{O}_{21}$ indicated above. From one side we see

$$uuyy: \begin{array}{ccc} u & \longrightarrow & u \\ \uparrow & & \downarrow \\ y & \longleftarrow & y \end{array} \quad yuuy: \begin{array}{ccc} y & \longrightarrow & u \\ \uparrow & & \downarrow \\ y & \longleftarrow & u \end{array} \quad yyuu: \begin{array}{ccc} y & \longrightarrow & y \\ \uparrow & & \downarrow \\ u & \longleftarrow & u \end{array} \quad uyyu: \begin{array}{ccc} u & \longrightarrow & y \\ \uparrow & & \downarrow \\ u & \longleftarrow & y \end{array},$$

whereas from the other, without distinction,

$$
yuuy:\begin{array}{ccc} y & \longleftarrow & u \\ \downarrow & & \uparrow \\ y & \longrightarrow & u \end{array}
\quad
yyuu:\begin{array}{ccc} y & \longleftarrow & y \\ \downarrow & & \uparrow \\ u & \longrightarrow & u \end{array}
\quad
uyyu:\begin{array}{ccc} u & \longleftarrow & y \\ \downarrow & & \uparrow \\ u & \longrightarrow & y \end{array}
\quad
uuyy:\begin{array}{ccc} u & \longleftarrow & u \\ \downarrow & & \uparrow \\ y & \longrightarrow & y \end{array}.
$$

Dihedral orbits for binary sequences in length of 3

Table (2.23) shows the orbits of $D_3 \simeq S_3$ acting on the binary sequences in length of 3 according to $s\tau^{-1}$.

τ	uuu	yyy	uyy	yuy	yyu	uuy	uyu	yuu
	1	2	3	4	5	6	7	8
1	1	2	3	4	5	6	7	8
(123)	1	2	4	5	3	8	6	7
(132)	1	2	5	3	4	7	8	6
(23)	1	2	3	5	4	7	6	8
(13)	1	2	5	4	3	8	7	6
(12)	1	2	4	3	5	6	8	7

(2.23)

The reader may want to verify that, in each one of the two nontrivial orbits, this action is equivalent to the action τi of S_3 on $i \in \{1, 2, 3\}$.

Ternary sequences in length of 4

Following with notation introduced earlier on in Section 1.8 on page 15 of Chapter 1, consider the mapping space $V = C^L$ representing the $4^3 = 81$ compositions

$$s:\ \{1, 2, 3, 4\} \mapsto \{\text{red } (\circ), \text{ blue } (\bullet), \text{ green } (\diamond)\}$$

of an urn with four numbered and colored marbles. Two urn compositions are defined as equivalent when they differ only by renumbering of the marbles. That is, when they share the same orbit under the action $s\tau^{-1}$, for $\tau \in S_4$ and $s \in V$. The elementary orbits, where S_4 acts transitively, are grouped into $m = 4$ classes $\lambda_1 = 40^2$, $\lambda_2 = 310$, $\lambda_3 = 2^20$ and $\lambda_4 = 21^2$, and

$$|V| = c^\ell = \sum_\lambda \frac{\ell!}{(a_1!)^{m_1}(a_2!)^{m_2}\ldots(a_k!)^{m_k}} \frac{c!}{m_1!m_2!\ldots m_k!} = \sum_\lambda \Omega_\lambda Q_\lambda$$

decomposes the Maxwell-Boltzmann count c^ℓ into the sum of the products $\Omega_\lambda Q_\lambda$ of the volumes Ω_λ of the elementary orbits and their multiplicities Q_λ. In the above decomposition, λ varies over the m integer partitions $(a_1^{m_1}, \ldots, a_k^{m_k})$ satisfying

$m_1a_1 + \cdots + m_ka_k = \ell$ and $m_1 + \cdots + m_k = c$. Moreover,

$$\sum_\lambda Q_\lambda = \binom{c+\ell-1}{\ell}$$

is a decomposition of the Bose-Einstein count $\binom{c+\ell-1}{\ell}$ into the sum of the number Q_λ of quantal states associated with the partition λ. Table (2.24) summarizes the correspondence among frames, orbits, volumes of elementary orbits, multiplicities, and urn compositions.

λ	urn composition	Ω_λ	Q_λ	$\Omega_\lambda Q_\lambda$
400	$\{\circ\ \circ\ \circ\ \circ\}$	1	3	3
310	$\{\circ\ \circ\ \circ\ \bullet\}$	4	6	24
220	$\{\circ\ \circ\ \bullet\ \bullet\}$	6	3	18
211	$\{\circ\ \circ\ \bullet\ \diamond\}$	12	3	36

(2.24)

We observe that $|V| = 3^4 = 81 = \sum_\lambda v(\lambda)$, and that $\sum_\lambda Q_\lambda = \binom{c+\ell-1}{\ell} = \binom{6}{4} = 15$.

Applying Propositions 2.3 and 2.4, we observe that the study of data indexed by ternary sequences in length of 4, when reduced by $s\tau^{-1}$, leads to Q_λ equivalent studies of transitive actions in each one of the orbits of type λ.

Translations

The actions $\sigma \mapsto \sigma\tau^{-1}$ and $\sigma \mapsto \tau\sigma$ are, respectively, the left and right translations. The argument supporting Proposition 2.4, when applied to translations, implies that $\tau \in \ker(\tau \mapsto \tau^*)$ if and only if $\tau = 1$, so that $\tau \mapsto \tau^*$ is a monomorphism from G to S_G. This leads to Cayley's Theorem:

Theorem 2.1 (Cayley, 1878). *Every group G is isomorphic to a subgroup of S_G. If G is finite with ℓ elements, then G is isomorphic to a subgroup of S_ℓ.*

2.6 Genotypic Classification

In genetics, many simple genotypic classifications follow from permutations in the genetic material. With two genes $\{A, a\}$ and two loci, the structure of interest is $V = \{AA, aa, Aa, aA\}$, with four chromosomes in it. The action

S_2	aa	AA	aA	Aa
1	aa	AA	aA	Aa
(12)	aa	AA	Aa	aA

of S_2 on the location of the two genes shows the presence of three

$$\underbrace{\{aa\},\ \{AA\}}_{\text{homozygous}},\ \underbrace{\{aA, Aa\}}_{\text{heterozygous}}$$

locus substitution invariant subsets, or orbits. The resulting classification, shown in Table (2.25), describes the genetic homogeneity in V and is characterized by number of integer partitions of 2.

partition	genotype	orbit size	number of orbits	partition size
20	homozygous	1	2	2
11	heterozygous	2	1	2

(2.25)

With 4 genes $\{A, A', a, a'\}$ and four loci, the corresponding mapping structure V has 256 (chromosomes) points and the genotypic variation is much larger. The genotypes are obtained by the action of S_4 on the loci. The resulting classification corresponds to the 5 integer partitions of 4, namely,

$$\text{tetraploid genotypes} = \begin{cases} \text{homogenic,} & 4000, \\ \text{digenic, simplex or triplex,} & 3100, \\ \text{digenic, duplex,} & 2200, \\ \text{trigenic,} & 2110, \\ \text{tetragenic,} & 1111. \end{cases} \tag{2.26}$$

Table (2.27) shows the resulting number of distinct genotypes. There are, for $c = 4$ genes and $\ell = 4$ loci,

$$\binom{c+\ell-1}{\ell} = 35$$

genotypes, corresponding to the total number of location-symmetry orbits.

partition	orbit volume	number of orbits	total
4000	1	4	4
3100	4	12	48
2200	6	6	36
2110	12	12	144
1111	24	1	24
		35	$256 = \lvert V \rvert$

(2.27)

For example, partition 3100 yields 12 genotypes, each one defined by a locus substitution orbit with 4 equivalent chromosomes, for a total of 48 chromosomes associated with that partition. Note that there are exactly six permutations in S_4 leaving each one of the 48 chromosome of the type V_{3100} invariant by location substitution, and that the number (12) or orbits in that partition is the average $(6 \times 48)/4!$ number of fixed points.

Following Propositions 2.3 and 2.4 on page 48, the study of data indexed by these 256 chromosomes, when reduced by loci symmetry, is equivalent to the study of the four (nontrivial) different group homomorphisms from S_4 to $S_{\mathcal{O}_\lambda}$, for $\lambda = 3100$, 2200, 2110, and 1111.

2.7 Actions on Binary Trees and Branches

The set $\mathcal{T}$ of all m-generation binary trees with nodes indexed by the elements in a finite group G can be obtained by defining $L_j = \{1, \ldots, 2^j\}$, $V_j = G^{L_j}$ so that $\mathcal{T} = V_0 \times V_1 \times \cdots \times V_m$ is the structure of interest. The trees, as labels for data, may be factored by the multiplicative action of G on their nodes.

To illustrate, consider the case in which $G = \{1, v, h, o\}$, as in Chapter 1, let x indicate a probability law in G, and define $P_t(u) = x(ut)$, for $t, u \in G$. Then

$$\sum_{u \in G} P_t(u) = \sum_{u \in G} x(ut) = \sum_{u \in G} x(u) = 1,$$

and consequently, P_t is a transition probability for each $t \in G$, and the matrix X with entries $X_{tu} = x(ut)$ determines a transition probability matrix for each fixed probability law x in G. Factorization of branching probabilities can then be obtained from the action of G on the tree structure $\mathcal{T}$. For example, the branch $o1h1$ has probability $P(o1h1|x) = x(1o)x(h1)x(1h) = x(o)x(h)x(h) = x(o)x^2(h)$, and is mapped into the branch $(ho)(h1)(hh)(h1) = vh1h$ under the product with $h \in G$. In turn,

$$P(vh1h|x) = x(hv)x(1h)x(h1) = x(o)x(h)x(h) = x(o)x^2(h)$$

so that these two branches are classified together. A similar argument, applied to phylogenetic trees, is found in the work of Evans and Speed (1993).

Data indexed by short branches or transitions

Table (2.28) summarizes the 16 different transitions among the elements of $G = \{1, v, h, o\}$ and their respective frequencies derived from three Sloan charts presented in Ferris III et al. (1993) and discussed earlier on page 3. Each frequency distribution was obtained as the sum of the transition frequencies over the 14 five-letter lines in each chart. Since each line has five letters, there are at most four left to right adjacent transitions in each line and hence at most 56 transitions in each chart.

transitions	1	1	1	1	v	v	v	v	h	h	h	h	o	o	o	o
	1	v	h	o	1	v	h	o	1	v	h	o	1	v	h	o
chart 1	56	17	22	28	18	4	4	9	22	8	7	13	27	7	7	14
chart 2	56	16	21	30	16	4	3	10	23	7	7	13	30	11	13	16
chart 3	56	18	23	28	17	1	2	8	22	6	5	12	27	10	10	13

(2.28)

Table (2.29) shows the componentwise action $(\eta, (\tau, \sigma)) \to (\eta\tau, \eta\sigma)$ of G on $G \times G$, the set of labels for the transitions among the symmetries.

$\eta = 1$	$\tau = 1$	1	1	1	v	v	v	v	h	h	h	h	o	o	o	o
	$\sigma = 1$	v	h	o	1	v	h	o	1	v	h	o	1	v	h	o
v	v	v	v	v	1	1	1	1	o	o	o	o	h	h	h	h
	v	1	o	h	v	1	o	h	v	1	o	h	v	1	o	h
h	h	h	h	h	o	o	o	o	1	1	1	1	v	v	v	v
	h	o	1	v	h	o	1	v	h	o	1	v	h	o	1	v
o	o	o	o	o	h	h	h	h	v	v	v	v	1	1	1	1
	o	h	v	1	o	h	v	1	o	h	v	1	o	h	v	1

(2.29)

Although G acts transitively on itself, its action on $G \times G$ identifies four orbits, namely,

$$\mathcal{O}_{11} = \{11, vv, hh, oo\}, \quad \mathcal{O}_{1v} = \{1v, v1, ho, oh\},$$
$$\mathcal{O}_{1h} = \{1h, h1, vo, ov\}, \quad \mathcal{O}_{1o} = \{1o, o1, vh, hv\}.$$

Following Propositions 2.3 and 2.4 on page 48, the study of transition data is reduced to the study of the distinct group homomorphisms from G to $S_{\mathcal{O}}$.

2.8 Counting Orbits in Linkage Analysis

The segregation products or inheritance vectors for a sibship of size 2 can be represented by the set V of all mappings $s : C = \{1, 2\} \to L_p \times L_m \equiv \{1, 2\} \times \{3, 4\}$, where each mapping represents a possible configuration of chromosome pairs for the two sibships. The paternal chromosomes are labeled here by $L_p = \{1, 2\}$ and the maternal chromosomes by $L_m = \{3, 4\}$. If $s(j) = (p(j), m(j)) \in L_p \times L_m$, $j = 1, 2$, indicates the components of each mapping s, then the *identity by descent* (IBD) of the inheritance vector s is given by $x(s) = \delta_{p(1)p(2)} + \delta_{m(1)m(2)}$, where δ denotes the Kronecker delta. Table (2.30) illustrates a number of inheritance vectors and their corresponding IBD values evaluated as $x(s)$.

$s(1)$	$s(2)$	s	$x(s) = \text{IBD}$
$a = (1, 3)$	$a = (1, 3)$	aa	2
$c = (2, 3)$	$a = (1, 3)$	ca	1
$c = (2, 3)$	$b = (1, 4)$	cb	0
$c = (2, 3)$	$d = (2, 4)$	cd	1

(2.30)

With the additional notation introduced in Table (2.30), we identify the space of inheritance vectors as the structure $V = \{a, b, c, d\}^{\{1,2\}}$ of all mappings defined in $\{1, 2\}$ with values in $\{a, b, c, d\}$. The space V has 16 labels or indices for possible IBD evaluations. Next, we consider the symmetries of interest, namely the permutations $\sigma_1 = (ac)(bd)$ for paternal chromosome symmetry; $\sigma_2 = (ab)(cd)$ for maternal chromosome symmetry, and $\sigma_3 = (bc)$, or equivalently $\sigma_3 = (ad)$, for parental origin symmetry. Together these permutations generate the dihedral group of the square

$$\begin{array}{ccc} a & = & b \\ \| & & \| \\ d & = & c \end{array}$$

of interest, with vertices the chromosome pairs $a = (1, 3)$, $b = (1, 4)$, $c = (2, 3)$, and $d = (2, 4)$. It acts on V according to the rule $(\sigma, s) \to \sigma s$, giving the orbits shown in Table (2.31).

$D_4 \backslash s$	aa	bb	cc	dd	ac	ca	ab	ba	cd	dc	bd	db	ad	da	bc	cb
1	1	2	3	4	5	6	7	8	9	10	11	12	13	14	15	16
$(abdc)$	2	4	1	3	8	7	11	12	5	6	10	9	15	16	14	13
$(ad)(bc)$	4	3	2	1	12	11	10	9	8	7	6	5	14	13	16	15
$(acdb)$	3	1	4	2	9	10	6	5	12	11	7	8	16	15	13	14
$(ac)(bd)$	3	4	1	2	6	5	9	10	7	8	12	11	16	15	14	13
$(ab)(cd)$	2	1	4	3	11	12	8	7	10	9	5	6	15	16	13	14
(ad)	4	2	3	1	10	9	12	11	6	5	8	7	14	13	15	16
(cb)	1	3	2	4	7	8	5	6	11	12	9	10	13	14	16	15
IBD	2	2	2	2	1	1	1	1	1	1	1	1	0	0	0	0

. (2.31)

In addition to the action of D_4, the sibship symmetry is generated by S_2, which then acts on the left of V according to the rule $(\tau, s) \to s\tau^{-1}$, $\tau \in S_2$. The orbits obtained from studying the joint action

$$\varphi((\tau, \sigma), s) = \sigma s \tau^{-1} \tag{2.32}$$

of $S_2 \times D_4$ coincide precisely with the classes of similar IBD-valued inheritance vectors obtained by Dudoit and Speed (1999).

2.9 Burnside's Lemma

The following result shows that the number of orbits generated by a group action is simply the average number of points fixed by the action. The reader may first refer to Table (2.19) on page 50, which summarizes the action $s\tau^{-1}$ of S_4 on the space V of all binary sequences in length of 4 and identifies, for each $\tau \in G$ the number $|\text{fix}(\tau)|$ of points in V fixed by the action on V, and the number $|G_s|$ of points in G that leave the element $s \in V$ fixed.

Lemma 2.1 (Burnside[4]). *If a finite group G acts on V, then*

$$\textit{Number of orbits in } V = \frac{1}{|G|} \sum_{\tau \in G} |\textit{fix}\ (\tau)|.$$

Proof. Indicate by φ the action of G on V and let $A = \{(\tau, s) \in G \times V; \varphi_\tau s = s\}$ so that $|A|$ evaluates as

$$|A| = \sum_{\tau} |\{s; \varphi_\tau s = s\}| = \sum_{\tau \in G} |\text{fix}\ (\tau)|.$$

Similarly, if m is the number of orbits in V and $s_1, \ldots, s_m$ are orbit representatives, we have

$$|A| = \sum_{s} |\{\tau; \varphi_s = s\}| = \sum_{s \in V} |G_s| = \sum_{i=1}^{m} |\mathcal{O}_i||G_{s_i}| = \sum_{i=1}^{m} |\mathcal{O}_i| \frac{|G|}{|\mathcal{O}_i|} = m \times |G|.$$

The result follows from equating the two evaluations of $|A|$. ■

Binary sequences in length of 4

To illustrate, consider the action $s\tau^{-1}$ on the binary sequences in length of 4, shown in Table (2.19). It then follows that

$$\text{Number of orbits of } V = \frac{1}{|\ G\ |} \sum_{G} |\ \text{fix}(\sigma)\ |= \frac{120}{24} = 5.$$

As introduced in Chapter 1, each orbit can be characterized by the number of symbols (u) in the sequences. Also note that $|G|/|G_s|$ is the number of elements in the orbit of which s is a representative.

[4] William Burnside was born July 2, 1852, in London, England and died August 21, 1927, in West Wickham, London, England. The proof of the lemma, however, is due to Frobenius in 1887, e.g. Rotman (1995, p. 58).

Further Reading

In genetics, the interplay among symmetry, classification, and the statistical analysis of experimental data has a long history, as evident from the early work of Fisher (1930) on natural selection. The laws of inheritance obtained by genetic studies, points Fisher, are the rules whereby, given the constitution of an organism, the kinds of gametes it can produce, and their relative frequencies, can be predicted (Fisher, 1947). Fisher was born in London, on February 17, 1890. His studies of errors in astronomical calculations, together with his interests in genetics and natural selection, led to his involvement in statistics. He died on July 29, 1962.

Exercises

Exercise 2.1. Following Section 1.8 on page 15, consider the space V of binary sequences in length of ℓ and let $P_k(s) = 1/\binom{\ell}{k}$ if $s \in \mathcal{O}_k$ and $P_k(s) = 0$ otherwise. Verify that

(1) $P = \sum_{k=0}^{\ell} \gamma_k P_k$, where $\sum_k \gamma_k = 1$, $\gamma_k > 0$, is a probability law in V;
(2) $P(s\tau^{-1}) = P(s)$ for all $\tau \in S_\ell$ and $s \in V$;
(3) Conversely, all probability laws in V satisfying property (2) are convex linear combinations of the elementary (within-orbit) uniform probability laws P_k.

Exercise 2.2. Following Exercise 2.1 with $\ell = 4$, show that there are only three types of position-symmetry orbits, $\mathcal{O}_\lambda$, namely $\mathcal{O}_{40}$, $\mathcal{O}_{31}$, and $\mathcal{O}_{22}$, corresponding to the integer partitions of four in length of 2. Consequently, verify that the elementary probability laws take values $1/16$, $1/4$, and $3/8$ in $\mathcal{O}_{40}$, $\mathcal{O}_{31}$, $\mathcal{O}_{22}$, respectively. Moreover, there are $\binom{2}{1,1} = 2$ orbits of type $\lambda = 40$ or $\lambda = 31$, and $\binom{2}{2,0} = 1$ orbit of type $\lambda = 22$.

Exercise 2.3. Show that the matrices Diag$(\pm 1, \pm 1)$ form a multiplicative group isomorphic with K_4. Study the groups Diag$(\pm 1, \ldots, \pm 1)$.

Exercise 2.4. Consider the action σs of S_2 on m copies of $\{uy, yu\}$ and show that it gives $2^m - 1$ orbits.

Exercise 2.5. Let $\mathcal{S}_2 = \{\emptyset, \{1\}, \{2\}, \{1, 2\}\}$ indicate the set of all subsets of $\{1, 2\}$, and define a multiplication $(*)$ in $\mathcal{S}_2$ by inclusion-exclusion so that, for example, $\{1\} * \{2\} = \{1, 2\}$ and $\{1\} * \{1, 2\} = \{2\}$. Show that $\mathcal{S}_2$ is isomorphic with K_4. The groups $(\mathcal{S}_n, *)$ appear in Fisher (1942).

Exercise 2.6. In analogy with the construction of the dihedral groups earlier on in this chapter, given two groups N and H and a homomorphism α from H to Aut(N), define for (τ, σ) and (τ_1, σ_1) in $N \times H$, the operation $(\tau, \sigma) \times_\alpha (\tau_1, \sigma_1) =$

$(\tau\alpha(\sigma)(\tau_1), \sigma\sigma_1)$. Show that $G = N \times H$, together with $\times_\alpha$, is a group, called the semidirect product of N and H under α, in which $(1_N, 1_H)$ is the identity in G and $(\alpha(\sigma^{-1})(\tau^{-1}), \sigma^{-1})$ is the inverse of (τ, σ) in G.

Exercise 2.7. Show that the multiplication table for $C_3 \times C_2$ is

$\times$	1	2	3	4	5	6
$1 = (1, 1)$	1	2	3	4	5	6
$2 = (\tau, 1)$	2	3	1	5	6	4
$3 = (\tau^2, 1)$	3	1	2	6	4	5
$4 = (1, \sigma)$	4	5	6	1	2	3
$5 = (\tau, \sigma)$	5	6	4	2	3	1
$6 = (\tau^2, \sigma)$	6	4	5	3	1	2

, (2.33)

where the elements of $C_3 \times C_2$ are indicated according to the first column of the table. Comparing the second and third rows of the table, we observe that they define the permutation (123)(456). Similarly, following with the remaining rows, we obtain the permutations $(\tau^2, 1) \simeq$ (132)(465), $(1, \sigma) \simeq$ (14)(25)(36), $(\tau, \sigma) \simeq$ (153426), and $(\tau^2, \sigma) \simeq$ (162435), which are isomorphic with those of C_6. We have then, $C_3 \times C_2 \simeq C_6$. In general, $C_m \times C_n \simeq C_{mn}$, provided that m and n are relative primes.

Exercise 2.8. Let $\alpha_t(\tau) = \tau$ if $t = 1$ and $\alpha_t(\tau) = \tau^{-1}$ if $t = (12)$. Show that the multiplication table for $C_3 \times_\alpha C_2$ is given by

$\times_\alpha$	1	2	3	4	5	6
$1 = (1, 1)$	1	2	3	4	5	6
$2 = (\tau, 1)$	2	3	1	5	6	4
$3 = (\tau^2, 1)$	3	1	2	6	4	5
$4 = (1, \sigma)$	4	6	5	1	3	2
$5 = (\tau, \sigma)$	5	4	6	2	1	3
$6 = (\tau^2, \sigma)$	6	5	4	3	2	1

.

Show that, relative to the first row in the above matrix,

$$(\tau, 1) \simeq (123)(456) \equiv (123), \quad (\tau^2, 1) \simeq (132)(465) \equiv (132),$$
$$(1, \sigma) \simeq (14)(26)(35) \equiv (12), \quad (\tau, \sigma) \simeq (15)(24)(36) \equiv (13),$$
$$(\tau^2, \sigma) \simeq (16)(25)(34) \equiv (23).$$

The subsets {1, 6}, {2, 4}, and {3, 5} constitute what is known as an imprimitive system.

Exercise 2.9. Show that $\left\{\begin{bmatrix} x & y \\ 0 & 1 \end{bmatrix}, \ x > 0\right\}$ is a group under matrix multiplication. It is called the proper affine group in the line.

Exercise 2.10. Show that the proper affine group in the line is isomorphic with the group of transformations $t \to xt + y$, $x > 0$, thus justifying its name. More specifically, show that if $F_i = \left\{\begin{bmatrix} x_i & y_i \\ 0 & 1 \end{bmatrix}, \ x_i > 0\right\}$, $f_i(t) = x_i t + y_i$, and $\xi(f_i) = F_i$, $i = 1, 2$, then $f_1 f_2(t) = f_1(f_2(t)) = x_1 x_2 t + x_1 y_2 + y_1$ so that $\xi(f_1 f_2) = \xi(f_1)\xi(f_2)$. Verify also that ξ maps the identity transformation $t \to t$ into the 2×2 identity matrix I. By showing that any point $t \to tx + y$ can be smoothly connected to the identity transformation, it follows that any point in the proper affine group can be smoothly connected to the identity matrix I. A (continuous) group in the neighborhood of I is called a Lie Group.

Exercise 2.11. Show that $\left\{\begin{bmatrix} x & y \\ 0 & x \end{bmatrix}, \ x > 0\right\}$ is a group under matrix multiplication. It is called the step transformation group.

Exercise 2.12. The convex hull of all permutation matrices in dimension of n is the set of all doubly stochastic matrices in dimension of n. Given a doubly stochastic matrix P and a vector y satisfying $y_1 \geq \cdots \geq y_n$ then the new vector $x = Py$ is said to majorize y in the sense that $x_1 \geq \cdots \geq x_n$ and

$$\sum_{j=1}^{n} x_j = \sum_{j=1}^{n} y_j, \quad \sum_{j=1}^{m} x_j \geq \sum_{j=1}^{m} y_j, \qquad m = 1, \ldots, n-1.$$

To indicate this we write $x \succ y$. Shown that $x \succ y \iff x = Py$. For a detailed discussion of majorization and further references, see Marshall and Olkin (1979).

Exercise 2.13. Apply Burnside's Lemma to evaluate the number of orbits derived from the group actions described in (2.20), (2.21), (2.29), and (2.31).

Exercise 2.14. Let V indicate the set of all 2×2 (incidence, or ralation) matrices with entries in $\{0, 1\}$ and indicate by W the set of all binary sequences in length of 2. Show that there is a one-to-one correspondence between V and $W \times W$, and study the orbits of V under different group actions on $W \times W$. Study the orbits of 3×3 incidence matrices under different group actions.

Exercise 2.15. For ω_1 and ω_2 probability laws in a finite group G, define the convolution $\omega_1 \star \omega_2$ of ω_1 and ω_2 by $\omega_1 \star \omega_2 : s \mapsto \sum_{t \in G} \omega_1(ts^{-1})\omega_2(t)$ for $s \in G$. Show that $\omega_1 \star \omega_2$ is a probability law in G. Describe the convolution laws for S_3, C_3, D_4, and K_4.

Exercise 2.16. Indicate by $\overline{x}(s) = \sum_{\tau \in G} x(s\tau^{-1})/|G|$ the (left) symmetrized version of a scalar function x in C^L. Let w be a probability law in C^L, which is permutation $(s\tau^{-1})$ symmetric. Then, relative to w, show that $E(\overline{x}) = E(x)$, when the action $s\tau^{-1}$ is transitive, and otherwise, $E(\overline{x}) = \sum_i E(x \mid \mathcal{O}_i)P_w(\mathcal{O}_i)$, where $\mathcal{O}_i$ are the distinct orbits generated by the action $s\tau^{-1}$.

Exercise 2.17. Show that the mapping $\tau^* : G \to G$ defined by $\tau^*(\sigma) = \tau\sigma\tau^{-1}$ is an isomorphism in G. Moreover, the mapping $\tau \mapsto \tau^*$ is a homomorphism in G taking values in Aut(G).

3

Connecting Symmetries and Data: Linear Representations

3.1 Introduction

This chapter gives continuation to the study of the remaining steps outlined earlier on in Section 1.9 of Chapter 1. We are interested in learning how to represent the symmetries of interest into the data vector space (step 5), which will lead us into the study of the linear representations of group actions in the corresponding data vector spaces. This will be complemented, in the next chapter, with the construction of the canonical projection matrices and the study and interpretation of the canonical invariants on the data (steps 6 and 7).

3.2 Representations

Permutation representations

The group homomorphisms $\tau \mapsto \varphi_\tau$ from G to S_V (see Proposition 2.4 on page 48) are called permutation representations of G on S_V, or, shortly, permutation representations of G on V. Its (permutation) matrix form ρ_τ, similarly to (2.5), is defined by

$$(\rho_\tau)_{sf} = 1 \iff \varphi_\tau s = f. \tag{3.1}$$

To illustrate, the restriction of the action

$\tau \backslash s$	uu	yy	uy	yu
1	uu	yy	uy	yu
(uy)	yy	uu	yu	uy

,

of $S_2 = \{1, (uy)\}$ on $V = \{uu, yy, uy, yu\}$ to each one of its orbits $\{uu, yy\}$ and $\{uy, yu\}$ gives a permutation representation

$$\rho_1 = \begin{bmatrix} 1 & 0 \\ 0 & 1 \end{bmatrix}, \quad \rho_{(uy)} = \begin{bmatrix} 0 & 1 \\ 1 & 0 \end{bmatrix}$$

of S_2 on $\mathbb{R}^2$, a subspace of the data space $\mathcal{V}$ for V.

Choose an ordering $1, 2, \ldots, v$, for the elements in V and assign them in correspondence with the canonical basis $\{e_1, e_2, \ldots, e_v\}$ for $\mathcal{V} = \mathbb{R}^v$. Then, equivalence (3.1) extends to

$$\varphi_\tau s = f \iff (\rho_\tau)_{sf} = 1 \iff e_s' \rho_\tau e_f = 1 \iff \rho_\tau e_f = e_s. \quad (3.2)$$

Applying (3.2) to the equality $\varphi_\tau(\varphi_{\tau^{-1}} h) = h$ gives $\rho_\tau e_{\varphi_{\tau^{-1}} h} = e_h$, or $e_{\varphi_{\tau^{-1}} h} = \rho_{\tau^{-1}} e_h$, for all $\tau \in G$. That is $\rho_\tau e_s = e_{\varphi_\tau s}$, for all $\tau \in G$ and $s \in V$. As a consequence, we have

$$(\rho_\tau \rho_\sigma) e_s = \rho_\tau(\rho_\sigma e_s) = \rho_\tau e_{\varphi_\sigma s} = e_{\varphi_\tau \varphi_\sigma} = e_{\varphi_{\tau\sigma} s} = \rho_{\tau\sigma} e_s,$$

for all $s \in V$. The homomorphism extends by linearity to all $x = \sum_{s \in V} x_s e_s \in \mathcal{V}$, where x_s are real scalars so that $\rho_\tau \rho_\sigma = \rho_{\tau\sigma}$ for all τ, σ in G.

The permutation representation of S_n. If in particular $V = \{1, \ldots, \ell\}$, $G = S_\ell$, and $\varphi_\tau i = \tau i$ then $(\rho_\tau)_{ij} = 1$ if and only if $\tau i = j$. We will refer to the corresponding representation simply as the permutation representation of S_n. The case $\ell = 3$ appeared earlier on page 36. The reference also applies to the subgroups of S_n.

Linear representations

More generally, group homomorphisms from G to $GL(\mathcal{V})$ are called linear representations of G on $\mathcal{V}$.

Linear representations map the identity of G into the identity matrix (or operator) of $GL(\mathcal{V})$, that is, $\rho_1 = I$. Also, the inverse τ^{-1} of τ is mapped to the inverse $\rho_{\tau^{-1}}$ of the matrix ρ_τ, that is $\rho_{\tau^{-1}} = \rho_\tau^{-1}$.

The dimension of a linear representation is defined as the dimension of the corresponding vector space. Often we refer to all linear representations simply as representations. The reader may want to verify that the transformations on page 42 are linear representations of D_n. For example, the matrices

$$\beta_{k,d} \simeq \begin{bmatrix} 1 & 0 \\ 0 & 1 \end{bmatrix}, \begin{bmatrix} -1 & -1 \\ 1 & 0 \end{bmatrix}, \begin{bmatrix} 0 & 1 \\ -1 & -1 \end{bmatrix}, \begin{bmatrix} -1 & -1 \\ 0 & 1 \end{bmatrix}, \begin{bmatrix} 1 & 0 \\ -1 & -1 \end{bmatrix}, \begin{bmatrix} 0 & 1 \\ 1 & 0 \end{bmatrix},$$

give a linear representation of D_3 (or S_3) on $\mathbb{R}^2$ and coincide with those in (2.14), for $k = 1, 2, 3$, $d = \pm 1$ after a change of basis.

Equivalent representations

If ρ is a linear representation of G on $\mathcal{V}$ and $B \in GL(\mathcal{V})$ then $\beta : \tau \mapsto B\rho_\tau B^{-1}$ is also a linear representation of G in $GL(\mathcal{V})$. Any two such linear representations, obtained one from another by a changing of basis, are called equivalent or isomorphic representations. It was apparent, in the construction of the permutation representations, that the identification of V with a basis for $\mathcal{V}$ could be done in as many different ways as the arbitrary orderings of the elements in V. The resulting representations, however, are all equivalent. The matrices (2.13) and (2.14) defined on page 42 are equivalent linear representations of the dihedral group D_n on $\mathbb{R}^2$.

One-dimensional representations

The unity or symmetric representation assigns the value $\rho_\tau = 1$ for all $\tau \in G$. The alternating or sign representation of S_ℓ is defined as

$$\mathrm{Sgn}_\tau = \begin{cases} +1 & \text{if the permutation } \tau \text{ is even;} \\ -1 & \text{if the permutation } \tau \text{ is odd.} \end{cases}$$

From Proposition 2.1, we know that $\mathrm{Sgn}_{\sigma\tau} = \mathrm{Sgn}_\sigma \mathrm{Sgn}_\tau$, for any two permutations σ, τ. When G is the subgroup C_n of S_n, the reader may verify that

$$\rho_k(\tau^j) = \omega^{jk}, \quad j = 0, \ldots, n-1, \quad k = 1, \ldots n,$$

where τ is a generator of C_n and ω is a primitive n-th root of 1, are n distinct one-dimensional representations of C_n.

The regular representation

The representation of the action $\varphi_\tau(\sigma) = \sigma\tau^{-1}$ of G on itself is called the (left) regular representation of G. Its dimension is the number of elements in G. Similarly, $\varphi_\tau(\sigma) = \tau\sigma$ defines the right regular representation. We will refer to either action as the regular action, the context indicating which one is at work. The regular representation

$$\phi_1 = \begin{bmatrix} 1 & 0 & 0 & 0 \\ 0 & 1 & 0 & 0 \\ 0 & 0 & 1 & 0 \\ 0 & 0 & 0 & 1 \end{bmatrix}, \quad \phi_v = \begin{bmatrix} 0 & 1 & 0 & 0 \\ 1 & 0 & 0 & 0 \\ 0 & 0 & 0 & 1 \\ 0 & 0 & 1 & 0 \end{bmatrix},$$

$$\phi_h = \begin{bmatrix} 0 & 0 & 1 & 0 \\ 0 & 0 & 0 & 1 \\ 1 & 0 & 0 & 0 \\ 0 & 1 & 0 & 0 \end{bmatrix}, \quad \phi_o = \begin{bmatrix} 0 & 0 & 0 & 1 \\ 0 & 0 & 1 & 0 \\ 0 & 1 & 0 & 0 \\ 1 & 0 & 0 & 0 \end{bmatrix} \qquad (3.3)$$

appeared earlier on in Chapter 1 under the action of $G = \{1, v, h, o\}$ on itself.

Data indexed by point groups. The symmetry transformations that leave the stable configuration of a molecule physically indistinguishable are generally known as *point groups*. The name indicates that at least one point in the molecular framework remains fixed. To illustrate, consider the following transformations in $\mathbb{R}^3$, represented with the standard notation used by chemists:

(1) E: the identity operator;
(2) C_2: a rotation by 180° around the z-axis;
(3) i: an *inversion* or point reflection through the origin (0, 0, 0);
(4) σ_h: a reflection on the xy-plane.

The transformations $\{E, C_2, i, \sigma_h\}$ in $\mathbb{R}^3$ define the point group $\mathcal{C}_{2h}$. Its multiplication table is given by

$$\mathcal{C}_{2h}: \begin{array}{|c|cccc|} * & E & C_2 & i & \sigma_h \\ \hline E & E & C_2 & i & \sigma_h \\ C_2 & C_2 & E & \sigma_h & i \\ i & i & \sigma_h & E & C_2 \\ \sigma_h & \sigma_h & i & C_2 & E \end{array}, \tag{3.4}$$

which is isomorphic with the multiplication table of $G = \{1, v, h, o\}$, on page 2. For example, the 180° rotation (C_2) around the z-axis followed by an inversion (i) through the origin is equivalent to a reflection on the xy-plane, that is, $iC_2 = \sigma_h$. The planar structure of a dichloroethene $C_2H_2Cl_2$-*trans* molecule is among the molecules characterized by the symmetries of $\mathcal{C}_{2h}$. The group acts transitively on the canonical basis for $\mathbb{R}^3$ giving a representation

$$\rho_E = \text{Diag}(1, 1, 1), \quad \rho_\sigma = \text{Diag}(1, 1, -1), \quad \rho_C = \text{Diag}(-1, -1, 1), \quad \rho_i = \text{Diag}(-1, -1, -1)$$

of $\mathcal{C}_{2h}$ on $\mathbb{R}^3$.

The molecular framework of the dichloroethene molecule can be used as a data structure. To see this, consider a rectangular parallelepiped with vertices $\{(\pm 2, \pm 1, \pm 1)\}$ expressed as the set of labels

$$V = \{abb,\ abB,\ aBb,\ aBB,\ Abb,\ AbB,\ ABb,\ ABB\}.$$

For example, AbB is the label for the point $(-2, 1, -1)$. Then $\mathcal{C}_{2h}$ acts on V according to

$$\begin{array}{|c|cccccccc|} & abb & abB & aBb & aBB & Abb & AbB & ABb & ABB \\ \hline E & abb & abB & aBb & aBB & Abb & AbB & ABb & ABB \\ \sigma_h & abB & abb & aBB & aBb & AbB & Abb & ABB & ABb \\ C_2 & ABb & ABB & Abb & AbB & aBb & aBB & abb & abB \\ i & ABB & ABb & AbB & Abb & aBB & aBb & abB & abb \end{array}, \tag{3.5}$$

and its restriction to the resulting two orbits

$$\{abb, abB, ABb, ABB\}, \quad \{aBb, aBB, Abb, AbB\}$$

of V gives two transitive actions, each one of which leading to (isomorphic copies of) a linear representation of $\mathcal{C}_{2h}$ on the data space $\mathbb{R}^4$.

3.3 Unitary Representations

The linear representations introduced in the previous section verify the property $(\rho_\tau x, \rho_\tau y) = (x, y)$, for all $\tau \in G$, under the standard inner product $(x, y) = x'y$ in $\mathbb{R}^v$. Equivalently then $\rho_\tau \rho'_\tau = I$ for all $\tau \in G$.

To define the corresponding property for complex linear spaces, recall that an inner product in a real or complex vector space $\mathcal{V}$ is a function $(.,.) : \mathcal{V}^2 \mapsto \mathbb{F}$ such that, for all $x, y, z \in \mathcal{V}$, and scalars a, b,

(1) $(x, y) = \overline{(y, x)}$, (Hermitian symmetric)
(2) $(ax + by, z) = a(x, y) + b(y, z)$, (conjugate bilinear)
(3) $(x, x) \geq 0$, $(x, x) = 0$ if and only if $x = 0$, (positive definite).

The vector space $\mathcal{V}$, together with $(.,.)$, is called an inner product space. An Euclidian (respectively unitary) space is a real (respectively complex) inner product space.

A linear representation ρ of G on the inner product space $\mathcal{V}$ is unitary if

$$(\rho_\tau x, \rho_\tau y) = (x, y)$$

for all $x, y \in \mathcal{V}$, and $\tau \in G$. If $(.|.)$ is an inner product in $\mathcal{V}$, direct evaluation shows that then

$$(x, y) = \frac{1}{|G|} \sum_{\tau \in G} (\rho_\tau x | \rho_\tau y) \tag{3.6}$$

is an inner product in $\mathcal{V}$, relative to which ρ is unitary.

Moreover, ρ is equivalent to a representation that is unitary in the original inner product space. To see this, indicate by $\{e_1, \ldots, e_v\}$ an orthonormal basis relative to $(.|.)$ and by $\{f_1, \ldots, f_v\}$ an orthonormal basis relative to the invariant inner product $(.,.)$, and let A be the linear transformation defined by $Ae_i = f_i$. Then $(Ae_i, Ae_j) = \delta_{ij} = (e_i|e_j)$, so that $(Ax, Ay) = (x|y)$. Define $r_\tau = A^{-1}\rho_\tau A$, $\tau \in G$. Then r and ρ are equivalent and, because

$$(r_\tau x | r_\tau y) = (A^{-1}\rho_\tau Ax | A^{-1}\rho_\tau Ay) = (\rho_\tau Ax, \rho_\tau Ay) = (Ax, Ay) = (x|y),$$

r is unitary in the original inner product space.

To illustrate, we will construct a representation unitarily equivalent to the two-dimensional representation of $S_3 \simeq D_3$ shown on page 64. The invariant scalar product derived from the Euclidean inner product $(\,|\,)$ in $\mathbb{R}^2$ is

$$(x, y) = \sum_\tau (\beta_\tau x | \beta_\tau y) = \sum_\tau x' \beta_\tau' \beta_\tau y = 4x' \begin{bmatrix} 2 & 1 \\ 1 & 2 \end{bmatrix} y \equiv x' F y.$$

Next, starting with the canonical basis $e_1 = (1, 0)$, $e_2 = (0, 1)$ for $\mathbb{R}^2$, use Gram-Schmidt to construct a basis $\{w_1, w_2\}$ that is orthonormal relative to the invariant inner product, such as $w_1' = (\sqrt{2}/4, 0)$ and $w_2' = (-\sqrt{6}/12, \sqrt{6}/6)$. The resulting new (unitarily equivalent) representation is then $b_\tau = A^{-1}\beta_\tau A$, where

$$A = \frac{\sqrt{2}}{12} \begin{bmatrix} 3 & -\sqrt{3} \\ 0 & 2\sqrt{3} \end{bmatrix}.$$

We obtain $b_1 = I_2$ and

$$\frac{1}{2}\begin{bmatrix} -1 & \sqrt{3} \\ -\sqrt{3} & -1 \end{bmatrix}, \quad \frac{1}{2}\begin{bmatrix} -1 & -\sqrt{3} \\ \sqrt{3} & -1 \end{bmatrix}, \quad \frac{1}{2}\begin{bmatrix} 1 & \sqrt{3} \\ \sqrt{3} & -1 \end{bmatrix}, \quad \begin{bmatrix} -1 & 0 \\ 0 & 1 \end{bmatrix}, \quad \frac{1}{2}\begin{bmatrix} 1 & -\sqrt{3} \\ -\sqrt{3} & -1 \end{bmatrix},$$

for $b_{(123)}, b_{(132)}, b_{(12)}, b_{(13)}$, and $b_{(23)}$, respectively. In each case we have $b'_\tau b_\tau = I_2$.

More generally, in the complex case, unitary representations satisfy $\rho^*_\tau \rho_\tau = I$, where ρ^*_τ indicates the conjugate transpose of ρ_τ, that is, $\rho^*_\tau = \overline{\rho}'_\tau$. To see this, first note that any linear operator ρ on a finite-dimensional vector space when applied to a basis $\{e_s; s \in V\}$ for $\mathcal{V}$ gives

$$\rho e_s = \sum_{h \in V} (\rho)_{hs} e_h, \quad s \in V.$$

Let then $\{e_s; s \in V\}$ be an orthonormal basis for $\mathcal{V}$ under the inner product $(\cdot, \cdot)$, relative to which ρ_τ (or an equivalent representation) is unitary. Then, writing ρ^τ_{sf} to indicate the entry sf of ρ_τ,

$$\begin{aligned} \delta_{sf} = (e_s, e_f) = (\rho_\tau e_s, \rho_\tau e_f) &= \sum_{h,h'} \rho^\tau_{hs} \, \overline{\rho}^\tau_{h'\,f} (e_h, e_{h'}) = \sum_h \rho^\tau_{hs} \overline{\rho}^\tau_{hf} \\ &= \sum_h \overline{\rho}^{\tau'}_{fh} \rho^\tau_{hs} = (\overline{\rho}'_\tau \rho_\tau)_{fs}, \end{aligned}$$

so that $\overline{\rho}'_\tau \rho_\tau = I$, or $\rho^*_\tau \rho_\tau = I$ for all $\tau \in G$.

3.4 Regular Representations and Group Algebras

Consider the (real or complex) vector space $\mathcal{V}$ in the dimension of the number g of elements in a finite group G and write the elements of $\mathcal{V}$ as symbolic linear combinations

$$x = \sum_{\sigma \in G} x_\sigma \sigma,$$

with scalar coefficients $x_\sigma \in G$. Then $\mathcal{V}$ has an operation of multiplication defined by

$$xy = \sum_{\sigma,\eta} x_\sigma y_\eta \sigma\eta = \sum_\tau \left[\sum_{\sigma\eta=\tau} x_\sigma y_\eta \right] \tau \in \mathcal{V} \tag{3.7}$$

so that for all $x, y, z \in \mathcal{V}$, and all scalars γ, we have $xy \in \mathcal{V}$, $x(y+z) = xy + xz$, $(x+y)z = xz + yz$, $\gamma(xy) = x(\gamma y) = (\gamma x)y$. Moreover, because the group operation is associative, we have $x(yz) = (xy)z = xyz$. In this case, we say that the vector space $\mathcal{V}$, along with the multiplication so defined, constitutes an associative group algebra. The subspaces, $\mathcal{B}$, corresponding to the (left) regular action of G satisfy $\tau\mathcal{B} \subset \mathcal{B}$ for all $\tau \in G$ and, by the linearity of the multiplication, are exactly those subalgebras $\mathcal{I}$ of $\mathcal{V}$ that satisfy $y\mathcal{I} \subset \mathcal{I}$ for all $y \in \mathcal{V}$. These subspaces are examples of stable or invariant subspaces and are studied in Section 3.7. Equivalently, in this example, they define left ideals of $\mathcal{V}$ so that the determination of the invariant subspaces of regular representations corresponds to searching for the ideals of $\mathcal{V}$.

3.5 Tensor Representations

If G acts on V and W giving linear representations ρ and η of G on $\mathcal{V}$ and $\mathcal{W}$, respectively, then the identification of $V \times W$ as a basis for $\mathcal{V} \times \mathcal{W}$ gives a linear representation of G on $\mathcal{V} \times \mathcal{W}$, indicated by $\rho \otimes \eta$, evaluated as

$$(\rho \otimes \eta)_\tau = \rho_\tau \otimes \eta_\tau,$$

and called the tensor product of the two representations ρ and η.

By expressing the data points in $\mathcal{V} \times \mathcal{W}$ as $x \otimes y$, we see that the these data then evaluate as $(\rho_\tau x) \otimes (\eta_\tau y)$.

Clearly, the same construction can be extended to define the tensor representation derived from more than two linear representations of G.

To illustrate, recall from Chapter 2 (page 53) that the set of characters or traits that are produced by two genes $\{A, a\}$ stems from S_2 acting on the location of the genes, with resulting orbits

$$\mathcal{O}_0 = \{aa\}, \quad \mathcal{O}_1 = \{Aa, aA\} \;\; \mathcal{O}_2 = \{AA\}.$$

In terms of their phenotype, if $\mathcal{O}_0$ is the recessive trait, then $\mathcal{O}_1$ and $\mathcal{O}_2$ are the dominant traits. Genotypically, $\mathcal{O}_1$ is the heterozygous pair, whereas the other two orbits contain the homozygous pairs. The basic aspects of the cell meiosis process can be represented by the subset of the structure V of all binary sequences in length of 4, introduced earlier on in Chapters 1 and 2. Under the position-symmetry action of S_4, the structure V decomposes as $V = V_{40} \cup V_{31} \cup V_{22}$, with $V_{40} = \mathcal{O}_0 \cup \mathcal{O}_4$, $V_{31} = \mathcal{O}_1 \cup \mathcal{O}_3$, and $V_{22} = \mathcal{O}_2$. The substructure of interest here is exactly

$$V_{22} = \{AAaa, AaAa, AaaA, aAAa, aAaA, aaAA\}$$

indicated, respectively, by the labels $\{13, 11, 7, 10, 6, 4\}$ shown in Table (2.19) on page 50.

Here, two organisms f and g in V_{22} are considered equivalent if they differ only by a location-symmetry transformation, that is, $f = g\tau^{-1}$ for some $\tau \in S_4$.

According to Hannan (1965), Fisher (1947) explored the fact that the trace $\chi_\rho(\tau)$ of the permutation matrix ρ_τ describes exactly the number of pairs of organisms that are self-similar under the given permutation, or the number of points fixed by the permutation. The number k of classes (or orbits) of equivalent organisms is then given by the average number

$$k = \frac{1}{4!} \sum_{\tau \in S_4} \chi_\rho(\tau)$$

of fixed points, an application of Burnside's Lemma. Since the action $s\tau^{-1}$ of S_4 on V_{22} is transitive, we must have $k = 1$. Indeed, a systematic reading of Table (2.19) on page 50 shows that $k = (6 + 9 \times 2)/4! = 1$, thus saying that all organisms are equivalent.

The classification problem is certainly more interesting when m loci are simultaneously considered. There are now 6^m labels in $V = V_{22} \times V_{22} \times \cdots V_{22}$, product of m copies of the original space V_{22}, where S_4 acts according to

$$\left(s\tau^{-1}, g\tau^{-1}, \ldots, h\tau^{-1}\right), \quad \tau \in S_4, \quad s, g, \ldots, h \in V_{22}.$$

The resulting representation of S_4 on $\mathbb{R}^{6m}$ is the m-fold tensor product $\rho \otimes \cdots \otimes \rho$ of the representation ρ.

Because the trace of the tensor product is simply the product of the component traces, it then follows that there are exactly $k = [6^m + 9.2^m]/4!$ orbits or classes of nonequivalent organisms or genotypes.

Tensor representations and direct products of groups. It is simple to verify that if ρ and η are linear representations of groups G and H on $\mathcal{V}$ and $\mathcal{W}$, respectively, then $\xi_{\tau,\sigma} = \rho_\tau \otimes \eta_\sigma$ is a linear representation of $G \times H$ on $\mathcal{V} \times \mathcal{W}$. To illustrate, the linear representation of $G = \{1, v, h, o\}$ on page 6 is isomorphic to a linear representation of $C_2 \times C_2 \simeq G$ on $\mathbb{R}^2 \times \mathbb{R}^2$ given by $\phi_{\tau,\sigma} = \rho_\tau \otimes \rho_\sigma$, where ρ is the linear representation of $C_2 = \{1, t\}$ on $\mathbb{R}^2$ given by

$$\rho_1 = I_2, \quad \rho_t = \begin{bmatrix} 0 & 1 \\ 1 & 0 \end{bmatrix}.$$

3.6 Matrices with Group Symmetry

The Lorentz group (Chapter 2, page 44) introduced a classical argument of commutativity and symmetry, which is useful for the analysis of data. To see this, note that Expression (2.18) says that $\mathcal{L}$ includes the subgroup G of all nonsingular orthogonal matrices that commute with η. The matrix η then is said to have the symmetry of G, or to be centralized by G.

Conversely, we may start with a given group G of symmetries and ask to characterize those matrices that commute with all elements of (a given representation of) G. To illustrate, consider the regular representation

$$\phi_E = \begin{bmatrix} 1 & 0 & 0 & 0 \\ 0 & 1 & 0 & 0 \\ 0 & 0 & 1 & 0 \\ 0 & 0 & 0 & 1 \end{bmatrix}, \quad \phi_{C_2} = \begin{bmatrix} 0 & 1 & 0 & 0 \\ 1 & 0 & 0 & 0 \\ 0 & 0 & 0 & 1 \\ 0 & 0 & 1 & 0 \end{bmatrix},$$
$$\phi_i = \begin{bmatrix} 0 & 0 & 1 & 0 \\ 0 & 0 & 0 & 1 \\ 1 & 0 & 0 & 0 \\ 0 & 1 & 0 & 0 \end{bmatrix}, \quad \phi_{\sigma_h} = \begin{bmatrix} 0 & 0 & 0 & 1 \\ 0 & 0 & 1 & 0 \\ 0 & 1 & 0 & 0 \\ 1 & 0 & 0 & 0 \end{bmatrix} \tag{3.8}$$

of $\mathcal{C}_{2h}$, derived from its multiplication table on page 66. If the data indexed by the components $\{E, C_2, i, \sigma_h\}$ of $\mathcal{C}_{2h}$ are indicated by (x, y, z, w), it then follows that

$$\frac{1}{4}\sum_{\tau\in G}\phi(\tau)\begin{bmatrix} x^2 & xy & xz & xw \\ xy & y^2 & yz & yw \\ xz & yz & z^2 & zw \\ xw & yw & zw & w^2 \end{bmatrix}\phi(\tau)' = \begin{bmatrix} F & B & C & D \\ B & F & D & C \\ C & D & F & B \\ D & C & B & F \end{bmatrix}$$

with

$$F = x^2 + y^2 + w^2 + z^2, \quad B = 2(xy + zw), \quad C = 2(xz + yw), \quad D = 2(xw + yz),$$

is the $\mathcal{C}_{2h}$-centralized version of the crossproduct matrix for the data.

Similar calculations appear, for example, in the determination of all covariance structures $\rho\Lambda\rho'$ that are consistent with the invariant property $\mathcal{L}(\rho Y) = \mathcal{L}(Y)$ of a probability law $\mathcal{L}$. See, for example, Andersson (1992) and Gao and Marden (2001). More specifically, we have,

Proposition 3.1. *Given a representation ρ of G on $\mathbb{R}^n$ and Λ a real $n \times n$ matrix, then*

$$\Sigma = \sum_{\tau\in G} \rho_\tau \Lambda \rho_{\tau^{-1}} \tag{3.9}$$

has the symmetry of (or is centralized by) ρ in the sense that $\rho_\sigma\Sigma = \Sigma\rho_\sigma$ for all $\sigma \in G$.

Proof. For any $\sigma \in G$, we have $\rho_\sigma \Sigma \rho_{\sigma^{-1}} = \sum_{\tau\in G} \rho_\sigma \rho_\tau \Lambda \rho_{\tau^{-1}} \rho_{\sigma^{-1}} = \sum_{\tau\in G} \rho_{\sigma\tau} \Lambda \rho_{\sigma\tau}^{-1} = \Sigma$, observing that $\sigma\tau$ spans G when $\tau \in G$, for any σ in G. ■

In this case, we also say that Σ commutes with the representation ρ. The set of all linear operators commuting with a representation ρ in $\mathcal{V}$ is a linear subspace of the space of linear operators in $\mathcal{V}$. Also in that subspace are the linear operators of the form $\widehat{x}(\rho) = \sum_{\tau\in G} x_\tau \rho_\tau$, where x_τ are scalar functions constant over the conjugacy classes of G. See also Naimark and Štern (1982, p. 55) and Simon (1996, p. 28).

Matrices with dihedral structure

In this section we apply Proposition 3.1 to determine the pattern of matrices centralized by the regular representation of the dihedral group D_4. First, we construct the regular representation of D_4, which will appear again in later chapters in data-analytic applications in which the data are indexed by those symmetries.

The regular representation of D_4. The reader may want to verify, from the multiplication table (2.12) of D_4, on page 42, and Algorithm 6 on page 223, that its regular representation can be expressed as

$$\phi_{k,d} = t^{(d+3)/2} \otimes (R^k H^{(d+3)/2}), \quad k = 1, 2, 3, 4\ d = \pm 1$$

where

$$t = \begin{bmatrix} 0 & 1 \\ 1 & 0 \end{bmatrix}, \quad R = \begin{bmatrix} 0 & 0 & 0 & 1 \\ 1 & 0 & 0 & 0 \\ 0 & 1 & 0 & 0 \\ 0 & 0 & 1 & 0 \end{bmatrix}, \quad H = \begin{bmatrix} 1 & 0 & 0 & 0 \\ 0 & 0 & 0 & 1 \\ 0 & 0 & 1 & 0 \\ 0 & 1 & 0 & 0 \end{bmatrix}.$$

Rotations are represented when $d = 1$ and reversals when $d = -1$. We now apply Proposition 3.1 to centralize the 8×8 matrix xx' of dihedral crossproducts, where x is the vector of (data indexed by the) dihedral rotations $x'_+ = (r_1, \ldots, r_4)$ and reversals $x'_- = (t_1, \ldots, t_4)$. Direct evaluation then shows that

$$\sum_{k,d} \phi_{k,d}(xx')\phi_{k,d}^{-1} = \begin{bmatrix} A & B & C & B & a & b & c & d \\ B & A & B & C & d & a & b & c \\ C & B & A & B & c & d & a & b \\ B & C & B & A & b & c & d & a \\ a & d & c & b & A & B & C & B \\ b & a & d & c & B & A & B & C \\ c & b & a & d & C & B & A & B \\ d & c & b & a & B & C & B & A \end{bmatrix}$$

where, writing γ to indicate the permutation matrix for the cycle (1234) and observing that $y'\gamma^k z$ is then the lag-k crossproduct of two vectors y, z in $\mathbb{R}^4$,

$$A = x'x, \quad B = x'_+\gamma x_+ + x'_-\gamma x_-, \quad C = x'_+\gamma^2 x_+ + x'_-\gamma^2 x_-,$$

$$a = 2x'_+x_-, \quad b = 2x'_+\gamma x_-, \quad c = 2x'_+\gamma^2 x_-, \quad d = 2x'_+\gamma^3 x_-.$$

Together, these coefficients can be generated as $x'(t^{(d+3)/2} \otimes \gamma^k)x$, for $= 0, 1, 2, 3$, and $d = \pm 1$.

Matrices with complex structure

The group defined by $G = \{1, i, -1, -i\}$, and its multiplication table

$*$	1	-1	i	$-i$
1	1	-1	i	$-i$
-1	-1	1	$-i$	i
i	i	$-i$	-1	1
$-i$	$-i$	i	1	-1

,

is called the complex group. The reader may want to verify that

$$\rho(i^k) = \begin{bmatrix} 0 & -1 \\ 1 & 0 \end{bmatrix}^k, \quad k = 0, 1, 2, 3.$$

is a representation of the complex group in $\mathbb{R}^2$, and that

$$\sum_{k=0}^{3} \rho(i^k) \begin{bmatrix} a & b \\ c & d \end{bmatrix} \rho(i^k)^{-1} = 2 \begin{bmatrix} d+a & -c+b \\ -b+c & d+a \end{bmatrix}.$$

Matrices of the form

$$M = \begin{bmatrix} A & B \\ -B & A \end{bmatrix}$$

where A and B are $n \times n$ real matrices are said to have *complex structure* and carry the symmetry of the complex group represented by ρ in the sense that $(\rho_\tau \otimes I_n)M = M(\rho_\tau \otimes I_n), \quad \text{for all } \tau \in G.$

Matrices with quaternionic structure

Following the comments on page 45, define

$$\rho(\pm 1) = \pm \begin{bmatrix} 1 & 0 & 0 & 0 \\ 0 & 1 & 0 & 0 \\ 0 & 0 & 1 & 0 \\ 0 & 0 & 0 & 1 \end{bmatrix}, \quad \rho(\pm \mathbf{k}) = \mp \begin{bmatrix} 0 & -1 & 0 & 0 \\ 1 & 0 & 0 & 0 \\ 0 & 0 & 0 & -1 \\ 0 & 0 & 1 & 0 \end{bmatrix},$$

$$\rho(\pm \mathbf{j}) = \mp \begin{bmatrix} 0 & 0 & -1 & 0 \\ 0 & 0 & 0 & 1 \\ 1 & 0 & 0 & 0 \\ 0 & -1 & 0 & 0 \end{bmatrix}, \quad \rho(\pm \mathbf{i}) = \mp \begin{bmatrix} 0 & 0 & 0 & -1 \\ 0 & 0 & -1 & 0 \\ 0 & 1 & 0 & 0 \\ 1 & 0 & 0 & 0 \end{bmatrix}.$$

Direct verification shows that ρ is a linear representation in $\mathbb{R}^4$ of the group of the quaternions. Given a real matrix

$$F = \begin{bmatrix} a & b & c & d \\ e & f & g & h \\ p & q & r & s \\ t & u & v & x \end{bmatrix},$$

it then follows that

$$\sum \rho_\tau F \rho_{\tau^{-1}} = 2 \begin{bmatrix} a+f+r+x & b-e-s+v & c+h-p-u & d-g+q-t \\ e-b-v+s & a+f+r+x & g-d+t-q & c+h-p-u \\ p+u-c-h & d-g+q-t & a+f+r+x & e-b-v+s \\ g-d+t-q & p+u-c-h & b-e-s+v & a+f+r+x \end{bmatrix},$$

so that matrices of the form

$$M = \begin{bmatrix} A & B1 & B2 & B3 \\ -B1 & A & -B3 & B2 \\ -B2 & B3 & A & -B1 \\ -B3 & -B2 & B1 & A \end{bmatrix},$$

where A, B_1, B_2, and B_3 are any $n \times n$ real matrices, are said to have a *quaternionic structure*. Those are exactly the matrices with the symmetry of the given representation, in the sense that

$$(\rho_\tau \otimes I_n)M = M(\rho_\tau \otimes I_n), \quad \text{for all } \tau \in Q.$$

3.7 Reducibility

Stable subspaces

If ρ is a representation of G on the vector space $\mathcal{V}$ and $\mathcal{W}$ is a subspace of $\mathcal{V}$ such that $\rho_\tau x \in \mathcal{W}$ for all $\tau \in G$ and $x \in \mathcal{W}$ then $\mathcal{W}$ is called a stable of invariant subspace. Clearly $\{0\}$ and $\mathcal{V}$ are stable subspaces of $\mathcal{V}$.

Note that if $x = \sum_{s \in W} x_s e_s$ is a point in the subspace $\mathcal{W}$ spanned by $\{e_s; s \in W\}$ and W is an orbit of V under the action φ of G on V then

$$\rho_\tau x = \sum_{s \in W} x_s \rho_\tau e_s = \sum_{s \in W} x_s e_{\varphi_\tau s} \in \mathcal{W},$$

for all $\tau \in G$, so that $\mathcal{W}$ is a stable subspace of the space generated by $\{e_s; s \in V\}$.

The Sym² and Alt² subspaces. Let ρ be the representation of S_2 shuffling the indices of the basis $\{e_1, e_2\}$ of $\mathcal{V} = \mathbb{R}^2$ and consider the tensor representation $\rho \otimes \rho$ of G on $\mathcal{V} \times \mathcal{V}$. Define also

$$v_1 = 2e_1 \otimes e_1, \quad v_2 = 2e_2 \otimes e_2, \quad v_3 = e_1 \otimes e_2 + e_2 \otimes e_1, \quad v_4 = e_1 \otimes e_2 - e_2 \otimes e_1.$$

It then follows that $\mathcal{V} \times \mathcal{V} =< v_1, v_2, v_3 > \oplus < v_4 >$ is a direct sum decomposition of $\mathcal{V}$, each component of which is a stable subspace of $\mathcal{V}$ under $\rho \otimes \rho$. The action of $\rho \otimes \rho$ on the bases of these subspaces, in turn, defines two representations, called respectively, the symmetric square (Sym²) and alternating square (Alt²) representations of S_2. More generally, if $\mathcal{V}$ is spanned by a basis indexed by the elements of a set V with v elements and ρ is a permutation representation of a group G acting on V, then

$$\mathcal{V} \times \mathcal{V} =< e_s \otimes e_f + e_f \otimes e_s; (s, f) \in D \cup U > \oplus < e_s \otimes e_f - e_f \otimes e_s; (s, f) \in U >,$$

where D indicates the main diagonal of $V \times V$, and U its upper triangular part. The component subspaces are stable subspaces of $\rho \otimes \rho$, the corresponding symmetric square representation is in dimension of $v(v+1)/2$ and the alternating square is in dimension of $v(v-1)/2$. We write $\rho \otimes \rho \simeq \mathrm{Sym}^2 \oplus \mathrm{Alt}^2$ to indicate the associated decomposition of $\rho \otimes \rho$.

The reader may want to verify that a decomposition $\mathcal{V}_1 \oplus \mathcal{V}_2$ of $\mathcal{V}$ in which $\mathcal{V}_1$ is a stable subspace under a representation ρ can be characterized by the matrix form

$$\rho_\tau = \begin{bmatrix} R_1(\tau) & 0 \\ M(\tau) & R_2(\tau) \end{bmatrix}$$

of ρ_τ so that R_1 and R_2 are representations of G, in the dimensions of the corresponding submatrices. In this case, we say that ρ is a *reducible* representation.

Irreducible representations

We say that a representation ρ of G in $GL(\mathcal{V})$ is irreducible when the only proper stable linear subspace of $\mathcal{V}$ is the null subspace.

Clearly, then, all one-dimensional representations are irreducible. To illustrate, consider the representation

$$\xi_1 = \begin{bmatrix} \begin{bmatrix} \boxed{1} \\ \boxed{0} \end{bmatrix} & \begin{matrix} 0 \\ \boxed{1} \end{matrix} \end{bmatrix}, \quad \xi_t = \begin{bmatrix} \begin{bmatrix} \boxed{-1} \\ \boxed{-1} \end{bmatrix} & \begin{matrix} 0 \\ \boxed{1} \end{matrix} \end{bmatrix},$$

of $S_2 = \{1, t\}$ given by its action (by multiplication) on the indices of the basis spanning the subspaces $< e_t - e_1 >$ and $< e_t >$ of $\mathcal{V} = \mathbb{R}^2$.

Here, $\mathcal{V} =< e_t - e_1 > \oplus < e_t >$ is a direct sum decomposition of $\mathcal{V}$ in which $< e_t - e_1 >$ is reduced by the alternating representation (page 65). Therefore, ρ is reducible. However, $< e_t >$ is not yet a stable complement of $< e_t - e_1 >$ in $\mathcal{V}$.

Consider, instead, the direct sum decomposition $\mathcal{V} =< e_t - e_1 > \oplus < e_t + e_1 >$, relative to which the representation is now

$$\beta_1 = \begin{bmatrix} \begin{bmatrix} \boxed{1} \\ 0 \end{bmatrix} & \begin{matrix} 0 \\ \boxed{1} \end{matrix} \end{bmatrix}, \quad \beta_t = \begin{bmatrix} \begin{bmatrix} \boxed{-1} \\ 0 \end{bmatrix} & \begin{matrix} 0 \\ \boxed{1} \end{matrix} \end{bmatrix}.$$

The new component reduces as the identity or symmetric representation, which is also one-dimensional and hence irreducible. In summary, ρ decomposes as the sum $1 \oplus \mathrm{Sgn}$ of two irreducible one-dimensional representations, $\mathcal{V}$ decomposes as the sum $\mathcal{V}_1 \oplus \mathcal{V}_{\mathrm{Sgn}}$ of two stable (and also irreducible) subspaces:

$$\mathcal{V} = \mathcal{V}_1 \oplus \mathcal{V}_{\mathrm{Sgn}}, \quad \rho \simeq \rho_1 \oplus \rho_{\mathrm{Sgn}}.$$

A reduction for binary sequences. The representation ρ for the action

$\tau \backslash s$	uu	yy	uy	yu
1	uu	yy	uy	yu
$t = (12)$	uu	yy	yu	uy

of S_2 on the binary sequences in length of 2 reduces as the identity in each one of the subspaces indexed by single-element orbits and as $1 \oplus \text{Sgn}$ in the subspace indexed by the two-element orbit. That is, $\rho \simeq 1 \oplus 1 \oplus 1 \oplus \text{Sgn}$.

A two-dimensional irreducible representation of S_3. Start with the subspace $W_1 =< e >$ generated by the sum $e' = (1, 1, 1)$ of the vectors in the canonical basis $\{e_1, e_2, e_3\}$ of $\mathcal{V} = \mathbb{R}^3$. If ρ indicates the permutation representation of S_3 (page 64) then clearly $\rho_\tau y \in W_1$ for all $y \in W_1$ and $\tau \in S_3$, so that W_1 is a stable subspace of ρ.

Let $W_0 = \{y \in \mathcal{V}; e'y = 0\}$ be the orthogonal complement of W_1 in $\mathcal{V}$ and $\mathcal{A} = ee'/3$ the projection on W_1 along W_0, that is, $\mathcal{V} = W_0 \oplus W_1$, $\mathcal{A}^2 = \mathcal{A}$ and $\mathcal{A}y = 0$ for all $y \in W_0$. Similarly, let

$$\mathcal{Q} = I - \mathcal{A} = \frac{1}{3}\begin{bmatrix} 2 & -1 & -1 \\ -1 & 2 & -1 \\ -1 & -1 & 2 \end{bmatrix} \tag{3.10}$$

indicate the projection on W_0 along W_1. The reader may want to verify that $\mathcal{Q}$ is centralized by ρ, that is, $\rho_\tau \mathcal{Q} = \mathcal{Q}\rho_\tau$ for all $\tau \in S_3$, and that, consequently, if $y \in W_0$ then $y \in \mathcal{Q}z$ for some $z \in \mathcal{V}$, and $\rho_\tau y = \rho_\tau \mathcal{Q}z = \mathcal{Q}\rho_\tau z \in W_0$, for all $\tau \in S_3$. That is, W_0 is a stable two-dimensional complement of W_1 in $\mathcal{V}$. To construct a two-dimensional representation (β) in W_0, note, from the corresponding projection in (3.10), that a basis $\{v_1, v_2\}$ for the image subspace of $\mathcal{Q}$ is $v_1 = 2e_1 - e_2 - e_3$, $v_2 = -e_1 + 2e_2 - e_3$. The resulting representation of $\tau = (12)$, for example, is obtained from the fact that

$$\tau v_1 = 2e_{\tau 1} - e_{\tau 2} - e_{\tau 3} = 2e_2 - e_1 - e_3 = v_2,$$
$$\tau v_2 = -e_{\tau 1} + 2e_{\tau 2} - e_{\tau 3} = -e_2 + 2e_1 - e_3 = v_1,$$

that is, $\beta_{(12)} = \begin{bmatrix} 0 & 1 \\ 1 & 0 \end{bmatrix}$. Similar calculations, noting that $-e_1 - e_2 + 2e_3 = -v_1 - v_2$, leads to the linear representation of S_3 shown on page 64.

The two-dimensional representation β is irreducible. In fact, if there were a proper one-dimensional stable subspace W, with generator y, then it would verify $\beta_{(12)}y = \lambda y$ for some scalar λ, which implies $y_2 = \lambda y_1$, $y_1 = \lambda y_2$. The nonzero eigenvalue solutions to $y_2 = \lambda^2 y_2$ are $\lambda = \pm 1$, that is, $y = (y_1, y_1)$ or $y = (y_1, -y_1)$. Since the subspace W must also be stable under $\beta_{(13)}$, then we would have

$$\beta_{(13)}y = \begin{bmatrix} -1 & -1 \\ 0 & 1 \end{bmatrix}\begin{bmatrix} y_1 \\ y_1 \end{bmatrix} = \begin{bmatrix} -2y_1 \\ y_1 \end{bmatrix} \in W$$
$$\iff y_1 = 0, \text{ using } y = (y_1, y_1) \text{ or } y = (y_1, -y_1) \implies W = \{0\}.$$

Because $\{0\}$ is the only proper stable subspace, β is irreducible.

Table (3.11) summarizes the irreducible representations of S_3. It includes the presently derived two-dimensional representation, along with the trivial and alternating (one-dimensional) ones. Since the trace Tr ρ_τ of a representation, indicated here by $\chi_\rho(\tau)$, is constant over conjugacy classes, it is sufficient to report it for representatives of these classes.

χ	1	(12)	(123)
χ_1	1	1	1
χ_β	2	0	-1
χ_{Sgn}	1	-1	1

. (3.11)

These tables completely describe the representations and will be studied later on in the text with greater detail.

A comment on irreducibility and the field of scalars. To appreciate the role of the field of scalars in the search for irreducible representations, consider the restriction

$$\gamma = \left\{ \begin{bmatrix} 1 & 0 \\ 0 & 1 \end{bmatrix}, \begin{bmatrix} -1 & -1 \\ 1 & 0 \end{bmatrix}, \begin{bmatrix} 0 & 1 \\ -1 & -1 \end{bmatrix} \right\}$$

of the two-dimensional representation of S_3, from page 64, to the subgroup $C_3 = \{1, (123), (132)\}$ of S_3. The equalities $\gamma_\tau y = \lambda y$, for $\tau \in C_3$, lead to the characteristic equations $(1-\lambda)^2 = 0$ and $\lambda^2 + \lambda + 1 = 0$, which identify three one-dimensional (irreducible) subspaces over the complex field. However, over the reals, γ is irreducible. Table (3.12) shows the three irreducible representations of C_3, indicated here by χ_1, χ_2, and χ_3.

χ	1	(123)	(132)
χ_1	1	1	1
χ_2	1	ω	ω^2
χ_3	1	ω^2	ω

. (3.12)

Note that χ_2 and χ_3 are complex conjugate to each other and that Tr $\gamma = \chi_2 + \chi_3$.

Planar rotations. Let $\mathcal{V}$ indicate the real vector space of all trigonometric Fourier series

$$f(t) = \sum_{m=0}^{\infty} a_m \cos(mt) + b_m \sin(mt), \quad -\pi \le t \le \pi, \quad a_m, b_m \in \mathbb{R},$$

spanned by $\{\cos(mt), \sin(mt) : m = 0, 1, 2, \ldots\}$, and consider the subspaces $\mathcal{W}_\ell$ of $\mathcal{V}$ spanned by $\mathcal{B}_\ell$, the set of the first ℓ elements in $\mathcal{B}$. Then, C_ℓ acts on $\mathcal{B}_\ell$ according to $\varphi(\tau^h, g)(t) = g\tau^{-h}(t) = g(t-h)$. In fact,

$$\begin{bmatrix} \cos(m(t-h)) \\ \sin(m(t-h)) \end{bmatrix} = \begin{bmatrix} \cos(mh) & -\sin(mh) \\ \sin(mh) & \cos(mh) \end{bmatrix} \begin{bmatrix} \cos(mt) \\ \sin(mt) \end{bmatrix},$$

so that $\varphi(\tau^h, \varphi(\tau^k, g) = \varphi(\tau^{h+k}, g))$. This linear relationship defines a family of representations

$$\rho_m(\tau^h) = \begin{bmatrix} \cos(mh) & -\sin(mh) \\ \sin(mh) & \cos(mh) \end{bmatrix}, \quad h, m = 0, 1, \ldots, \ell - 1,$$

in dimension of 2, the direct sum $\rho = \rho_0 \oplus \rho_1 \oplus \ldots \oplus \rho_{\ell-1}$ of which gives a real representation of C_ℓ in dimension of 2ℓ. Moreover, each component ρ_m is irreducible (over $\mathbb{R}$) for $m > 0$ and ρ_0 further reduces as $1 \oplus 1$. For example, when $\ell = 3$, $\rho \simeq 1 \oplus 1 \oplus \rho_1 \oplus \rho_2$.

The construction of the linear representation ρ, described in terms of three-fold planar rotations in the present example ($\ell = 3$), extends naturally to $\ell = \infty$.

This example also outlines the general structure of the representation obtained when the planar rotations are replaced by spherical rotations. In that case, the invariant subspaces under the action of the full three-dimensional rotation group are spanned by the spherical harmonics $Y_{\ell m}(\theta, \phi)$. To each basis indexed by ℓ there corresponds a $2\ell + 1$-dimensional subspace. See also Riley et al. (2002, p. 930).

Theorem 3.1. *Let $\rho : G \to GL(\mathcal{V})$ be a linear representation of G in $\mathcal{V}$ and let W_1 be a vector subspace of $\mathcal{V}$ stable under G. Then there is a complement W_0 of W_1 in $\mathcal{V}$ which is also stable under G.*

Proof. Let $\mathcal{P}_1$ be a projection on W_1 along some vector space complement of W_1 in $\mathcal{V}$. Form the average

$$\overline{\mathcal{P}}_1 = \frac{1}{|G|} \sum_{\tau \in G} \rho_\tau \mathcal{P}_1 \rho_{\tau^{-1}}$$

of projections on W_1 along that vector space complement. Then, $\operatorname{im} \overline{\mathcal{P}}_1 = \{\overline{\mathcal{P}}_1 z; z \in \mathcal{V}\} = W_1$. To see this, first note that for $z \in \mathcal{V}$ we have $\mathcal{P}_1 \rho_{\tau^{-1}} z \in W_1$, and because W_1 is a stable subspace, $\rho_\tau[\mathcal{P}_1 \rho_{\tau^{-1}} z] \in W_1$, so that $\overline{\mathcal{P}}_1 z \in W_1$, that is, $\operatorname{im} \overline{\mathcal{P}}_1 \subseteq W_1$. Second, if $z \in W_1$, which is stable, we have $\rho_{\tau^{-1}} z \in W_1$ for all $\tau \in G$, so that $\mathcal{P}_1 \rho_{\tau^{-1}} z = \rho_{\tau^{-1}} z$. This implies

$$\overline{\mathcal{P}}_1 z = \frac{1}{|G|} \sum_{\tau \in G} \rho_\tau \mathcal{P}_1 \rho_{\tau^{-1}} z = \frac{1}{|G|} \sum_{\tau \in G} \rho_\tau \rho_{\tau^{-1}} z = z,$$

that is, if $z \in W_1$ then $z = \overline{\mathcal{P}}_1 z \in \operatorname{im} \overline{\mathcal{P}}_1$, and hence $W_1 \subseteq \operatorname{im} \overline{\mathcal{P}}_1$. Therefore, $W_1 = \operatorname{im} \overline{\mathcal{P}}_1$. Let then $W_0 = \ker \overline{\mathcal{P}}_1 = \{z \in \mathcal{V}; \overline{\mathcal{P}}_1 z = 0\}$, so that $\mathcal{V} = W_1 \oplus W_0$. To conclude the proof, we must show that W_0 is G-stable: In fact, for all $\tau \in G$,

$$\rho_\tau \overline{\mathcal{P}}_1 \rho_{\tau^{-1}} = \frac{1}{|G|} \sum_{\sigma \in G} \rho_\tau \rho_\sigma \mathcal{P}_1 \rho_{\sigma^{-1}} \rho_{\tau^{-1}} = \frac{1}{|G|} \sum_{\sigma \in G} \rho_{\tau\sigma} \mathcal{P}_1 \rho_{(\tau\sigma)^{-1}} = \frac{1}{|G|} \sum_{\sigma \in G} \rho_\sigma \mathcal{P}_1 \rho_{\sigma^{-1}} = \overline{\mathcal{P}}_1,$$

so that $y \in W_0 = \ker \overline{\mathcal{P}}_1$ implies $\overline{\mathcal{P}}_1 y = 0$ and hence $\overline{\mathcal{P}}_1 \rho_\tau y = \rho_\tau \overline{\mathcal{P}}_1 y = 0$, thus showing that $\rho_\tau y \in W_0$, for all $\tau \in G$. Consequently, W_0 is a stable subspace of $\mathcal{V}$ under G. ∎

The invariant subspaces of a group algebra

With the definitions and notation introduced on page 68, let $\mathcal{I}_1$ indicate a left ideal of the group algebra $\mathcal{A}$. Theorem 3.1 implies that $\mathcal{A}$ decomposes as the direct sum $\mathcal{A} = \mathcal{I}_1 \oplus \mathcal{I}_2$ of $\mathcal{I}_1$ and a complementary ideal $\mathcal{I}_2$. If $x \in \mathcal{A}$ then $x = x_1 + x_2$ with x_1, x_2 in $\mathcal{I}_1, \mathcal{I}_2$ respectively. In particular, the identify $1 \in \mathcal{A}$ can be expressed as $1 = e_1 + e_2$, so that $x = xe_1 + xe_2$, for all $x \in \mathcal{A}$, thus showing that the subspaces $\mathcal{I}_1, \mathcal{I}_2$ are spanned by e_1 and e_2, respectively. When $x \in \mathcal{I}_1$, because $\mathcal{I}_1$ is a left ideal, $xe_1 \in \mathcal{I}_1$ and $x = x(e_1 + e_2) = xe_1$. In particular, for $x = e_1$, $e_1 = e_1^2$. Similarly, $e_2 = e_2^2$. In addition, $e_1 = e_1(e_1 + e_2) = e_1^2 + e_1 e_2 = e_1 + e_1 e_2$, so that $e_1 e_2 = 0$. Similarly, $e_2 e_1 = 0$. Repeating the argument in each component, we obtain a final decomposition of the form

$$\mathcal{A} = \mathcal{I}_1 \oplus \mathcal{I}_2 \oplus \cdots \oplus \mathcal{I}_h, \quad 1 = e_1 + e_2 + \cdots + e_h,$$

with $e_i^2 = e_i$ and $e_i e_j = 0$ for $i \neq j$, and such that each ideal cannot be further reduced as a sum of two left ideals. The irreducible left ideals are called the *primitive idempotents* of the group algebra.

Theorem 3.2. *Every representation is a direct sum of irreducible representations.*

Proof. Given a linear representation in $GL(\mathcal{V})$, if $\mathcal{V}$ is irreducible, the proof is complete. Suppose, otherwise, that $\dim \mathcal{V} = n + 1$ and that the result holds for every representation in $\dim \mathcal{V} \leq n$. Since $\mathcal{V}$ is not irreducible, then, from Theorem 3.1, there are stable subspaces V' and V'' such that $\mathcal{V} = V' \oplus V'' \oplus$ with $\dim V' \leq n$ and $\dim V'' \leq n$. By the induction hypothesis, V' and V'' are direct sum of irreducible representations, and then so is $\mathcal{V}$. ∎

3.8 Schur's Lemma

Lemma 3.1 (Schur). *Let $\rho_i : G \to \mathcal{V}_i$ be irreducible (complex) representations of $G, i = 1, 2$, and let $f : \mathcal{V}_1 \to \mathcal{V}_2$ be a nonzero linear mapping satisfying $f\rho_1(\tau) = \rho_2(\tau) f$ for all $\tau \in G$. Then ρ_1 and ρ_2 are isomorphic. If, in addition, $\mathcal{V}_1 = \mathcal{V}_2$ and $\rho_1 = \rho_2$ then f is a scalar multiple of the identity mapping.*

Proof. Let $W_1 = \ker f = \{x; f(x) = 0\}$. If $x \in W_1$ then $f(x) = 0$ and $f\rho_2(\tau)x = \rho_1(\tau) f(x) = 0$, which implies $\rho_1(\tau)x \in W_1$, for all $\tau \in G$. That is, W_1 is a stable subspace. Since ρ is irreducible, we must have $W_1 = \{0\}$ or $W_1 = \mathcal{V}_1$. If $W_1 = \mathcal{V}_1$

then $f = 0$, contrary to the hypothesis, hence $W_1 = \{0\}$. Similarly, we obtain im f is stable and equal to $\mathcal{V}_2$. Hence, f is an isomorphism, and the two representations are equivalent or isomorphic. To verify the second statement, let λ be an eigenvalue (possibly complex) of f and define $f' = f - \lambda$, understanding that $\lambda \equiv \lambda I$. If $f(x) = \lambda x$ then $(f - \lambda)x = 0$, so that ker $(f - \lambda) \neq \{0\}$, and equivalently, $f - \lambda$ is not an isomorphism. Moreover, $(f - \lambda)\rho_\tau = f\rho_\tau - \lambda\rho_\tau = \rho_\tau f - \rho_\tau\lambda = \rho_\tau(f - \lambda)$, for all $\tau \in G$. From the first part of the Lemma, it follows that $f - \lambda = 0$, or $f = \lambda I$. ■

Proposition 3.2. *For every nonequivalent irreducible representations ρ_1, ρ_2 of G and every linear mapping $H : \mathcal{V}_1 \to \mathcal{V}_2$, it holds that $\sum_{\tau \in G} \rho_1(\tau)H\rho_2(\tau^{-1}) = 0$.*

Proof. Note that $H_0 = \sum_{\tau \in G} \rho_1(\tau)H\rho_2(\tau^{-1})$ is a linear mapping from $\mathcal{V}_1$ into $\mathcal{V}_2$ which intertwines with $\rho_1(\tau)$ and $\rho_2(\tau)$ for all $\tau \in G$, that is, $\rho_1(\tau)H_0 = H_0\rho_2(\tau)$ for all $\tau \in G$. From Schur's Lemma (the representations are nonequivalent irreducible) it follows that $H_0 = 0$. ■

Proposition 3.3. *Let ρ be an irreducible representation of G into $GL(\mathcal{V})$ with dim $\rho = n$. Then, for any linear mapping H in $\mathcal{V}$,*

$$\frac{1}{|G|}\sum_{\tau \in G} \rho_\tau H \rho_{\tau^{-1}} = \frac{Tr\ H}{n} I_n.$$

Proof. Schur's Lemma implies that $H_0 = \sum_{\tau \in G} \rho_\tau H \rho_{\tau^{-1}}/|G| = \lambda I_n$ for some scalar λ. Taking the trace on both sides (and using its invariance under similarity) the result $\lambda = \text{Tr } H/n$ obtains. ■

Consider again the irreducible representations 1, Sgn and β of S_3 (page 64). Let $H = (h_{11}, h_{12})$ be any linear mapping from $\mathbb{R}^2$ into $\mathbb{R}$. From Schur's Lemma we know that

$$\sum_{\tau \in G} \text{Sgn}_\tau H\beta_{\tau^{-1}} = 0,$$

so that the linear forms

$$\sum_{\tau \in G} \text{Sgn}_\tau \left[h_{11}\beta_{11}(\tau^{-1}) + h_{12}\beta_{21}(\tau^{-1})\right], \quad \sum_{\tau \in G} \text{Sgn}_\tau \left[h_{11}\beta_{12}(\tau^{-1}) + h_{12}\beta_{22}(\tau^{-1})\right]$$

in h_{11} and h_{12} must vanish for all values of h_{11} and h_{12}. Therefore, the corresponding coefficients must be zero, that is,

$$\sum_{\tau \in G} \text{Sgn}_\tau \beta_{ij}(\tau^{-1}) = 0, \quad i, j = 1, 2. \tag{3.13}$$

The reader may want to verify the relations in (3.13) from Table (3.14).

τ	1	Sgn(τ)	$\beta_{11}(\tau)$	$\beta_{21}(\tau)$	$\beta_{12}(\tau)$	$\beta_{22}(\tau)$
1	1	1	1	0	0	1
(123)	1	1	−1	1	−1	0
(132)	1	1	0	−1	1	−1
(12)	1	−1	−1	0	−1	1
(13)	1	−1	1	−1	0	−1
(22)	1	−1	0	1	1	0

(3.14)

This is the argument that proves

Corollary 3.1. *For any two nonequivalent irreducible representations ρ, β of G, the relation*

$$\sum_{\tau \in G} \rho_{ij}(\tau)\beta_{k\ell}(\tau^{-1}) = 0$$

holds for all i, j, k, ℓ indexing the entries of these representations.

Consider again the irreducible two-dimensional representation β of S_3. From Proposition 3.2, we know that

$$\frac{1}{|G|} \sum_{\tau \in G} \beta_\tau \begin{bmatrix} h_{11} & h_{12} \\ h_{21} & h_{22} \end{bmatrix} \beta_{\tau^{-1}} = \frac{\operatorname{Tr} H}{2} I_2,$$

implying that, for all scalars $h_{11}, h_{12}, h_{21}, h_{22}$, we must have

$$\frac{1}{|G|} \sum_{\tau \in G} \sum_{j,k=1}^{2} \beta_{ij}(\tau) h_{jk} \beta_{ki}(\tau^{-1}) = \frac{1}{2} h_{11} + \frac{1}{2} h_{22}, \quad i = 1, 2.$$

Consequently, equating the coefficients of the linear forms, the equality $\sum_{\tau \in G} \beta_{ij}(\tau)\beta_{k\ell}(\tau^{-1}) = \delta_{i\ell}\delta_{jk}/2$ must obtain. This is the argument proving the following result:

Proposition 3.4. *For any n-dimensional irreducible representation, ρ, of G we have*

$$\sum_{\tau \in G} \rho_{ij}(\tau)\rho_{k\ell}(\tau^{-1}) = \delta_{i\ell}\delta_{jk}|G|/n.$$

The reader may refer to Table (3.14) and verify the relations obtained from Proposition 3.4.

We conclude this chapter with an outline of the steps leading to the explicit contruction of the irreducible representations of the symmetric group.

3.9 Constructing the Irreducible Representations of S_n

Fix, to illustrate, $n = 3$ and the partition $\lambda = 21$. Indicate by $ij.k$ the assignment

$$\begin{array}{|c|}\hline i \\ \hline \end{array}\ \begin{array}{|c|}\hline j \\ \hline \end{array}$$
$$\begin{array}{|c|}\hline k \\ \hline \end{array}$$

of the permutation $\binom{1\,2\,3}{i\,j\,k}$ to the Young frame (page 33). The Young tableaux or diagrams are the $n!$

12.3, 21.3, 13.2, 31.2, 23.1, 32.1

possible assignments of permutations to Young frames. Two diagrams are equivalent if and only if one is obtained from the other by a permutation of S_3 fixing their row elements. The resulting classes of diagrams are then indexed by the collection of sets defining the different rows. In the present example, we can identify the classes $\{1, 2\}$, $\{1, 3\}$, and $\{2, 3\}$ of diagrams, corresponding to $\{12.3, 21.3\}$, $\{13.2, 31.2\}$, and $\{23.1, 32.1\}$ There are $3 = 3!/2!1!$ classes for $\lambda = 21$.

The symmetric group S_3 acts transitively on these classes according to $\varphi :$ $\{\tau i, \tau j\}$, specifically,

	$\{1, 2\}$	$\{1, 3\}$	$\{2, 3\}$
1	$\{1, 2\}$	$\{1, 3\}$	$\{2, 3\}$
(123)	$\{2, 3\}$	$\{1, 2\}$	$\{1, 3\}$
(132)	$\{1, 3\}$	$\{2, 3\}$	$\{1, 2\}$
(12)	$\{1, 2\}$	$\{2, 3\}$	$\{1, 3\}$
(13)	$\{2, 3\}$	$\{1, 3\}$	$\{1, 2\}$
(23)	$\{1, 3\}$	$\{1, 2\}$	$\{2, 3\}$

giving a permutation representation, indicated here by ρ, of S_3 on M^λ (the subspace of $\mathbb{R}^n$ in dimension equal to the number of classes). This representation is reducible but contains the irreducible representation of interest, which will be identified in the following step.

Define the column stabilizer of a given diagram as the subgroup S_X of S_n leaving all columns of the diagram fixed. Note that S_X is determined by a given diagram and not by the classes of diagrams. However, different classes of diagrams lead to different (sets of) column stabilizers. In our example, $S_{\{1,3\}}$, $S_{\{1,2\}}$, and $S_{\{2,3\}}$ are the stabilizers. Then, for each S_X, construct the projections

$$\mathcal{P}_X = \frac{1}{|S_X|} \sum_{\tau \in S_X} \text{Sgn}_\tau \rho_\tau .$$

Specifically, in the present case,

$$\mathcal{P}_{\{1,3\}} = \frac{1}{2}[\rho_1 - \rho_{(13)}] = \frac{1}{2}\begin{bmatrix} 1 & 0 & -1 \\ 0 & 0 & 0 \\ -1 & 0 & 1 \end{bmatrix}, \quad \mathcal{P}_{\{1,2\}} = \frac{1}{2}[\rho_1 - \rho_{(12)}] = \frac{1}{2}\begin{bmatrix} 0 & 0 & 0 \\ 0 & 1 & -1 \\ 0 & -1 & 1 \end{bmatrix},$$

$$\mathcal{P}_{\{2,3\}} = \frac{1}{2}[\rho_1 - \rho_{(23)}] = \frac{1}{2}\begin{bmatrix} 1 & -1 & 0 \\ -1 & 1 & 0 \\ 0 & 0 & 0 \end{bmatrix},$$

which suggest defining $u = \{1,2\} - \{2,3\}$, $v = \{1,3\} - \{2,3\}$, so that $< u >$, $< v >$, $< u - v >$ are the subspaces spanned by $\mathcal{P}_{\{1,3\}}$, $\mathcal{P}_{\{1,2\}}$ and $\mathcal{P}_{\{2,3\}}$ respectively, and $S^\lambda =< u, v >$ is the resulting subspace spanned by the distinct $\mathcal{P}_X$. It gives a representation

β :

	u	v
1	u	v
(123)	$-v$	$u - v$
(132)	$-u + v$	$-u$
(12)	$u - v$	$-v$
(13)	$-u$	$-u + v$
(23)	v	u

or

$$\beta_1 = \begin{bmatrix} 1 & 0 \\ 0 & 1 \end{bmatrix}, \quad \beta_{(123)} = \begin{bmatrix} 0 & -1 \\ 1 & -1 \end{bmatrix}, \quad \beta_{(132)} \begin{bmatrix} -1 & 1 \\ -1 & 0 \end{bmatrix},$$

$$\beta_{(12)} = \begin{bmatrix} 1 & -1 \\ 0 & -1 \end{bmatrix}, \quad \beta_{(13)} = \begin{bmatrix} -1 & 0 \\ -1 & 1 \end{bmatrix}, \quad \beta_{(23)} = \begin{bmatrix} 0 & 1 \\ 1 & 0 \end{bmatrix},$$

which is the irreducible representation we are looking for.

Similarly, if $\lambda = 111$, there are $3!/1!1!1! = 6$ classes of diagrams of the type $i.j.k$. The action $\tau i.\tau j.\tau k$ of S_3 gives its regular representation ϕ. The single column stabilizer is $S_{\{1,2,3\}}$ and the resulting projection $\sum_{\tau \in S_3} \mathrm{Sgn}_\tau \phi_\tau / 6$ identifies the subspace $\mathcal{V}_{\mathrm{Sgn}} =< r_1 + r_2 + r_3 - t_1 - t_2 - t_3 >$ to appear later on page 105. It gives, of course, the Sign (irreducible) representation of S_3. Similarly, $\lambda = 300$ gives the trivial representation.

Proposition 3.5. *For all* $\tau \in S_n$, $\rho_{\tau^{-1}} \mathcal{P}_X \rho_\tau = \mathcal{P}_{\tau S_X \tau^{-1}}$.

Proof.

$$\begin{aligned} \rho_{\tau^{-1}} \mathcal{P}_X \rho_\tau &= \frac{1}{|S_X|} \sum_{\sigma \in S_X} \mathrm{Sgn}_{\tau^{-1}\sigma\tau} \rho_{\tau^{-1}\sigma\tau} \\ &= \frac{1}{|S_X|} \sum_{\gamma \in \tau S_X \tau^{-1}} \mathrm{Sgn}_\gamma \rho_\gamma = \frac{1}{|\tau S_X \tau^{-1}|} \sum_{\gamma \in \tau S_X \tau^{-1}} \mathrm{Sgn}_\gamma \rho_\gamma = \mathcal{P}_{\tau S_X \tau^{-1}}. \end{aligned}$$

■

Because diagrams have the same cycle structure, when a permutation $\tau \in S_n$ acts on a diagram then its column stabilizer S_X moves to $\tau^{-1} S_X \tau$. For example,

$$\begin{array}{ccc} \begin{matrix} 1 & 2 & 3 \\ 4 & 5 & \end{matrix} & \longrightarrow & S_X = \{1, (14), (25), (14)(25)\} \\ \tau = (1234) \downarrow & & \tau^{-1} S_X \tau \downarrow \\ \begin{matrix} 2 & 3 & 4 \\ 1 & 5 & \end{matrix} & \longrightarrow & S_{X'} = \{1, (12), (35), (12)(35)\} \end{array} \tag{3.15}$$

Now let $U \subset S^\lambda \subseteq M^\lambda$ be an invariant subspace of ρ. Because $U \subset S^\lambda$, U is then spanned by all points $\mathcal{P}_X y$ where X, the index of S_X, is determined by the different diagrams and y is in $\mathcal{V} \subset \mathbb{R}^n$ in dimension equal to the number of classes of diagrams in that partition. Given any $y \in \mathcal{V}$, suppose first that we can choose a diagram such that $\mathcal{P}_X y \neq 0$. Since U is invariant then $\rho_\tau \mathcal{P}_X y \in U$ so that, from Proposition 3.5,

$$\rho_\tau \mathcal{P}_X y = \mathcal{P}_{\tau S_X \tau^{-1}} \rho_\tau y \in U,$$

for all $\tau \in S_n$. As τ varies in S_n, $\mathcal{P}_{\tau S_X \tau^{-1}} \rho_\tau y$ covers a basis for S^λ so that then $S^\lambda \subseteq M^\lambda$.

If otherwise $\mathcal{P}_X y = 0$ for all X (i.e., for all diagrams), let $<, >$ indicate an inner product in $\mathcal{V}$ relative to which ρ is invariant. Direct calculation shows that

$$< \mathcal{P}_X y, z > = < y, \mathcal{P}_X z > \tag{3.16}$$

for all $y, z \in \mathcal{V}$. Consequently, $< y, \mathcal{P}_X z > = 0$ for all X, and hence y is in the orthogonal complement $S^{\lambda\perp}$ of S^λ. In summary.

Proposition 3.6. *Let U be an invariant subspace of M^λ. Then either $S^\lambda \subset U$ or $U \subset S^{\lambda\perp}$. In particular, S^λ is irreducible.*

Proposition 3.6 (Diaconis (1988, p. 133, Submodule Theorem)) gives an irreducible representation for each partition λ and since there are as many irreducible representations as the number of partitions, it only remains to show that these representations, indexed by λ, are all nonequivalent.

Given two partitions λ and μ of n, let

$$\mathcal{P}_\lambda^\mu = \frac{1}{|S_{X_\lambda}|} \sum_{\tau \in S_{X_\lambda}} \text{Sgn}_\tau \rho_\tau^\mu.$$

Then, if any pair of distinct points $\{i, j\} \subset X_\lambda$ share a common row of μ then $\mathcal{P}_\lambda^\mu = 0$. To see this, write $S_{X_\lambda} = \sigma_1 H + \cdots + \sigma_k H$ where $\sigma_1, \ldots, \sigma_k$ are representative of the cosets determined by $H = \{1, (ij)\}$ in S_{X_λ}, so that then

$$\sum_{\tau \in S_{X_\lambda}} \mathrm{Sgn}_\tau \rho_\tau^\mu = \sum_k \sum_{\eta \in H} \mathrm{Sgn}_{\sigma_k \eta} \rho_{\sigma_k \eta}^\mu = \sum_k \mathrm{Sgn}_{\sigma_k} (\rho_1^\mu - \rho_{(ij)}^\mu) = 0.$$

Consequently, if $\mathcal{P}_\lambda^\mu \neq 0$ then the numbers in any given row of μ are all in different columns of λ.

Proposition 3.7. *If the numbers in any given row of μ are all in different columns of λ then $\mu \preceq \lambda$.*

Proof. If the numbers in the first row of $\mu = (\mu_1, \ldots, \mu_n)$ are all in different columns of $\lambda = (\lambda_1, \ldots, \lambda_n)$ then clearly $\mu_1 \leq \lambda_1$. If the numbers in the second row of μ are all in different columns of λ then no column of λ can have more than two of the numbers in the first or second row of μ so that after sliding those entries up to the top of their columns in λ we have that $\mu_1 + \mu_2 \leq \lambda_1 + \lambda_2$. The same argument applies to the subsequent rows. ■

Combining Proposition 3.7 and the remark preceding it, we have

Proposition 3.8. *If $\mathcal{P}_\lambda^\mu \neq 0$ then $\mu \preceq \lambda$.*

Suppose now that there is a nonnull linear mapping $F : M^\mu \mapsto M^\lambda$ intertwining ρ^λ and ρ^μ, that is, $F\rho_\tau^\mu = \rho_\tau^\lambda F$ for all $\tau \in S_n$. Summing in S_{X_λ} gives

$$\frac{|S_{X_\mu}|}{|S_{X_\lambda}|} F\, \mathcal{P}_\lambda^\mu = \mathcal{P}_{X_\lambda}\, F,$$

so that either (i) $F \subset \ker \mathcal{P}_{X_\lambda}$ for all X_λ indexed by λ or (ii) $\mathcal{P}_{X_\lambda}\, F \neq 0$ for at least one X_λ, and hence $\mathcal{P}_\lambda^\mu \neq 0$, in which case $\mu \preceq \lambda$, from Proposition 3.8. In case of (i), we have $\mathcal{P}_{X_\lambda} F y = 0$ for all X indexed by λ, so that then $F \subset S^{\lambda\perp}$.

As a result, given a mapping $S^\mu \mapsto S^\lambda$ intertwining with ρ, extend it linearly to a mapping $F : S^\mu \mapsto S^\lambda$ by defining it to be zero on $S^{\lambda\perp}$ and if $F \neq 0$ then $\mu \preceq \lambda$. Similar argument applied to $S^\mu \mapsto S^\lambda$ then gives $\lambda = \mu$. This proves

Theorem 3.3. *The S^λ are all of the irreducible representations of S_n.*

An irreducible representation of S_4

We will obtain an irreducible representation of S_4 indexed by the partition $\lambda = 3100$. Associated with the Young frame

$$\begin{array}{ccc} \boxed{i} & \boxed{j} & \boxed{k} \\ \boxed{\ell} & & \end{array}$$

there are $4! = 24$ Young diagrams, such as 123.4, and $4 = 4!/3!1!$ classes

$$a = \{1, 2, 3\}, \quad b = \{1, 2, 4\}, \quad c = \{1, 3, 4\}, \quad d = \{2, 3, 4\}$$

of (row permutation) equivalent diagrams, upon which S_4 acts transitively according to $\varphi : \{\tau i, \tau j, \tau k\}$, giving a representation ρ. Since the column stabilizers are

$$S_X = S_{\{1,2\}}, \quad S_{\{1,3\}}, \quad S_{\{1,4\}}, \quad S_{\{2,3\}}, \quad S_{\{2,4\}}, \quad S_{\{3,4\}},$$

we need to evaluate φ only at the permutations defined by these subgroups. It gives

τ	a	b	c	d
1	a	b	c	d
(12)	a	b	d	c
(13)	a	d	c	b
(14)	d	b	c	a
(23)	a	c	b	d
(24)	c	b	a	d
(34)	b	a	c	d

,

from which the corresponding ρ_τ are obtained. The evaluation of

$$\mathcal{P}_X = \frac{1}{|S_X|} \sum_{\tau \in S_X} \mathrm{Sgn}_\tau \rho_\tau,$$

for each S_X gives six projections in the dimension of 1, jointly defining S^λ. For example,

$$\mathcal{P}_{\{1,2\}} = \frac{1}{2}(\rho_1 - \rho_{(12)}) = \frac{1}{2} \begin{bmatrix} 0 & 0 & 0 & 0 \\ 0 & 0 & 0 & 0 \\ 0 & 0 & 1 & -1 \\ 0 & 0 & -1 & 1 \end{bmatrix},$$

which identifies the invariant $\pm(c - d)$. Together, the six projections identify the basis $S^\lambda =< u, v, w >$ for the irreducible representation, where

$$u = c - d = \{1, 3, 4\} - \{2, 3, 4\}, \quad v = b - c = \{1, 2, 4\} - \{1, 3, 4\},$$
$$w = a - c = \{1, 2, 3\} - \{1, 3, 4\}.$$

Table (3.17) gives the action of the elementary generators of S_4 on S^λ. The full representation is shown in Table (3.19), corresponding to the permutations given by Table (3.18).

	u	v	w
1	u	v	w
(12)	$-u$	$v+u$	$u+w$
(13)	$-v$	$-u$	w
(14)	$-w$	v	$-u$
(23)	$v+u$	$-v$	$-v+w$
(24)	$u+w$	$v-w$	$-w$
(34)	u	w	v
123)	v	$-v-u$	$-v+w$
(234)	$v+u$	$-v+w$	$-v$
(134)	$-v$	w	$-u$
(124)	w	$v-w$	$-u-w$
(1234)	v	$-v+w$	$-v-u$
(1324)	$-v+w$	$-w$	$-u-w$
(1243)	w	$-u-w$	$v-w$

(3.17)

1	(12)	(13)	(14)	(23)	(24)
(34)	(123)	(132)	(234)	(243)	(134)
(143)	(124)	(142)	(1234)	(13)(24)	(1432)
(1324)	(12)(34)	(1423)	(1243)	(14)(23)	(1342)

. (3.18)

1	0	0	−1	0	0	0	−1	0	0	0	−1	1	1	0	1	0	1
0	1	0	1	1	0	−1	0	0	0	1	0	0	−1	0	0	1	−1
0	0	1	1	0	1	0	0	1	−1	0	0	0	−1	1	0	0	−1
1	0	0	0	1	0	−1	−1	0	1	1	0	1	0	1	0	−1	0
0	0	1	−1	−1	0	1	0	0	0	−1	1	0	0	−1	0	0	1
0	1	0	0	−1	1	1	0	1	0	−1	0	0	1	−1	−1	0	0
0	0	−1	0	0	1	−1	0	−1	0	1	0	0	−1	1	−1	0	−1
−1	0	0	0	1	−1	1	1	0	0	−1	1	−1	0	−1	1	0	0
0	1	0	−1	0	−1	1	0	0	−1	−1	0	0	0	−1	1	1	0
0	−1	1	−1	0	0	0	1	−1	0	0	1	0	1	−1	−1	−1	0
0	0	−1	1	0	1	−1	−1	0	−1	0	−1	0	−1	0	1	0	1
−1	0	−1	1	1	0	0	−1	0	0	1	−1	−1	−1	0	1	0	0

. (3.19)

Further Reading

The reader interested in a concise and yet comprehensive historical account of the first 100 years of representations of finite groups will enjoy reading Lam (1998a) and Lam (1998b). The theory of the representations of the symmetric groups, developed in James (1978), is elegantly summarized in Diaconis (1988, Chapter 7).

Exercises

Exercise 3.1. Show that any linear operator ρ on a finite-dimensional vector space when applied to a basis $\{e_s; s \in V\}$ for $\mathcal{V}$ gives $\rho e_s = \sum_h (\rho)_{hs} e_h$.

Exercise 3.2. Show that if ρ is a linear representation of S_ℓ in dimension of v then $\tau \mapsto \text{Sgn}_\tau \rho_\tau$ is also a linear representation of S_ℓ in dimension of v.

Exercise 3.3. Following with the notation of Section 3.7 on page 74, show that

$$\mathcal{V} \times \mathcal{V} = < v_1 + v_2, v_3 > \oplus < v_1 - v_2, v_4 >,$$

and that each component of $\mathcal{V} \times \mathcal{V}$ is a stable subspace under $\rho \otimes \rho$.

Exercise 3.4. Show that

$$\rho_i = \begin{bmatrix} 0 & -1 \\ 1 & 0 \end{bmatrix}, \quad \rho_{-1} = \begin{bmatrix} -1 & 0 \\ 0 & -1 \end{bmatrix}, \quad \rho_{-i} = \begin{bmatrix} 0 & 1 \\ -1 & 0 \end{bmatrix}, \quad \rho_1 = \begin{bmatrix} 1 & 0 \\ 0 & 1 \end{bmatrix}$$

is a two-dimensional representation of the complex group.

Exercise 3.5. Indicate by $\mathcal{F}(G)$ the vector space of all complex-valued functions defined on a given group G and let $\varphi_\sigma(x)(\tau) = x(\sigma\tau\sigma^{-1})$, where for $\sigma, \tau \in G$ and $x \in \mathcal{F}(G)$. Show that

(1) φ is a group action of G on $\mathcal{F}(G)$;
(2) $\varphi_\sigma \in GL(\mathcal{F}(G))$;
(3) $\sigma \in G \mapsto \varphi_\sigma \in GL(\mathcal{F}(G))$ is a group homomorphism.

The result is a linear representation of G in $GL(\mathcal{F}(G))$.

Exercise 3.6. Data indexed on a subset (fraction) of $\{0, 1, 2\}^3$. Let $L = \{0, 1, 2\}$ and consider the subset

$$V = \begin{array}{|c|ccc|} \hline & 0 & 1 & 2 \\ \hline 0 & 0 & 1 & 2 \\ 1 & 2 & 0 & 1 \\ 2 & 1 & 2 & 0 \\ \hline \end{array} \subset L^3,$$

or,

$$V = \left[0\ 0\ 0 \middle| 0\ 1\ 1 \middle| 0\ 2\ 2 \middle| 1\ 0\ 2 \middle| 1\ 1\ 0 \middle| 1\ 2\ 1 \middle| 2\ 0\ 1 \middle| 2\ 1\ 2 \middle| 2\ 2\ 0\right],$$

as the set of labels of interest. Show that the group $C_3 \times C_3$ acts on V by permutation according to

$$((\sigma, \tau), (i, j, k)) = (\sigma i, \tau j, \sigma^{-1}\tau k), \tag{3.20}$$

giving a representation

σ	τ	0	0	0	0	1	1	0	2	2	1	0	2	1	1	0	1	2	1	2	0	1	2	1	2	2	2	0
0	0	0	0	0	0	1	1	0	2	2	1	0	2	1	1	0	1	2	1	2	0	1	2	1	2	2	2	0
0	1	0	1	1	0	2	2	0	0	0	1	1	0	1	2	1	1	0	2	2	1	2	2	2	0	2	0	1
0	2	0	2	2	0	0	0	0	1	1	1	2	1	1	0	2	1	1	0	2	2	0	2	0	1	2	1	2
1	0	1	0	2	1	1	0	1	2	1	2	0	1	2	1	2	2	2	0	0	0	0	0	1	1	0	2	2
1	1	1	1	0	1	2	1	1	0	2	2	1	2	2	2	0	2	0	1	0	1	1	0	2	2	0	0	0
1	2	1	2	1	1	0	2	1	1	0	2	2	0	2	0	1	2	1	2	0	2	2	0	0	0	0	1	1
2	0	2	0	1	2	1	2	2	2	0	0	0	0	0	1	1	0	2	2	1	0	2	1	1	0	1	2	1
2	1	2	1	2	2	2	0	2	0	1	0	1	1	0	2	2	0	0	0	1	1	0	1	2	1	1	0	2
2	2	2	2	0	2	0	1	2	1	2	0	2	2	0	0	0	0	1	1	1	2	1	1	0	2	1	1	0

or, equivalently,

σ	τ	1	2	3	4	5	6	7	8	9
0	0	1	2	3	4	5	6	7	8	9
0	1	2	3	1	5	6	4	8	9	7
0	2	3	1	2	6	4	5	9	7	8
1	0	4	5	6	7	8	9	1	2	3
1	1	5	6	4	8	9	7	2	3	1
1	2	6	4	5	9	7	8	3	1	2
2	0	7	8	9	1	2	3	4	5	6
2	1	8	9	7	2	3	1	5	6	4
2	2	9	7	8	3	1	2	6	4	5

Argue that this example suggests the construction of group actions θ on $V \times V$ starting with an action φ on V, such as $\theta_\tau(s, f) = (s, \varphi_\tau f)$.

Exercise 3.7. Let ρ be a representation of G (with g elements) on a finite-dimensional vector space $\mathcal{V}$, in which a inner product $(\,|\,)$ is defined (e.g., page 66). Show that $(x, y) = \sum_{\tau \in G}(\rho_\tau x | \rho_\tau y)/g$ is a inner product in $\mathcal{V}$ and that it satisfies $(\rho_\tau x, \rho_\tau y) = (x, y)$ for all $\tau \in G$ and all $x, y \in \mathcal{V}$.

Exercise 3.8. [Contributed by K.S. Mallesh] Show that the matrices centralized by the regular representation of S_3 have the pattern

and identify their components.

4

Data Reduction and Inference: The Canonical Projections and Their Invariants

4.1 Introduction

In the previous chapter we introduced the linear representations as a connection between the symmetries applied to the data labels and the data vector space. We now follow it with the construction of the canonical projection matrices and the study and interpretation of the canonical invariants on the data. These are the concluding steps outlined on page 23 of Chapter 1.

4.2 Characters of a Linear Representation

Given a linear representation ρ in $GL(\mathcal{V})$, the complex-valued function $\chi_\rho(\tau) = \text{Tr } \rho_\tau$ is called the character of the representation. Since ρ_1 is the indentity matrix in $\mathcal{V}$, we see that $\chi_\rho(1) = \dim \mathcal{V}$.

If λ is an eigenvalue of ρ, then, relative to the invariant inner product defined in (3.6), we have

$$(y, y) = (\rho_\tau y, \rho_\tau y) = (\lambda y, \lambda y) = \lambda\overline{\lambda}(y, y),$$

so that $\lambda\overline{\lambda} = 1$. Let $\lambda_1, \ldots, \lambda_m$ indicate the (eventually complex) eigenvalues of ρ_τ. Then

$$\chi_\rho(\tau^{-1}) = \text{Tr } \rho_{\tau^{-1}} = \text{Tr } \rho_\tau^{-1} = \sum \lambda_i^{-1} = \sum \overline{\lambda}_i = \text{Tr } \overline{\rho}_\tau = \overline{\chi}_\rho(\tau). \qquad (4.1)$$

Also note, since the trace operations is invariant under similarity, that $\chi_\rho(\tau\sigma\tau^{-1}) = \chi_\rho(\sigma)$, for all $\tau, \sigma \in G$. It is also simple to verify that if ρ and θ are two linear representations of G with corresponding characters χ_ρ and χ_θ, then $\chi_{\rho\oplus\theta} = \chi_\rho + \chi_\theta$ and $\chi_{\rho\otimes\theta} = \chi_\rho\chi_\theta$.

Recall, from page 74, that $\rho \otimes \rho \simeq \text{Sym}^2 \oplus \text{Alt}^2$. Direct evaluation of the action of C_4 on the components of

$$D \cup U = \{(1, 1), (1, 2), (1, 3), (1, 4), (2, 2), (2, 3), (2, 4), (3, 3), (3, 4), (4, 4)\},$$

$$U = \{(1, 2), (1, 3), (1, 4), (2, 3), (2, 4), (3, 4)\},$$

leads to the following table of characters:

C_4	$\chi_\rho(\tau)$	$\chi_{\mathrm{Sym}^2}(\tau)$	$\chi_{\mathrm{Alt}^2}(\tau)$
1	4	10	6
(1234)	0	0	0
(13)(24)	0	2	−2
(1432)	0	0	0

Note that, for all $\tau \in C_4$, $\chi_\rho^2(\tau) = \chi_{\rho\otimes\rho}(\tau) = \chi_{\mathrm{Sym}^2}(\tau) + \chi_{\mathrm{Alt}^2}(\tau)$, and

$$\chi_{\mathrm{Sym}^2}(\tau) = \frac{1}{2}\left(\chi_\rho^2(\tau) + \chi_\rho(\tau^2)\right), \quad \chi_{\mathrm{Alt}^2}(\tau) = \frac{1}{2}\left(\chi_\rho^2(\tau) - \chi_\rho(\tau^2)\right).$$

It can be shown that these two equalities hold in general for any linear representation ρ of G.

Characters and direct products of groups. Recall, from page 70, that if ρ and η are linear representations of groups G and H with characters χ_ρ and χ_η, respectively, then $\xi_{\tau,\sigma} = \rho_\tau \otimes \eta_\sigma$ is a linear representation of $G \times H$ on $\mathcal{V} \times \mathcal{W}$ and $\chi_\xi(\tau, \sigma) = \chi_\rho(\tau)\chi_\eta(\sigma)$ is its character.

4.3 Orthogonality Relations for Characters

Following Section 3.3, we observe that

$$(f \mid g) = \frac{1}{|G|}\sum_{\tau\in G} f(\tau)\overline{g}(\tau) \tag{4.2}$$

is a inner product in the vector space $\mathcal{F}(G)$ of complex-valued functions defined in G. In particular, if χ_1 and χ_2 are characters of a representation of G, then $\chi_1, \chi_2 \in \mathcal{F}(G)$, and because $\chi(\tau^{-1}) = \overline{\chi}(\tau)$, we have

$$(\chi_1 \mid \chi_2) = \frac{1}{|G|}\sum_{\tau\in G} \chi_1(\tau)\overline{\chi}_2(\tau) = \frac{1}{|G|}\sum_{\tau\in G} \chi_1(\tau)\chi_2(\tau^{-1}).$$

From Section 3.3 we may assume that the representation ρ is unitary so that Proposition 3.4 can then be expressed as

$$\sum_{\tau\in G} \rho_{ij}(\tau)\rho_{k\ell}(\tau^{-1}) = \sum_{\tau\in G} \rho_{ij}(\tau)\rho_{k\ell}^*(\tau) = |G|(\rho_{ij} \mid \rho_{\ell k}) = \delta_{i\ell}\delta_{jk}|G|/n. \tag{4.3}$$

Similarly, Corollary 3.1 becomes

$$(\rho_{ij} \mid \beta_{k\ell}) = 0, \quad \text{for all } i, j, k, \ell, \tag{4.4}$$

where ρ and β are two nonequivalent irreducible representations of G.

Irreducible characters

We often refer to the character of an irreducible representation as an irreducible character and indicate the set of irreducible characters of a group G by $\widehat{G}$. The same notation will be used to indicate a set of nonequivalent irreducible representations of G.

Theorem 4.1. $(\chi_i \mid \chi_j) = \delta_{ij}$ *for any* $\chi_i, \chi_j \in \widehat{G}$.

Proof. From Expression (4.3), we have

$$(\chi \mid \chi) = \frac{1}{|G|}\sum_{\tau\in G}\left(\sum_{i=1}^{n}\rho_{ii}(\tau)\Big|\sum_{j=1}^{n}\rho_{jj}(\tau)\right) = \sum_{i=1}^{n}(\rho_{ii} \mid \rho_{ii}) = \sum_{i=1}^{n}\frac{1}{n} = 1,$$

whereas, from Expression (4.4), similarly, we obtain $(\chi_1 \mid \chi_2) = 0$. ■

The irreducible characters of S_3. Table (4.5) shows three irreducible characters of S_3, corresponding to the irreducible representations 1, Sgn, and β discussed earlier on page 77.

τ	χ_ρ	χ_1	χ_{Sgn}	χ_β
1	3	1	1	2
(123)	0	1	1	−1
(132)	0	1	1	−1
(12)	1	1	−1	0
(13)	1	1	−1	0
(23)	1	1	−1	0

(4.5)

It also shows the character χ_ρ of the representation ρ of S_3. The reader may want to verify, from Table (4.5) that $(\chi_1|\chi_\rho) = (\chi_\beta|\chi_\rho) = (\chi_{\text{Sgn}}|\chi_\rho) = 1$. We will argue, later in the chapter, that these are all the irreducible representations of S_3.

A remark on notation. Given a linear representation ρ, the notation $\rho \simeq m_1\rho_1 \oplus \ldots \oplus m_h\rho_h$ indicates that there is a basis in $\mathcal{V}$ relative to which

$$\rho_\tau = \text{Diag}\,(I_{m_1} \otimes \rho_1(\tau), \ldots, I_{m_h} \otimes \rho_h(\tau)), \quad \tau \in G,$$

where I_m is the $m \times m$ identity matrix. If $\rho_1, \rho_2, \ldots \in \widehat{G}$ then, from Theorem 4.1, it follows that $m_i = (\chi_i|\chi)$. That is:

Proposition 4.1. *If ρ is a linear representation of G with character χ and $\xi \in \widehat{G}$, then $(\xi|\chi)$ is the number of representations in any decomposition of ρ that are isomorphic to ξ.*

To illustrate, consider the action $s\tau^{-1}$ of $S_2 = \{1, t\}$ on the position of the symbols in $\{uu, yy, uy, yu\}$, giving the representation

$$\rho_1 = I_4, \quad \rho_t = \text{Diag}\left(1, 1, \begin{bmatrix} 0 & 1 \\ 1 & 0 \end{bmatrix}\right),$$

or, relative to a basis for $\mathcal{V}$ indexed by $\{uu, yy, uy + yu, uy - yu\}$,

$$\rho_1 = I_4, \quad \rho_t = \text{Diag}(1, 1, 1, \text{Sgn}),$$

that is $\rho \simeq 1 \oplus 1 \oplus 1 \oplus \text{Sgn}$. In fact, from the character table

τ	χ_ρ	χ_1	χ_{Sgn}
1	4	1	1
t	2	1	-1

we obtain $(\chi_1 \mid \chi_\rho) = [\chi_1(1)\chi_\rho(1) + \chi_1(t)\chi_\rho(t)]/2 = (4 + 2)/2 = 3$, which is the multiplicity with which the symmetric representation appears in ρ. The alternating representation appears with multiplicity

$$(\chi_{\text{Sgn}} \mid \chi_\rho) = \frac{1}{2}(1 \times 4 + (-1) \times 2) = 1.$$

In correspondence, $\mathcal{V}$ decomposes as the direct sum $3\mathcal{V}_1 \oplus \mathcal{V}_{\text{Sgn}}$ of irreducible subspaces.

Similarly, the action σs of S_2 V gives the representation

$$\rho_1 = I_4, \quad \rho_t = \text{Diag}\left(\begin{bmatrix} 0 & 1 \\ 1 & 0 \end{bmatrix}, \begin{bmatrix} 0 & 1 \\ 1 & 0 \end{bmatrix}\right) \simeq \text{Diag}(1, \text{Sgn}, 1, \text{Sgn}),$$

so that $\rho \simeq 1 \oplus 1 \oplus \text{Sgn} \oplus \text{Sgn}$.

Note that the multiplicity of a given irreducible component does not depend on the underlying choice of basis. Moreover, two representations with the same set of characters are isomorphic, because they contain each irreducible component with exactly the same multiplicity. These arguments reflect the importance of characters in the study of linear representations. It is in that sense that irreducible representations are the building blocks of generic representations.

In summary, if ρ is a linear representation of G on $\mathcal{V}$, we may then restrict our attention to the set $\chi_1, \ldots, \chi_h$ of distinct irreducible characters of G, and write

$$\rho \simeq m_1\rho_1 \oplus \ldots \oplus m_h\rho_h, \quad \chi_\rho = m_1\chi_1 + \cdots + m_h\chi_h, \quad \mathcal{V} = m_1\mathcal{V}_1 \oplus \ldots \oplus m_h\mathcal{V}_h, \quad (4.6)$$

where the multiplicities m_i are given by the integers $(\chi_\rho \mid \chi_i) \geq 0$ and $\mathcal{V}_i$ are subspaces of $\mathcal{V}$ in the dimension equal to the dimension of ρ_i, $i = 1, \ldots, h$. In the previous example, under location symmetry, $\chi_\rho = 3\chi_1 + \chi_{\text{Sgn}}$.

Moreover, the orthogonality relations among the irreducible components imply that $(\chi_\rho \mid \chi_\rho) = \sum_{i=1}^h m_i^2$. The following result is a useful characterization of the irreducible representations.

Theorem 4.2. $(\chi_\rho \mid \chi_\rho) = 1$ *if and only if* ρ *is irreducible.*

Proof. We have $(\chi_\rho \mid \chi_\rho) = \sum_{i=1}^h m_i^2 = 1$ if and only if exactly one of the m_i's is equal to 1 and all the others are equal to 0, in which case ρ is isomorphic to that irreducible component. ■

To illustrate, consider the irreducible representations 1, β and Sgn of S_3 (page 77), along with the representation $\beta \otimes \beta$. Table (4.7) shows the corresponding characters, from which it can be verified that $(\chi_\beta \mid \chi_\beta) = (\chi_1 \mid \chi_1) = (\chi_{\text{Sgn}} \mid \chi_{\text{Sgn}}) = 1$.

τ	β	$\beta \otimes \beta$	1	Sgn
1	2	4	1	1
(123)	−1	1	1	1
(132)	−1	1	1	1
(12)	0	0	1	−1
(13)	0	0	1	−1
(23)	0	0	1	−1

(4.7)

Since $(\chi_{\beta\otimes\beta} \mid \chi_{\beta\otimes\beta}) = 18/6 = 3$, the representation $\beta \otimes \beta$ must be reducible. On the other hand,

$$(\chi_{\beta\otimes\beta} \mid \chi_\beta) = (\chi_{\beta\otimes\beta} \mid \chi_1) = (\chi_{\beta\otimes\beta} \mid \chi_{\text{Sgn}}) = 1,$$

so that these representations appear in its decomposition with single multiplicity. That is, $\beta \otimes \beta \simeq 1 \oplus \beta \oplus \text{Sgn}$, with the corresponding character decomposition.

The characters of the regular representation

Of particular interest in the study of structured data is the regular representation (ϕ), introduced on page 65. Its dimension is the number $|G|$ of elements in the group G. For example, the regular representation of D_4 is shown on page 72.

Since, for all $\sigma \in G$, $\varphi(\tau, \sigma) = \varphi(\eta, \sigma)$ if and only if $\tau = \sigma$, and $\varphi(\tau, 1) = \tau$ for all $\tau \in G$, it follows that its character is given by

$$\chi_\phi(\tau) = \begin{cases} 0 & \text{if } \tau \neq 1; \\ |G| & \text{if } \tau = 1. \end{cases}$$

Consequently, for any irreducible representation ρ of G with character χ_ρ, we have

$$(\chi_\phi, \chi_\rho) = \frac{1}{|G|} \sum_{\tau \in G} \chi_\phi(\tau)\chi_\rho(\tau^{-1}) = \chi_\rho(1) = \dim \rho, \tag{4.8}$$

that is, from (4.6), every irreducible representation is contained in the regular representation with multiplicity $(\chi_\phi|\chi_\rho)$ equal to its dimension.

Proposition 4.2. *The dimensions $n_1, \ldots, n_h$ of the distinct irreducible representations of G, satisfy the relation $|G| = \sum_{i=1}^h n_i^2$.*

Proof. From (4.6) and (4.8) we have $\chi_\phi(\tau) = \sum_{i=1}^h m_i \chi_i(\tau)$, with $m_i = (\chi_\phi|\chi_i) = \dim \rho_i = n_i$. Taking $\tau = 1$, the proposed equality obtains. ■

Note that for $\tau \neq 1$, the defining property of χ_ϕ implies that $\sum_{i=1}^h n_i \chi_i(\tau) = 0$. This equality, together with Proposition 4.2, shows that $\sum_{i=1}^h n_i \chi_i(\sigma^{-1}\tau) = |G|\delta_{\sigma\tau}$.

To illustrate, the irreducible nonequivalent representations 1, β and Sgn of S_3 are contained in the regular representation with multiplicities 1, 2, 1, respectively. Because $|S_3| = 6 = 1^2 + 2^2 + 1^2$, these must be all the distinct irreducible nonequivalent representations of S_3.

The irreducible characters of S_4. Table (4.9) gives the character of the irreducible representation of S_4, from page 87, indicated here by χ_θ. The first row shows representatives of the conjugacy classes of S_4, with the corresponding number of elements in each class in the second row. In addition to the symmetric χ_1 and alternating χ_{Sgn} characters, it also includes the irreducible character $\chi_{\text{Sgn}}\chi_\theta$ (applying Theorem 4.2 on the previous page) in dimension of 3. From Proposition 4.2, and knowing that the number of conjugacy classes of S_4 is five, the remaining character, indicated here by χ_β must be in dimension of 2 so that $24 = 1 + 1 + 3^2 + 3^2 + 2^2$. This gives $\chi_\beta(1) = 2$. Using the orthogonality relations (Theorem 4.1 on page 93) with the existing (four) characters determines the remaining (four) entries of χ_β, thus completing the table.

	1	(12)	(123)	(12)(34)	(1234)
	1	6	8	3	6
1	1	1	1	1	1
χ_{Sgn}	1	−1	1	1	−1
χ_θ	3	1	0	−1	−1
$\chi_\theta\chi_{\text{Sgn}}$	3	−1	0	−1	1
χ_β	2	0	−1	2	0

(4.9)

The irreducible characters of D_n. Assume first that n is even and refer to the representation $\beta_{k,d}$ of D_n shown on page 42. Using its canonical space realization, it is simple to verify that $\beta^m_{k,d}$, for $m = 1, \ldots, n/2 - 1$ is also a representations of D_n, in dimension of 2. Indicating by χ_m its character, then, clearly $\chi_m \neq \chi_\ell$ for $m \neq \ell$, so that the corresponding representations are nonisomorphic. In addition,

$$(\chi_m \mid \chi_\ell) = \frac{1}{2n} \sum_{d=\pm 1} \sum_{k=0}^{n-1} (1+d)^2 \cos(mk\phi)\cos(\ell k\phi) = \frac{2}{n} \sum_{k=0}^{n-1} \cos(mk\phi)\cos(\ell k\phi) = \delta_{m\ell},$$

for $m, \ell = 1, \ldots, n/2 - 1$, so that (from Theorem 4.2) these are nonequivalent irreducible characters, in dimension of 2. Together, their square dimention adds to $2^2(n/2 - 1) = 2n - 4$. The remaining four dimensions (Proposition 4.2) are found by assigning four one-dimensional characters as follows: χ_1 is the trivial character; $\chi_{\text{Sgn}}(k, d) = d$; $\chi_\theta(k, d) = (-1)^k$ and finally $\chi_{\text{Sgn}}(k, d)\chi_\theta(k, d)$.

When n is odd, the construction above gives, similarly, $(n - 1)/2$ irreducible representations in dimension of 2, namely, $\beta^m_{k,d}$, for $m = 1, \ldots, (n - 1)/2$. They account for $2^2(n - 1)/2 = 2n - 2 = |G| - 2$ dimensions. The remaining two are the symmetric and the alternating representations.

Irreducible characters and direct product of groups. If χ_ρ and χ_η are irreducible characters of groups G and H then $\chi_\xi(\tau, \sigma) = \chi_\rho(\tau)\chi_\eta(\sigma)$ is an irreducible character of $G \times H$. Indeed, applying Theorem 4.2,

$$(\chi_\xi \mid \chi_\xi) = \frac{1}{|G||H|} \sum_{(\tau,\sigma)\in G\times H} \chi_\xi(\tau, \sigma)^2 = (\chi_\rho \mid \chi_\rho)(\chi_\eta \mid \chi_\eta) = 1.$$

Moreover, every irreducible character of $G \times H$ is a product of irreducible characters of G and H. To see this, note, from Proposition 4.2 on the facing page that

$$\sum_{\rho,\eta} [\chi_\rho(1)\chi_\eta(1)]^2 = \sum_\rho \chi^2_\rho(1) \sum_\eta \chi^2_\eta(1) = |G||H|,$$

where the sum is over all irreducible representations of G and H. The same argument extends similarly to the product of more than two groups.

To illustrate, let $G = \{1, v, h, o\} \simeq C_2 \times C_2$. The irreducible characters of $C_2 = \{1, t\}$ are the characters

$$\chi_1 : (1, t) \mapsto (1, 1), \quad \chi_{\text{Sgn}} : (1, t) \mapsto (1, -1)$$

of the symmetric and the alternating representations. It then follows that

τ	σ	$\chi_1(\tau)\chi_1(\sigma)$	$\chi_1(\tau)\chi_{\mathrm{Sgn}}(\sigma)$	$\chi_{\mathrm{Sgn}}(\tau)\chi_1(\sigma)$	$\chi_{\mathrm{Sgn}}(\tau)\chi_{\mathrm{Sgn}}(\sigma)$
1	1	1	1	1	1
1	t	1	-1	1	-1
t	1	1	1	-1	-1
t	t	1	-1	-1	1

(4.10)

are the irreducible characters of $C_2 \times C_2$. Note that $C_2 \times C_2$ is Abelian and that all characters are in dimension of 1.

Class functions

A scalar-valued function x defined on G and satisfying $x(\tau\sigma\tau^{-1}) = x(\sigma)$, for all $\sigma, \tau \in G$ is called a class function.

Clearly, class functions are constant within each conjugacy class of G. We indicate by $\mathcal{C}$ the set of class functions on G. Note that $\mathcal{C}$ is a linear subspace of the vector space $\mathcal{F}(G)$ of scalar-valued functions defined on G. All characters belong to $\mathcal{C}$. From Exercise 3.5 on page 88, we observe that $\mathcal{C}$ is a stable subspace of $\mathcal{F}(G)$ under the linear representation $\sigma \xrightarrow{\varphi} \varphi_\sigma$, that is, $x \in \mathcal{C}$ implies $\varphi_\sigma(\sigma)x = x$, for all $\sigma \in G$. More precisely, $\mathcal{C}$ is the subspace of $\mathcal{F}(G)$ of functions invariant under this conjugation action. For each class function x and any representation ρ, define the linear mapping

$$\widehat{x}(\rho) = \sum_{\tau \in G} x_\tau \rho_\tau. \tag{4.11}$$

Note that $\widehat{x}(\rho)$ commutes with ρ_τ for all $\tau \in G$. In fact,

$$\begin{aligned} \rho_\tau \widehat{x}(\rho)\rho_{\tau^{-1}} &= \rho_\tau \sum_\sigma x_\sigma \rho_\sigma \rho_{\tau^{-1}} = \sum_\sigma x_\sigma \rho_\tau \rho_\sigma \rho_{\tau^{-1}} = \sum_\sigma x_\sigma \rho_{\tau\sigma\tau^{-1}} \\ &= \sum_\sigma x_{\tau\sigma\tau^{-1}} \rho_{\tau\sigma\tau^{-1}} = \sum_\sigma x_\sigma \rho_\sigma = \widehat{x}(\rho). \end{aligned}$$

Therefore, if ρ is an irreducible representation, it follows from Schur's Lemma that $\widehat{x}(\rho) = \lambda I$. To evaluate λ we take the trace in each side of the above equality, to obtain

$$\operatorname{Tr} \widehat{x}(\rho) = \sum_{\tau \in G} x_\tau \operatorname{Tr} \rho_\tau = \sum_{\tau \in G} x_\tau \chi_\rho(\tau) = \sum_{\tau \in G} x_\tau \overline{\chi}_\rho(\tau^{-1}) = |G|(x, \overline{\chi}_\rho) = \operatorname{Tr} \lambda I_n = n\lambda,$$

so that $\lambda = |G|(x, \overline{\chi}_\rho)/n$. This proves

Proposition 4.3. *If ρ is an n-dimensional irreducible representation of G and x is a class function, then*

$$\widehat{x}(\rho) = \frac{|G|}{n}(x|\overline{\chi}_\rho)I_n.$$

Theorem 4.3. *The distinct irreducible characters form an orthonormal basis for $\mathcal{C}$.*

Proof. From Theorem 4.1 we know that the set of distinct irreducible characters form an orthonormal set of functions in $\mathcal{C}$. We need to show that this set generates $\mathcal{C}$. Suppose that $x \in \mathcal{C}$ and that x is orthogonal to $\overline{\chi}_1, \ldots, \overline{\chi}_h$. Therefore, for any irreducible n-dimensional representation ρ of G, we have

$$\widehat{x}(\rho) = \frac{|G|}{n}(x \mid \overline{\chi}_\rho)I_n = 0, \quad x \in \mathcal{C}.$$

Because every representation decomposes isomorphically as a sum of irreducible components, it follows that $\widehat{x}(\rho) = 0$ for every representation ρ of G. In particular, if ρ is the regular representation we have: $\widehat{x}(\rho) = 0$, $\{e_\tau : \tau \in G\}$ is a basis for $\mathcal{V}$, and

$$0 = \widehat{x}(\rho)e_1 = \sum_{\tau \in G} x_\tau \rho_\tau e_1 = \sum_{\tau \in G} x_\tau e_\tau, \quad x \in \mathcal{C}$$

which implies $x(\tau) = 0$ for all $\tau \in G$. That is, $x = 0$. ■

Note that the dimension of the subspace $\mathcal{C}$ of class functions is then the number of distinct irreducible representations of G or, equivalently the number of orbits or conjugacy classes of G under the action $\sigma\tau\sigma^{-1}$, in which the class functions can be arbitrarily defined. Consequently, the number of distinct irreducible representations of G coincides with the number of its conjugacy classes. If $G = S_n$, this number is also equal to its distinct cycle structures, as discussed earlier on page 32.

Irreducible representations of C_n. If G is a commutative group, then G has $|G|$ conjugacy classes and hence $|G|$ distinct irreducible representations $\rho_1, \rho_2, \ldots$. Moreover, because $|G| = \sum_j \dim^2 \rho_j$, we conclude that these representations are all one-dimensional. In particular, the irreducible representations of C_n are given by $\rho_j(\tau^k) = e^{2\pi ijk/n}$, $j, k = 0, 1, \ldots, n-1$, which clearly coincide with the irreducible characters of C_n.

Proposition 4.4. *If $\chi_1, \ldots, \chi_h$ are the distinct irreducible characters of group G and $\mathcal{O}_\tau = \{\sigma\tau\sigma^{-1},\ \sigma \in G\}$, then*

$$\frac{|\mathcal{O}_\tau|}{|G|}\sum_i \overline{\chi}_i(\eta)\chi_i(\tau) = \begin{cases} 1 & \text{if } \eta \in \mathcal{O}_\tau; \\ 0 & \text{if } \eta \notin \mathcal{O}_\tau. \end{cases}$$

Proof. Define $x_\tau(\eta) = 1$ if $\eta \in \mathcal{O}_\tau$ and equal to zero otherwise. Then x_τ is a class function and, consequently, can be expressed as a linear combination $\sum_i c_i \chi_i$ of the distinct irreducible characters $\chi_1, \ldots, \chi_h$ of G. The reader may verify that, in this case, $c_i = (x_\tau \mid \chi_i) = |\mathcal{O}_\tau| \, \overline{\chi}_i(\tau)/|G|$, so that

$$x_\tau(\eta) = \frac{|\mathcal{O}_\tau|}{|G|} \sum_i \overline{\chi}_i(\tau)\chi_i(\eta) = \begin{cases} 1 & \text{if } \eta \in \mathcal{O}_\tau, \\ 0 & \text{if } \eta \notin \mathcal{O}_\tau, \end{cases}$$

from which the result follows. ■

To illustrate, Table (4.12) shows the irreducible characters $\chi_1, \chi_{\text{Sgn}}, \chi_\beta$ of S_3, along with the characters $\chi_\rho, \chi_{\beta\otimes\beta}, \chi_\phi$ of the permutation, $\beta \otimes \beta$ and the regular representation of S_3, respectively.

τ	χ_ρ	χ_1	χ_{Sgn}	χ_β	$\chi_{\beta\otimes\beta}$	χ_ϕ
1	3	1	1	2	4	6
(123)	0	1	1	−1	1	0
(132)	0	1	1	−1	1	0
(12)	1	1	−1	0	0	0
(13)	1	1	−1	0	0	0
(23)	1	1	−1	0	0	0

(4.12)

From the conjugacy orbits $\mathcal{O}_1 = \{1\}$, $\mathcal{O}_t = \{(12), (13), (23)\}$ and $\mathcal{O}_c = \{(123), (132)\}$ of S_3 we obtain

$$\overline{\chi}_1(\tau)\chi_1(\tau) + \overline{\chi}_{\text{Sgn}}(\tau)\chi_{\text{Sgn}}(\tau) + \overline{\chi}_\beta(\tau)\chi_\beta(\tau) = \begin{cases} 4+1+1 = 6 = |G|/|\mathcal{O}_1|, & \text{if } \tau \in \mathcal{O}_1; \\ 0+1+1 = 2 = |G|/|\mathcal{O}_t|, & \text{if } \tau \in \mathcal{O}_t; \\ 1+1+1 = 3 = |G|/|\mathcal{O}_c|, & \text{if } \tau \in \mathcal{O}_c, \end{cases}$$

whereas $x_\tau(\eta) = 0$ if $\tau = 1$, $\eta = (12)$, $\tau = 1$, $\eta = (123)$ or $\tau = (12)$, $\eta = (123)$.

To decompose, say, the character of $\beta \otimes \beta$, we write $\chi_{\beta\otimes\beta} = c_1\chi_1 + c_{\text{Sgn}}\chi_{\text{Sgn}} + c_\beta\chi_\beta$, in which the coefficients are determined by $c_1 = (\chi_{\beta\otimes\beta} \mid \chi_1) = 6/6 = 1$, $c_{\text{Sgn}} = (\chi_{\beta\otimes\beta} \mid \chi_{\text{Sgn}}) = 6/6 = 1$ and $c_\beta = (\chi_{\beta\otimes\beta} \mid \chi_\beta) = 6/6 = 1$. In fact, $\chi_{\beta\otimes\beta} = \chi_1 + \chi_{\text{Sgn}} + \chi_\beta$.

Reducing the conjugacy action on S_3

Table (4.13) shows the conjugacy action $\varphi_\tau\sigma = \tau\sigma\tau^{-1}$ of S_3 on itself, followed by the character (χ_ρ) of the resulting representation (ρ), and the irreducible characters of S_3 (page 77). The resulting orbits are exactly the three conjugacy classes of S_3.

φ	r_1	r_2	r_3	r_1	t_1	t_2	χ	χ_{Sgn}	χ_β	χ_1
$1 = r_1$	r_1	r_2	r_3	t_1	t_2	t_3	6	1	2	1
$(123) = r_2$	r_1	r_2	r_3	t_3	t_1	t_2	3	1	−1	1
$(132) = r_3$	r_1	r_2	r_3	t_2	t_3	t_1	3	1	−1	1
$(12) = t_1$	r_1	r_3	r_2	t_1	t_3	t_2	2	−1	0	1
$(13) = t_2$	r_1	r_3	r_2	t_3	t_2	t_1	2	−1	0	1
$(23) = t_3$	r_1	r_3	r_2	t_2	t_1	t_3	2	−1	0	1

(4.13)

Applying Proposition 4.1, we obtain $\rho = 1 \oplus 1 \oplus 1 \oplus \text{Sgn} \oplus \beta$, whereas the restriction of ρ to each one of the orbits

$$\mathcal{O}_1 = \{1\}, \quad \mathcal{O}_t = \{t_1, t_2, t_3\}, \quad \mathcal{O}_r = \{r_2, r_3\}$$

reduces according to $1, 1 \oplus \beta$ and $1 \oplus \text{Sgn}$, respectively. This separation is obtained by restricting the representation to the orbits where it acts transitively. This fact will be important in the next section where we construct the canonical projections.

4.4 The Canonical Projections

In the previous chapters we have emphasized that, for data-analytic purposes, our interest is focused on permutation representations associated with transitive actions. This was remarked earlier (Chapter 1, on page 23) in the identification of the classes and multiplicities of the elementary orbits where G acts transitively. The next three examples will explain why such identification is sufficient, although not always necessary, for data-analytic purposes. These examples also illustrate the algument used in the proof of the canonical decomposition theorem.

The canonical projections for the Sloan Fonts study

Returning to our starting point in Chapter 1 (page 5), consider again the group $G = \{1, v, h, o\}$ and its representation ρ described in (1.9). Note that $\rho_1 = \phi_1 \otimes \phi_1$, $\rho_v = \phi_1 \otimes \phi_t$, $\rho_h = \phi_t \otimes \phi_1$ and $\rho_o = \phi_t \otimes \phi_t$, where ϕ is the regular representation of $C_2 \simeq \{1, t\}$. The irreducible characters of $G \simeq C_2 \times C_2$ are given by Table (4.10).

The projection matrices $\mathcal{P}_1, \ldots, \mathcal{P}_4$ shown on page 6 were obtained from linear combinations

$$\frac{1}{4} \sum_{\tau \in G} \chi_\tau \rho_\tau,$$

of the ρ_τ in which the scalar coefficients are indexed by the irreducible characters χ_τ of G. These matrices are the canonical projections associated with the representation $\rho = \phi \otimes \phi$ of $C_2 \times C_2$ on $\mathbb{R}^4$. There are, in this case then, four canonical projections, one for each irreducible character of G.

To understand the role of these projections in decomposing the data space $\mathcal{V} = \mathbb{R}^4$, let $\{1, v, h, o\}$ indicate a basis for $\mathcal{V}$ indexed by the elements of G (the double notation should be self-evident) so that

$$\mathcal{V} =< 1 + v + h + o > \oplus < 1 + v - h - o > \oplus < 1 + h - v - o >$$
$$\oplus < 1 + o - v - h > \equiv \mathcal{V}_1 \oplus \ldots \oplus \mathcal{V}_4,$$

is a direct sum of stable subspaces in dimension of 1. Correspondingly, ρ reduces isomorphically as the direct sum $\xi_1 \oplus \xi_2 \oplus \xi_3 \oplus \xi_4$ of the irreducible representations of G.

Observing that $\xi = \text{Tr}\,\xi = \chi_\xi$ when ξ is a representation in dimension of 1, and applying Proposition 4.3 with $x = \overline{\chi} \in \widehat{G}$ we obtain, in the new basis,

$$P_i = \frac{1}{4}\sum_{\tau \in G} \xi_i(\tau) B \rho_\tau B^{-1} = 4\widehat{\chi}(\phi) = \text{Diag}((\xi_i|\overline{\xi}_1)I_1, (\xi_i|\overline{\xi}_2)I_1, (\xi_i|\overline{\xi}_3)I_1, (\xi_i|\overline{\xi}_4)I_1),$$

where

$$B = \begin{bmatrix} 1 & 1 & 1 & 1 \\ 1 & 1 & -1 & -1 \\ 1 & -1 & 1 & -1 \\ 1 & -1 & -1 & 1 \end{bmatrix}$$

is the matrix connecting the two bases. Consequently, from Proposition 4.1,

$$P_1 = \text{Diag}(1, 0, 0, 0), \quad P_2 = \text{Diag}(0, 1, 0, 0),$$
$$P_3 = \text{Diag}(0, 0, 1, 0), \quad P_4 = \text{Diag}(0, 0, 0, 1).$$

Clearly, these are algebraically orthogonal ($P_i P_j = P_j P_i = 0,\ i \neq j$) projection matrices ($P_i^2 = P_i, i = 1, 2, 3, 4$) such that

$$I_4 = P_1 + P_2 + P_3 + P_4,$$

in correspondence with the decomposition $\mathcal{V}_1 \oplus \ldots \oplus \mathcal{V}_4$ of $\mathcal{V}$. Specifically, P_i is a projection from $\mathcal{V}$ to $\mathcal{V}_i$, for $i = 1, 2, 3, 4$.

Equivalently, the matrices $\mathcal{P}_i$ on page 6 are the canonical projections $B^{-1} P_i B$ on the original basis $\{1, v, h, o\}$. These matrices remain a set of algebraically orthogonal projections decomposing the identity matrix in $\mathcal{V}$. In that sense of equivalence, therefore, we have derived the canonical projections of the given representation of G.

Canonical invariants. The invariants in (1.10) are precisely $\mathcal{I} = Bx$, where x is the structured data vector.

This derivation illustrates the case of an Abelian group acting on itself so that the action was transitive (single orbit) and in addition all irreducible representations were in dimension of 1. In the next derivation the group action is not transitive.

The canonical projections for the binary sequences study

Consider the permutation representations of $S_2 \simeq \{1, t\}$ acting on the space $V = \{uu, yy, uy, yu\}$ of binary sequences in length of 2, first according to position symmetry and then according to letter symmetry (page 49). In each case the data space for the structured data is $\mathcal{V} = \mathbb{R}^4$. Also recall that here $\widehat{G} = \{1, \text{Sgn}\}$.

In the position symmetry representation, indicated here by ρ, the orbits of G suggest the decomposition

$$\mathcal{V} =< uu > \oplus < yy > \oplus < uy, yu >$$

of $\mathcal{V}$ into a direct sum of stable subspaces, in which $< uy, yu >$ further reduces as $< uy + yu > \oplus < uy - yu >$, so that

$$\mathcal{V} =< uu > \oplus < yy > \oplus < uy + yu > \oplus < uy - yu >$$

is a direct sum decomposition of $\mathcal{V}$ into irreducible subspaces. In $< uu > \oplus < yy > \oplus < uy + yu >$, ρ reduces as three copies of the symmetric representation, whereas in $< uy - yu >$ as a single copy of the alternating representation, that is,

$$\rho \simeq 1 \oplus 1 \oplus 1 \oplus \text{Sgn}.$$

Clearly, the multiplicities of the symmetric and the sign representations come directly from Proposition 4.1 on page 93. Therefore, without repeating the details of the previous illustration, we obtain the canonical projections

$$P_1 = \text{Diag}(1, 1, 1, 0), \quad P_{\text{Sgn}} = \text{Diag}(0, 0, 0, 1),$$

corresponding to the decomposition $\mathcal{V}_1 \oplus \mathcal{V}_{\text{Sgn}}$ of $\mathcal{V}$. Specifically, P_1 is a projection from $\mathcal{V}$ onto $\mathcal{V}_1$, in dimension of 3, whereas P_{Sgn} is a projection on $\mathcal{V}_{\text{Sgn}}$, in dimension of 1. Equivalently, the matrices

$$B^{-1} P_1 B = \text{Diag}(1, 1, \mathcal{A}), \quad B^{-1} P_{\text{Sgn}} B = \text{Diag}(0, 0, \mathcal{Q})$$

are the canonical projections on the original basis $\{uu, yy, uy, yu\}$, respectively, and $\mathcal{I} = Bx$ are the canonical invariants on the original data, where

$$B = \text{Diag}\left(1, 1, \begin{bmatrix} 1 & 1 \\ 1 & -1 \end{bmatrix}\right), \quad \mathcal{A} = \frac{1}{2}\begin{bmatrix} 1 & 1 \\ 1 & 1 \end{bmatrix}, \quad \mathcal{Q} = \frac{1}{2}\begin{bmatrix} 1 & -1 \\ -1 & 1 \end{bmatrix}.$$

Remark. This example shows an important fact: that the canonical projections are not in correspondence with the stable subspaces of the representation on $\mathcal{V}$ (equivalently, the orbits on V). Here, although $< uu > \oplus < yy >$ and $< uy + yu > \oplus < uy - yu >$ are stable subspaces, the canonical projections are on $\mathcal{V}_1 =< uu > \oplus < yy > \oplus < uy + yu >$ and $\mathcal{V}_{\text{Sgn}} =< uy - yu >$.

The following is the summary of the results for the symbol symmetry decomposition, where the same remark clearly applies:

(1) $\mathcal{V} \simeq < uu + yy > \oplus < uy + yu > \oplus < uu - yy > \oplus < uy - yu >$;
(2) $\rho \simeq 1 \oplus 1 \oplus \text{Sgn} \oplus \text{Sgn}$;
(3) $P_1 = \text{Diag}(1, 1, 0, 0), \quad P_{\text{Sgn}} = \text{Diag}(0, 0, 1, 1)$;
(4) $B^{-1} P_1 B = \text{Diag}(\mathcal{A}, \mathcal{A}), \quad B^{-1} P_{\text{Sgn}} B = \text{Diag}(\mathcal{Q}, \mathcal{Q}), \mathcal{I} = Bx$,

where

$$B = \text{Diag}\left(\begin{bmatrix} 1 & 1 \\ 1 & -1 \end{bmatrix}, \begin{bmatrix} 1 & 1 \\ 1 & -1 \end{bmatrix}\right).$$

Our third illustration is similar to the first one in that the set of labels coincides with the symmetries of interest ($V = G$), and hence the action is transitive action. This prototypic study of data indexed by S_3 illustrates many of the basic features of that type of symmetry study.

The canonical projections for the regular representation of S_3

Table (4.14) shows the three irreducible characters of S_3, corresponding to the irreducible representations 1, Sgn, and β discussed earlier on page 77. The data space $\mathcal{V} = \mathbb{R}^6$ has a natural basis indexed by the elements $\{r_1, r_2, r_3, t_1, t_2, t_3\}$ of S_3, in dimension of 6.

	τ	χ_1	χ_{Sgn}	χ_β
r_1	1	1	1	2
r_2	(123)	1	1	-1
r_3	(132)	1	1	-1
t_1	(12)	1	-1	0
t_2	(13)	1	-1	0
t_3	(23)	1	-1	0

(4.14)

The regular representation (ϕ) of S_3 follows directly from its multiplicative action on itself and can be derived from Table (2.4) on page 35 using Algorithm 1. We reproduce the multiplication table here,

r_1	r_2	r_3	t_1	t_2	t_3
r_2	r_3	r_1	t_3	t_1	t_2
r_3	r_1	r_2	t_2	t_3	t_1
t_1	t_2	t_3	r_1	r_2	r_3
t_2	t_3	t_1	r_3	r_1	r_2
t_3	t_1	t_2	r_2	r_3	r_1

with the notation introduced above. From (4.8) on page 96, it follows that $\phi \simeq 1 \oplus \text{Sgn} \oplus 2\beta$. It is clear that the subspace

$$\mathcal{V}_1 = < r_1 + r_2 + r_3 + t_1 + t_2 + t_3 >$$

reduces as the symmetric representation whereas

$$\mathcal{V}_{\text{Sgn}} = < r_1 + r_2 + r_3 - t_1 - t_2 - t_3 >$$

reduces as the alternating representation. With a little more of effort, the reader can verify that the subspace

$$\mathcal{V}_\beta = < r_1 - r_2 - t_1 + t_2,\quad -r_2 + r_3 - t_1 + t_3 > \oplus < r_2 - r_3 - t_2 + t_3,\quad r_1 - r_3 + t_1 - t_2 >$$

reduces as two copies of β. More specifically, writing

$$\{r_1 - r_2 - t_1 + t_2,\quad -r_2 + r_3 - t_1 + t_3\} = \{v_1, v_2\} = \mathcal{B}_v,$$
$$\{r_2 - r_3 - t_2 + t_3,\quad r_1 - r_3 + t_1 - t_2\} = \{w_1, w_2\} = \mathcal{B}_w$$

and using the multiplication table of S_3 above, we see that the basis $\mathcal{B}_v$ transforms according to

$\mathcal{B}_v \xrightarrow{\beta^v} \tau * \mathcal{B}_v :$

	v_1	v_2
r_1	$r_1 - r_2 - t_1 + t_2$	$-r_2 + r_3 - t_1 + t_3$
r_2	$r_2 + t_1 - r_3 - t_3$	$r_1 + t_2 - r_3 - t_3$
r_3	$r_3 + t_3 - r_1 - t_2$	$r_2 + t_1 - r_1 - t_2$
t_1	$r_2 + t_1 - r_1 - t_2$	$r_3 + t_3 - r_1 - t_2$
t_2	$r_1 + t_2 - r_3 - t_3$	$r_2 + t_1 - r_3 - t_3$
t_3	$-r_2 + r_3 - t_1 + t_3$	$r_1 - r_2 - t_1 + t_2$

or,

$$\beta^v : \tau \rightarrow \begin{bmatrix} 1 & 0 \\ 0 & 1 \end{bmatrix}, \begin{bmatrix} 0 & -1 \\ 1 & -1 \end{bmatrix}, \begin{bmatrix} -1 & 1 \\ -1 & 0 \end{bmatrix}, \begin{bmatrix} -1 & 0 \\ -1 & 1 \end{bmatrix}, \begin{bmatrix} 1 & -1 \\ 0 & -1 \end{bmatrix}, \begin{bmatrix} 0 & 1 \\ 1 & 0 \end{bmatrix}, \tag{4.15}$$

$\tau = r_1, r_2, r_3, t_1, t_2, t_3$, respectively. It then follows that $\beta^v_{\tau * \sigma} = \beta^v_\tau \beta^v_\sigma$, thus giving a two-dimensional representation of S_3. Its character coincides with that of β, so it is irreducible, and hence isomorphic to (the unique two-dimensional) β. Similarly, working with the other basis, we obtain

$\mathcal{B}_w \xrightarrow{\beta^w} \tau * \mathcal{B}_w :$

	w_1	w_2
r_1	$r_2 - r_3 - t_2 + t_3$	$r_1 - r_3 + t_1 - t_2$
r_2	$r_3 + t_2 - r_1 - t_1$	$r_2 + t_3 - r_1 - t_1$
r_3	$r_1 + t_1 - r_2 - t_3$	$r_3 + t_2 - r_2 - t_3$
t_1	$r_3 + t_2 - r_2 - t_3$	$r_1 + t_1 - r_2 - t_3$
t_2	$r_2 + t_3 - r_1 - t_1$	$r_3 + t_2 - r_1 - t_1$
t_3	$r_1 - r_3 + t_1 - t_2$	$r_2 - r_3 - t_2 + t_3$

and $\beta^w = \beta^v$. Comparing with

$$\beta_{k,d} \simeq \begin{bmatrix} 1 & 0 \\ 0 & 1 \end{bmatrix}, \begin{bmatrix} -1 & -1 \\ 1 & 0 \end{bmatrix}, \begin{bmatrix} 0 & 1 \\ -1 & -1 \end{bmatrix}, \begin{bmatrix} -1 & -1 \\ 0 & 1 \end{bmatrix}, \begin{bmatrix} 1 & 0 \\ -1 & -1 \end{bmatrix}, \begin{bmatrix} 0 & 1 \\ 1 & 0 \end{bmatrix},$$
$$k = 1, 2, 3,\ d = \pm 1,$$

from page 64, we see that $\beta^v = \beta'^{-1}$.

Relative to this new basis, then,

$$\phi = \text{Diag}(1, \text{Sgn}, I_2 \otimes \beta).$$

The canonical projections are the linear combinations

$$\mathcal{P}_\chi = \frac{n_\chi}{6} \sum_{\tau \in G} \overline{\chi}_\tau \phi_\tau,$$

indexed by the irreducible characters of S_3, where n_χ is the dimension of the corresponding irreducible representation and $|S_3| = 6$. Therefore, in the new basis

$$P_\chi = \frac{n_\chi}{6} \sum_\tau \text{Diag}(\overline{\chi}_\tau 1_\tau, \overline{\chi}_\tau \text{Sgn}_\tau, I_2 \otimes \overline{\chi}_\tau \beta_\tau).$$

From Proposition 4.3,

$$\sum_\tau \overline{\chi}_\tau \xi_\tau = \frac{6}{n_\chi} (\chi_\xi | \chi) I_{n_\chi}, \quad \xi \in \{1, \text{Sgn}, \beta\} = \widehat{G},$$

so that, applying Theorem 4.1, we obtain

$$P_1 = \text{Diag}(1, 0, 0, 0, 0, 0), \quad P_{\text{Sgn}} = \text{Diag}(0, 1, 0, 0, 0, 0), \quad P_\beta = \text{Diag}(0, 0, I_2 \otimes I_2).$$

The matrix connecting the two bases is

$$B = \begin{bmatrix} 1 & 1 & 1 & 1 & 1 & 1 \\ 1 & 1 & 1 & -1 & -1 & -1 \\ 1 & -1 & 0 & -1 & 1 & 0 \\ 0 & -1 & 1 & -1 & 0 & 1 \\ 0 & 1 & -1 & 0 & -1 & 1 \\ 1 & 0 & -1 & 1 & -1 & 0 \end{bmatrix},$$

and the canonical invariants $\mathcal{I} = Bx$ on the data coincide with the bases for the irreducible subspaces. In particular, $\mathcal{B}_v$ and $\mathcal{B}_w$. Moreover, direct calculation shows that in the original basis

$$\mathcal{P}_1 = B^{-1} P_1 B = \mathcal{A}_2 \otimes \mathcal{A}_3 = 1/6 \begin{bmatrix} 1 & 1 & 1 & 1 & 1 & 1 \\ 1 & 1 & 1 & 1 & 1 & 1 \\ 1 & 1 & 1 & 1 & 1 & 1 \\ 1 & 1 & 1 & 1 & 1 & 1 \\ 1 & 1 & 1 & 1 & 1 & 1 \\ 1 & 1 & 1 & 1 & 1 & 1 \end{bmatrix},$$

$$\mathcal{P}_{\mathrm{Sgn}} = B^{-1} P_{\mathrm{Sgn}} B = I_2 \otimes \mathcal{Q}_3 = 1/6 \begin{bmatrix} 1 & 1 & 1 & -1 & -1 & -1 \\ 1 & 1 & 1 & -1 & -1 & -1 \\ 1 & 1 & 1 & -1 & -1 & -1 \\ -1 & -1 & -1 & 1 & 1 & 1 \\ -1 & -1 & -1 & 1 & 1 & 1 \\ -1 & -1 & -1 & 1 & 1 & 1 \end{bmatrix},$$

$$\mathcal{P}_{\beta} = B^{-1} P_{\beta} B = \mathcal{Q}_2 \otimes \mathcal{A}_3 = 1/3 \begin{bmatrix} 2 & -1 & -1 & 0 & 0 & 0 \\ -1 & 2 & -1 & 0 & 0 & 0 \\ -1 & -1 & 2 & 0 & 0 & 0 \\ 0 & 0 & 0 & 2 & -1 & -1 \\ 0 & 0 & 0 & -1 & 2 & -1 \\ 0 & 0 & 0 & -1 & -1 & 2 \end{bmatrix}.$$

Invariant plots

In contrast to the previous examples, here we obtained two invariants in dimension of 2 given by the components of $\mathcal{B}_v$ and $\mathcal{B}_w$ appearing in $\mathcal{I} = Bx$, each one reducing as β. A graphical display of data in an invariant subspace is generally called an invariant plot. Consequently, any symmetry study of data indexed by S_3, or D_3, generates essentially two non trivial invariant plots. One in the alternating subspace

$$\mathcal{V}_{\mathrm{Sgn}} =< r_1 + r_2 + r_3 - t_1 - t_2 - t_3 >,$$

in dimension of 1, and one (type of) bivariate plot for the each one of the two copies of $\mathcal{V}_\beta$.

The canonical projections for the regular representation of D_4

Table (4.16) shows the multiplication table of D_4, isomorphic with Table (2.12) shown on page 42, in the present notation.

1	r_1	r_2	r_3	r_4	t_1	t_2	t_3	t_4
(1432)	r_2	r_3	r_4	r_1	t_4	t_1	t_2	t_3
(13)(24)	r_3	r_4	r_1	r_2	t_3	t_4	t_1	t_2
(1234)	r_4	r_1	r_2	r_3	t_2	t_3	t_4	t_1
(13)	t_1	t_2	t_3	t_4	r_1	r_2	r_3	r_4
(14)(23)	t_2	t_3	t_4	t_1	r_4	r_1	r_2	r_3
(24)	t_3	t_4	t_1	t_2	r_3	r_4	r_1	r_2
(12)(34)	t_4	t_1	t_2	t_3	r_2	r_3	r_4	r_1

(4.16)

The irreducible characters of D_4 are shown in Table (4.17). The representation α in dimension of 1 is suggested by the two-coset structure of D_4 and the fact that rotations and reversals have opposing parity, whereas $\gamma = \text{Sgn} \times \alpha$.

χ	r_1	r_2	r_3	r_4	t_1	t_2	t_3	t_4
χ_1	1	1	1	1	1	1	1	1
χ_{Sgn}	1	−1	1	−1	−1	1	−1	1
χ_α	1	1	1	1	−1	−1	−1	−1
χ_γ	1	−1	1	−1	1	−1	1	−1
χ_β	2	0	−2	0	0	0	0	0

(4.17)

The two-dimensional representation β was introduced earlier on page 96. Since $1+1+1+1+2^2 = 8$ these are all the irreducible representations of D_4, and consequently, its regular representation reduces as the direct sum $1 \oplus \text{Sgn} \oplus \alpha \oplus \gamma \oplus \beta$.

If ξ is in dimension of 1 then $\xi = \chi_\xi$ and we may write its corresponding subspace as

$$\mathcal{V}_\xi = < \sum_{\tau \in G} x_\tau \xi_\tau >$$

Remarkably, it turns out that this construction extends to higher-order representations and can be used to determine the canonical bases of interest. In fact, the bases $\mathcal{B}_v$ and $\mathcal{B}_w$ for $\mathcal{V}_\beta$ in the previous example are precisely the rows of the 2×2 matrix $\sum_{\tau \in G} x_\tau \beta_\tau$. We will return to this argument later on in the chapter. In the present case, $\mathcal{V}_\beta = < \mathcal{B}_v > \oplus < \mathcal{B}_w >$, where

$$\mathcal{B}_v = \{v_1, v_2\} = \{r_1 - r_3 + t_1 - t_3,\ -r_2 + r_4 + t_2 - t_4\},$$
$$\mathcal{B}_w = \{r_2 - r_4 + t_2 - t_4,\ r_1 - r_3 - t_1 + t_3\}.$$

Together, then, $\mathcal{V} = \mathcal{V}_1 \oplus \mathcal{V}_{\text{Sgn}} \oplus \mathcal{V}_\alpha \oplus \mathcal{V}_\gamma \oplus \mathcal{V}_\beta$. Referring to the multiplication table of D_4 shown above, we see that the basis $\mathcal{B}_v$ transforms according to

$\mathcal{B}_v \xrightarrow{\beta^v} \tau * \mathcal{B}_v$:

	v_1	v_2
r_1	$r_1 - r_3 + t_1 - t_3$	$-r_2 + r_4 + t_2 - t_4$
r_2	$r_2 - r_4 + t_4 - t_2$	$r_1 - r_3 + t_1 - t_3$
r_3	$r_3 - r_1 + t_3 - t_1$	$r_2 - r_4 + t_4 - t_2$
r_4	$-r_2 + r_4 + t_2 - t_4$	$r_3 - r_1 + t_3 - t_1$
t_1	$r_1 - r_3 + t_1 - t_3$	$r_2 - r_4 + t_4 - t_2$
t_2	$-r_2 + r_4 + t_2 - t_4$	$r_1 - r_3 + t_1 - t_3$
t_3	$r_3 - r_1 + t_3 - t_1$	$-r_2 + r_4 + t_2 - t_4$
t_4	$r_2 - r_4 + t_4 - t_2$	$r_3 - r_1 + t_3 - t_1$

or,

$$\beta^v : \tau \rightarrow \begin{bmatrix} 1 & 0 \\ 0 & 1 \end{bmatrix}, \begin{bmatrix} 0 & -1 \\ 1 & 0 \end{bmatrix}, \begin{bmatrix} -1 & 0 \\ 0 & -1 \end{bmatrix}, \begin{bmatrix} 0 & 1 \\ -1 & 0 \end{bmatrix}, \begin{bmatrix} 1 & 0 \\ 0 & -1 \end{bmatrix}, \begin{bmatrix} 0 & 1 \\ 1 & 0 \end{bmatrix}, \begin{bmatrix} -1 & 0 \\ 0 & 1 \end{bmatrix}, \begin{bmatrix} 0 & -1 \\ -1 & 0 \end{bmatrix},$$

$\tau = r_1, r_2, r_3, r_4, t_1, t_2, t_3, t_4$, respectively, which is precisely the realization of β shown on page 43. Similarly, working with $\mathcal{B}_w$, we obtain

$\mathcal{B}_w \xrightarrow{\beta^w} \tau * \mathcal{B}_w$:

	w_1	w_2
r_1	$r_2 - r_4 + t_2 - t_4$	$r_1 - r_3 - t_1 + t_3$
r_2	$t_1 - t_3 + r_3 - r_1$	$r_2 - r_4 + t_2 - t_4$
r_3	$t_4 - t_2 - r_2 + r_4$	$t_1 - t_3 + r_3 - r_1$
r_4	$r_1 - r_3 - t_1 + t_3$	$t_4 - t_2 - r_2 + r_4$
t_1	$r_2 - r_4 + t_2 - t_4$	$t_1 - t_3 + r_3 - r_1$
t_2	$r_1 - r_3 - t_1 + t_3$	$r_2 - r_4 + t_2 - t_4$
t_3	$t_4 - t_2 - r_2 + r_4$	$r_1 - r_3 - t_1 + t_3$
t_4	$t_1 - t_3 + r_3 - r_1$	$t_4 - t_2 - r_2 + r_4$

and $\beta^w = \beta^v$.

For D_4, the matrix connecting the two bases is

$$B = \begin{bmatrix} 1 & 1 & 1 & 1 & 1 & 1 & 1 & 1 \\ 1 & -1 & 1 & -1 & -1 & 1 & -1 & 1 \\ 1 & 1 & 1 & 1 & -1 & -1 & -1 & -1 \\ 1 & -1 & 1 & -1 & 1 & -1 & 1 & -1 \\ 1 & 0 & -1 & 0 & 1 & 0 & -1 & 0 \\ 0 & -1 & 0 & 1 & 0 & 1 & 0 & -1 \\ 0 & 1 & 0 & -1 & 0 & 1 & 0 & -1 \\ 1 & 0 & -1 & 0 & -1 & 0 & 1 & 0 \end{bmatrix},$$

and again, the canonical invariants $\mathcal{I} = Bx$ on the data are defined as the bases for the irreducible subspaces.

In the new basis, then,

$$\phi = \text{Diag}(1, \text{Sgn}, \alpha, \gamma, I_2 \otimes \beta).$$

Similarly to D_3, the canonical projections are the linear combinations

$$\mathcal{P}_\chi = \frac{n_\chi}{8} \sum_{\tau \in G} \overline{\chi}_\tau \phi_\tau,$$

indexed by the irreducible charactes of D_4, where n_χ is the dimension of the corresponding irreducible representation and $8 = |D_4|$. In that basis

$$P_\chi = \frac{n_\chi}{8} \sum_\tau \text{Diag}(\overline{\chi}_\tau 1_\tau, \overline{\chi}_\tau \text{Sgn}_\tau, \overline{\chi}_\tau \alpha_\tau, \overline{\chi}_\tau \gamma_\tau, I_2 \otimes \overline{\chi}_\tau \beta_\tau).$$

From Proposition 4.3 and Theorem 4.1, we obtain

$$P_1 = \text{Diag}(1, 0, 0, 0, 0, 0, 0, 0), \quad P_{\text{Sgn}} = \text{Diag}(0, 1, 0, 0, 0, 0, 0, 0),$$
$$P_\alpha = \text{Diag}(0, 0, 1, 0, 0, 0, 0, 0), \quad P_\gamma = \text{Diag}(0, 0, 0, 1, 0, 0, 0, 0),$$
$$P_\beta = \text{Diag}(0, 0, 0, 0, I_2 \otimes I_2).$$

Direct calculation shows that in the original basis $\mathcal{P}_1 = B^{-1}P_1B = \mathcal{A}_8$,

$$\mathcal{P}_{\text{Sgn}} = B^{-1}P_{\text{Sgn}}B = 1/8\begin{bmatrix} 1 & -1 & 1 & -1 & -1 & 1 & -1 & 1 \\ -1 & 1 & -1 & 1 & 1 & -1 & 1 & -1 \\ 1 & -1 & 1 & -1 & -1 & 1 & -1 & 1 \\ -1 & 1 & -1 & 1 & 1 & -1 & 1 & -1 \\ -1 & 1 & -1 & 1 & 1 & -1 & 1 & -1 \\ 1 & -1 & 1 & -1 & -1 & 1 & -1 & 1 \\ -1 & 1 & -1 & 1 & 1 & -1 & 1 & -1 \\ 1 & -1 & 1 & -1 & -1 & 1 & -1 & 1 \end{bmatrix} = \mathcal{Q}_2 \otimes \mathcal{A}_2 \otimes \mathcal{Q}_2,$$

$$\mathcal{P}_\alpha = B^{-1}P_\alpha B = 1/8\begin{bmatrix} 1 & 1 & 1 & 1 & -1 & -1 & -1 & -1 \\ 1 & 1 & 1 & 1 & -1 & -1 & -1 & -1 \\ 1 & 1 & 1 & 1 & -1 & -1 & -1 & -1 \\ 1 & 1 & 1 & 1 & -1 & -1 & -1 & -1 \\ -1 & -1 & -1 & -1 & 1 & 1 & 1 & 1 \\ -1 & -1 & -1 & -1 & 1 & 1 & 1 & 1 \\ -1 & -1 & -1 & -1 & 1 & 1 & 1 & 1 \\ -1 & -1 & -1 & -1 & 1 & 1 & 1 & 1 \end{bmatrix} = \mathcal{Q}_2 \otimes \mathcal{A}_2 \otimes \mathcal{A}_2,$$

$$\mathcal{P}_\gamma = B^{-1}P_\gamma B = 1/8\begin{bmatrix} 1 & -1 & 1 & -1 & 1 & -1 & 1 & -1 \\ -1 & 1 & -1 & 1 & -1 & 1 & -1 & 1 \\ 1 & -1 & 1 & -1 & 1 & -1 & 1 & -1 \\ -1 & 1 & -1 & 1 & -1 & 1 & -1 & 1 \\ 1 & -1 & 1 & -1 & 1 & -1 & 1 & -1 \\ -1 & 1 & -1 & 1 & -1 & 1 & -1 & 1 \\ 1 & -1 & 1 & -1 & 1 & -1 & 1 & -1 \\ -1 & 1 & -1 & 1 & -1 & 1 & -1 & 1 \end{bmatrix} = \mathcal{A}_2 \otimes \mathcal{A}_2 \otimes \mathcal{Q}_2,$$

$$\mathcal{P}_\beta = B^{-1}P_\beta B = 1/2\begin{bmatrix} 1 & 0 & -1 & 0 & 0 & 0 & 0 & 0 \\ 0 & 1 & 0 & -1 & 0 & 0 & 0 & 0 \\ -1 & 0 & 1 & 0 & 0 & 0 & 0 & 0 \\ 0 & -1 & 0 & 1 & 0 & 0 & 0 & 0 \\ 0 & 0 & 0 & 0 & 1 & 0 & -1 & 0 \\ 0 & 0 & 0 & 0 & 0 & 1 & 0 & -1 \\ 0 & 0 & 0 & 0 & -1 & 0 & 1 & 0 \\ 0 & 0 & 0 & 0 & 0 & -1 & 0 & 1 \end{bmatrix} = I_2 \otimes \mathcal{Q}_2 \otimes I_2.$$

Invariant plots

The invariant plots are defined, similarly, by the bases of the irreducible subspaces, or $\mathcal{I} = Bx$. Any symmetry study of data indexed by D_4 generates four nontrivial invariant plots, corresponding to $\mathcal{V}_{\text{Sgn}}$, $\mathcal{V}_\alpha$, and $\mathcal{V}_\gamma$, and two bivariate plots for the two copies of $\mathcal{V}_\beta$.

With these illustrations in mind, we now prove the following theorem.

Theorem 4.4 (Canonical Decomposition). *If ρ is a linear representation of G on $GL(\mathcal{V})$ then, for each irreducible representation ξ of G in dimension of n_ξ and character χ_ξ,*

$$\mathcal{P}_\xi = \frac{n_\xi}{|G|} \sum_{\tau \in G} \overline{\chi}_\xi(\tau)\rho_\tau,$$

is a projection of $\mathcal{V}$ onto a subspace $\mathcal{V}_\xi$, sum of m_ξ isomorphic copies of irreducible subspaces of ξ. Moreover: $\mathcal{P}_\xi\mathcal{P}_{\xi'} = \mathcal{P}_{\xi'}\mathcal{P}_\xi = 0$, for any two distinct irreducible representations ξ, ξ' of G; $\mathcal{P}_\xi^2 = \mathcal{P}_\xi$ and $\sum_{\xi \in \widehat{G}} \mathcal{P}_\xi = I_v$, where $v = \dim \mathcal{V} = \sum_{\xi \in \widehat{G}} m_\xi n_\xi$.

Proof. Indicate by ξ_i the distinct irreducible representations of G, with corresponding characters and dimensions χ_i and n_i, respectively, $i = 1, \ldots, h$. From Proposition 4.1 we know that $\rho \simeq \sum_{j=1}^h m_j \rho_j$, where $\xi_1, \ldots, \xi_h$ are the distinct irreducible representations of G. That is, there is a basis B in $\mathcal{V}$ relative to which

$$\rho = \text{Diag}(I_{m_1} \otimes \xi_1, \ldots, I_{m_h} \otimes \xi_h).$$

Therefore,

$$P_i = B\mathcal{P}_i B^{-1} = \frac{n_i}{|G|} \sum_\tau \text{Diag}(I_{m_1} \otimes \overline{\chi}_i(\tau)\xi_1(\tau), \ldots, I_{m_h} \otimes \overline{\chi}_i(\tau)\xi_h(\tau)).$$

Applying Proposition 4.3 with $x = \overline{\chi}_i$, so that $\widehat{x}(\xi_j) = \sum_\tau \overline{\chi}_i(\tau)\xi_j(\tau)$, we have

$$\sum_\tau \overline{\chi}_i(\tau)\xi_j(\tau) = \frac{|G|}{n_i}(\chi_j|\chi_i)I_{n_i} = \frac{|G|}{n_i}\delta_{ij}I_{n_i},$$

and consequently

$$P_i = \text{Diag}(\delta_{i1}I_{m_1} \otimes I_{n_1}, \ldots, \delta_{ih}I_{m_h} \otimes I_{n_h}).$$

It is then clear that $P_i^2 = P_i$, so that P_i is a projection of $\mathcal{V}$ into the subspace $\mathcal{V}_i$ direct sum of m_i copies of the irreducible subspaces associated with ξ_i, $i = 1, \ldots, h$. It is also clear that, in addition, $P_i P_j = 0$ for $j \neq i$ and that

$$\sum_{i=1}^h P_i = \text{Diag}(I_{m_1} \otimes I_{n_1}, \ldots, I_{m_h} \otimes I_{n_h}) = I_v,$$

all of these statements holding true for the original components $\mathcal{P}_i = B^{-1}P_iB$ of the decomposition, concluding the proof. ■

Note that $\text{Tr}\,\mathcal{P}_i = n_i m_i = \dim \mathcal{V}_i$. From now on we will refer to $\sum_i \mathcal{P}_i = I$, under the conditions of Theorem 4.4, simply as the canonical reduction of the identity operator I in the data space $\mathcal{V}$.

Data reduction by conjugacy action in S_3

Returning to the conjugacy action of S_3 on itself, defined by Table (4.13) on page 101, direct evaluation shows that the resulting canonical projections are given by

$$\mathcal{P}_1 = \text{Diag}(1, \mathcal{A}_2, \mathcal{A}_3), \quad \mathcal{P}_{\text{Sgn}} = (0, \mathcal{Q}_2, 0), \quad \mathcal{P}_\beta = (0, 0, \mathcal{Q}_3),$$

in dimensions of 3, 1, 2, respectively, where $\mathcal{A}$ and $\mathcal{Q}$ are the matrices introduced in (1.14) on page 10 with the appropriate dimensions. The reader may want to identify the resulting canonical reduction appearing in each one of the three conjugacy orbits of S_3.

4.5 The Standard Decomposition

In this section we will derive the canonical decomposition for the permutation representation of S_n, indicated here by ρ, which is naturally associated with data that are indexed by $\{1, 2, \ldots, n\}$, such as in uniform sampling.

With the notation $\mathcal{A} = ee'/n$, where ee' is the $n \times n$ matrix of ones and $\mathcal{Q} = I - \mathcal{A}$ in mind, note that the reduction $I = \mathcal{A} + \mathcal{Q}$ satisfies $\mathcal{A}^2 = \mathcal{A}$, $\mathcal{Q}^2 = \mathcal{Q}$ and $\mathcal{A}\mathcal{Q} = \mathcal{Q}\mathcal{A} = 0$. Moreover, $\mathcal{A}$ projects $\mathcal{V} = \mathbb{R}^n$ into a subspace $\mathcal{V}_a$ of dimension $\dim \mathcal{V}_a = \text{Tr}\,\mathcal{A} = 1$ generated by $e = e_1 + \cdots + e_n = (1, 1, \ldots, 1) \in \mathcal{V}$, whereas $\mathcal{Q}$ projects $\mathcal{V}$ into the subspace $\mathcal{V}_q$ in dimension $n - 1$, the orthogonal complement of $\mathcal{V}_a$ in $\mathcal{V}$. We will show that the reduction $\mathcal{V} = \mathcal{V}_a \oplus \mathcal{V}_q$ is exactly the canonical reduction determined by ρ. We refer to this decomposition as the standard decomposition or standard reduction.

To illustrate the argument, consider first the case $n = 3$. The joint character table for ρ and the irreducible representations of S_3 is

χ	1	(12)	(123)
χ_ρ	3	1	0
χ_1	1	1	1
χ_β	2	0	−1
χ_{Sgn}	1	−1	1

where β is the two-dimensional irreducible representation derived earlier on page 77. Recall also that there are 3 elements in the conjugacy class of (12) and two elements in the class of (123). It then follows that $(\chi_1|\chi_\rho) = 1$, $(\chi_{\text{Sgn}}|\chi_\rho) = 0$ and $(\chi_\beta|\chi_\rho) = 1$, so that $\rho \simeq 1 \oplus \beta$ and $\chi_\beta = \chi_\rho - 1$. In general, we have

Proposition 4.5. *$\chi_\beta = \chi_\rho - 1$ is an irreducible character of S_n. Its dimension is $n-1$.*

Proof. Write $\chi_\rho = \chi$ to indicate the character of the permutation representation of S_n. To evaluate

$$(\chi_\beta|\chi_\beta) = \frac{1}{|G|}\sum_\tau(\chi(\tau)-1)^2 = \frac{1}{|G|}\sum_\tau(\chi^2(\tau) - 2\chi(\tau) + 1)$$

and verify the irreducibility criteria $(\chi_\beta|\chi_\beta) = 1$ of Proposition 4.2, we need the first two moments

$$\sum_\tau \chi(\tau)/|G|, \quad \sum_\tau \chi^2(\tau)/|G|,$$

of χ. The argument is as follows: Consider the action $(\tau i, \tau j)$ of S_n on the product space V^2. Its character is χ^2 and the number of orbits is clearly 2, namely

$$\mathcal{O}_0 = \{(i,i), i = 1,\ldots,n\}, \quad \mathcal{O}_1 = \{(i,j), i, j = 1,\ldots,n, i \neq j\}.$$

Now apply Burnside's Lemma to write

$$2 = \text{Number of orbits in } V^2 = \frac{1}{|G|}\sum_\tau \chi^2(\tau).$$

Similarly, since S_n acts transitively on $\{1,\ldots,n\}$, we have

$$1 = \text{Number of orbits in } V = \frac{1}{|G|}\sum_\tau \chi(\tau).$$

Consequently,

$$(\chi_\beta|\chi_\beta) = \frac{1}{|G|}\sum_\tau(\chi^2(\tau) - 2\chi(\tau) + 1) = (2 - 2 + 1) = 1,$$

thus showing that χ_β is an irreducible character of S_n. Its dimension is $\chi_\beta(1) = \chi(1) - 1 = n - 1$, concluding the proof. ■

Because $(\chi_1|\chi_\rho) = (\chi_\beta|\chi_\rho) = 1$ and the dimension of ρ is n we conclude that $\rho \simeq 1 \oplus \beta$ is an irreducible decomposition of ρ.

The implication for the canonical decomposition of ρ is as follows: Because $\rho \simeq 1 \oplus \beta$ there are only two (nonnull) projections, namely $\mathcal{P}_1$ associated with the symmetric character, and $\mathcal{P}_\beta$ associated with the irreducible character χ_β of dimension $n-1$. Clearly,

$$\mathcal{P}_1 = \frac{1}{|G|}\sum_\tau \rho_\tau = \mathcal{A}.$$

Moreover, $I = \mathcal{A} + \mathcal{Q} = \mathcal{A} + \mathcal{P}_\beta$ so we must have $\mathcal{P}_\beta = \mathcal{Q}$. That is, $\mathcal{A}$ and $\mathcal{Q}$ are the only nonnull canonical projections associated with ρ, which is the characterization we had in mind.

The results summarized in the following proposition are useful in obtaining new canonical decompositions from existing ones. Its proof is by direct verification that in each case the appropriate identity matrix decomposes as a sum of pairwise algebraically orthogonal projections.

Proposition 4.6. *If $I_m = \sum_i \mathcal{P}_i$ and $I_n = \sum_j \mathcal{T}_j$ are canonical reductions, then $I_{mn} = \sum_{i,j} \mathcal{P}_i \otimes \mathcal{T}_j$ is a canonical reduction. In particular, for any component $\mathcal{P} \otimes \mathcal{T}$ in the decomposition,*

$$(\mu \otimes e)'(\mathcal{P} \otimes \mathcal{T})(\mu \otimes e) = \begin{cases} n\mu'\mathcal{P}\mu & \text{if } \mathcal{T} = \mathcal{A} \\ 0 & \text{if } \mathcal{T} = \mathcal{Q} \end{cases}.$$

If $L_1, \ldots, L_h$ is a disjoint partition of L then

$$I_m = \left(\sum_{i\in L_1} \mathcal{P}_i\right) + \cdots + \left(\sum_{i\in L_h} \mathcal{P}_i\right)$$

is also a canonical reduction. If, in addition, $m = n$ and the components $\mathcal{P}_i$ and $\mathcal{T}_j$ all commute, then $I_n = \sum_{i,j} \mathcal{P}_i\mathcal{T}_j$ is a canonical reduction. In particular, the components $\mathcal{A}$ and $\mathcal{Q}$ of the standard reduction commute with every symmetric matrix of same dimension.

Sampling considerations

An important application of Proposition 4.6 is in obtaining the sampling or error component in the analysis of variance based on a canonical decomposition. Specifically, if a sample of size n is obtained in each point $s \in V$ (with v points) where the action of a group G leads to a canonical decomposition

$$I_v = \sum_j \mathcal{P}_j,$$

and $I_n = \mathcal{A}_n + \mathcal{Q}_n$ is the standard decomposition for S_n (shuffling the labels for the sample) then,

$$I_{nv} = I_v \otimes I_n = \left(\sum_j \mathcal{P}_j\right) \otimes (\mathcal{A}_n + \mathcal{Q}_n)$$

is the canonical decomposition for the data $x \in \mathbb{R}^{vn}$. The error term $x'(I_v \otimes \mathcal{Q}_n)x$ is obtained by collecting together the components $\sum_j x'(\mathcal{P}_j \otimes \mathcal{Q}_n)x$ of I_{nv}.

Matrices with the symmetry of S_n

The following result describes the matrices that are centralized by the permutation representation of S_n. In multivariate analysis, these matrices play a significant role in describing the (intraclass) covariance structure of permutation symmetric random variables. It also complements the content of Section 3.6 on page 70.

Proposition 4.7. *If ρ indicates the permutation representation of S_n, then, for every real or complex $n \times n$ matrix H,*

$$\frac{1}{n!}\sum_{\tau \in S_n} \rho_\tau H \rho_{\tau^{-1}} = a_0 ee' + a_1 I_n,$$

where the coefficients a_0 and a_1 are scalars defined by the relations $n(a_0 + a_1) = \mathrm{Tr}\, H$ and $n(n-1)a_0 = e'He - \mathrm{Tr}\, H$, in which $e'He$ is the sum of the entries in H.

Proof. Let $M = \sum_{\tau \in S_n} \rho_\tau H \rho_{\tau^{-1}}/n!$ and $J = BHB^{-1}$ where

$$B = \begin{bmatrix} 1 & 1 & \dots & 1 & 1 \\ n-1 & -1 & \dots & -1 & -1 \\ \vdots & \vdots & \vdots & \vdots & \vdots \\ -1 & -1 & \dots & n-1 & -1 \end{bmatrix}. \tag{4.18}$$

It is simple to verify that the irreducible decomposition $\rho \simeq 1 \oplus \beta$ of ρ is realized by $B\rho B^{-1}$. Consequently, applying Proposition 3.3, we have

$$PMP^{-1} = \frac{1}{n!}\sum_\tau (B\rho_\tau B^{-1}) J (B\rho_{\tau^{-1}} B^{-1}) = \mathrm{Diag}\left(J_{11}, \frac{\mathrm{Tr}\, J_{22}}{n-1} I_{n-1}\right),$$

from which we obtain

$$M = B^{-1}\mathrm{Diag}\left(J_{11}, \frac{\mathrm{Tr}\, J_{22}}{n-1} I_{n-1}\right) B.$$

Direct evaluation, using the definition of the matrix B, shows that M is the matrix with entries

$$M_{ij} = \begin{cases} \mathrm{Tr}\, H/n & \text{if } i = j, \\ (e'He - \mathrm{Tr}\, H)/(n-1) & \text{if } i \neq j, \end{cases}$$

which is the proposed result. ■

Linear representations of order statistics and ranks

Let x indicate the vector of the (distinct) ranks derived from the components of a random vector $y \in \mathbb{R}^n$ when these components are ordered from the smallest to the largest value. Let also ρ_τ indicate the permutation matrix arranging the observed ranks in increasing order.

By virtue of the randomness in the ranks of y, the matrices $U = \rho_\tau$ are then random permutation matrices. In particular, $U'y$ gives the ordered version of y. For example, if the vector of ranks of an observed vector y is $x' = (3, 4, 2, 1)$, then

$$\rho_\tau = \begin{bmatrix} 0 & 0 & 1 & 0 \\ 0 & 0 & 0 & 1 \\ 0 & 1 & 0 & 0 \\ 1 & 0 & 0 & 0 \end{bmatrix}.$$

If we indicate by $\mathcal{Y}$ the ordered version of y, then the random permutation matrix $U = \rho_\tau$ is such that $x = Ur$ and $\mathcal{Y} = U'y$. If (y, z) are concomitant random vectors of corresponding dimensions, then the action $U'z$ of U on z generates the vector of concomitants of (or induced) order statistics. If the probability distribution of y is permutation symmetric, in the sense that y and $\rho_\tau y$ are equally distributed for all $\tau \in S_n$, then the probability law of U is uniform in S_n and the covariance structure of ranks, order statistics, and concomitants may then be expressed as multiplicative actions of uniformly distributed random permutations. See, for example, Lee and Viana (1999).

The covariance structure. In general, to obtain the covariance structure of vectors resulting from actions such as $x = Ur$, we need to evaluate the uniform expected value $\sum_\tau \rho_\tau r/n!$ of $U'r$, indicated here by $E(Ur)$, and the uniform expected value $\sum_\tau \rho_\tau (rr')\rho_\tau'/n!$ of $U'rr'U$, denoted by $E(Urr'U')$. The reader may want to verify that

$$E(Ur) = E(U)r = \frac{ee'}{n}r = \frac{n(n+1)}{2n}e = \frac{n+1}{2}e,$$

whereas the covariance of Ur is given by $E(Urr'U') - E(Ur)E(Ur)'$. To evaluate $E(Urr'U')$, following Proposition 4.7, we note that Tr $rr' = n(n+1)(2n+1)/6$ and that $e'rr'e = [n(n+1)/2]^2$. Therefore,

$$n(a_0 + a_1) = \frac{n(n+1)(2n+1)}{6},$$

$$n(n-1)a_0 = \left[\frac{n^2(n+1)^2}{4} - \frac{n(n+1)(2n+1)}{6}\right] = \frac{1}{12}n(n+1)(n-1)(3n+2),$$

from which we obtain

$$\mathrm{Cov}(Ur) = \left[a_0 - \frac{(n+1)^2}{4}\right]ee' + a_1 I = \frac{-(n+1)}{12}ee' + \frac{n(n+1)}{12}I.$$

These results show that the common variance among the ranks is $(n+1)(n-1)/12$, and the common covariance between any two ranks is $-(n+1)/12$, so that the resulting common correlation is $-1/(n-1)$. The mean and variance of the usual Wilcoxson rank-sum statistics W follow from fixing, say, the first m components of R. Writing $f' = (1, \ldots, 1, 0, \ldots, 0)$ with $1 \leq m < n$ components equal to 1, then $W = f'R$ and $EW = f'E(Ur) = m(n+1)/2$, whereas $var(W) = f'\text{Cov}(Ur)f = m(n-m)(n+1)/12$. Similar arguments can be applied to rank correlations.

Concomitants of order statistics. Suppose that y and z are related by $z = Ty + F$, where F and y are jointly independent and T is a constant matrix with the symmetry of S_n. That is, T commutes with every permutation matrix in dimension of n. Then, the covariance structure between Uy, the ordered version of y, and the induced order Uz is given by

$$\text{Cov}[(Uy, Uz)] = \begin{bmatrix} \Gamma & \Gamma T \\ & \\ T\Gamma & \Sigma_{22} + T(\Gamma - \Sigma_{11})T \end{bmatrix},$$

where $\Gamma = \text{Cov}(Uy)$, $\Sigma_{11} = \text{Cov}(y)$ and $\Sigma_{22} = \text{Cov}(z)$. To see this, write

$$\begin{bmatrix} y \\ z \end{bmatrix} = A \begin{bmatrix} y \\ F \end{bmatrix}, \text{ with } A = \begin{bmatrix} I & 0 \\ T & I \end{bmatrix}.$$

Then, because T commutes with U, $Uz = U(Ty + F) = TUy + UF$, and consequently,

$$\begin{bmatrix} Uy \\ Uz \end{bmatrix} = A \begin{bmatrix} Uy \\ UF \end{bmatrix}.$$

Moreover, because y and F are jointly independent, $\text{Cov}(F) = \text{Cov}(UF) = \Sigma_{22} - T\Sigma_{11}T$ so that

$$\text{Cov}[(Uy, Uz)] = A \begin{bmatrix} \Gamma & 0 \\ 0 & \Sigma_{22} - T\Sigma_{11}T \end{bmatrix} A',$$

from which the result follows. The covariance structure of ordered observations from symmetrically dependent observations is described with detail in Viana (1998), Viana and Olkin (1997), and Olkin and Viana (1995).

4.6 Harmonic Analysis

Earlier in Section 4.3 (page 92) we introduced the linear transforms

$$\widehat{x}(\rho) = \sum_{\tau \in G} x_\tau \rho_\tau,$$

defined for class functions x and linear representations ρ. If ρ is an irreducible representation of G and x is any scalar-valued function defined on G, the evaluation $\widehat{x}(\rho)$ is called the Fourier transform of x at ρ. In analogy to the Fourier transform

$$\widehat{f}(r) = \int_{\mathbb{R}} f(t)\omega^{rt} dt$$

in the real line we think of each irreducible representation as the spectral frequencies. Similarly, the discrete Fourier transform of a scalar-valued function x defined in $C_n = \{1, \omega, \ldots, \omega^{n-1}\}$, evaluated at $\omega^j \in \widehat{G}$, is

$$\widehat{x}(j) = \sum_k x_k \omega^{jk}, \quad k, j = 0, 1, \ldots, n-1.$$

Its inverse Fourier transform is

$$x_k = \frac{1}{n} \sum_j \widehat{x}(j) \omega^{-jk}.$$

By observing that the irreducible representations of C_n have dimension $n_j = 1$, we may rewrite the above formula as

$$x_k = \sum_j \frac{n_j}{g} \text{Tr}\,[\omega^{-jk}\widehat{x}(j)],$$

which suggests the formula for arbitrary finite groups. This formula will be derived next, starting with the canonical projections of the regular representation of the group.

Theorem 4.5. *If x is a data vector indexed by the finite group G with g elements and $\widehat{x}(\beta) = \sum_{\tau \in G} x_\tau \beta_\tau$ is its Fourier transform at the irreducible representation β then, conversely,*

$$x_\tau = \sum_{\beta \in \widehat{G}} n_\beta Tr\,[\beta_{\tau^{-1}} \widehat{x}(\beta)]/g,$$

where the sum is over all irreducible representations of G.

Proof. Indicate by ϕ the (left) regular representation of G and by x the vector of observations indexed by G so that the σ-th component of $\phi_\tau x$ is $e'_\sigma \phi_\tau x = x_{\sigma\tau^{-1}}$. Evaluating the canonical projections for ϕ, we obtain

$$g e'_\sigma P_\beta x / n_\beta = \sum_{\tau \in G} \chi_\beta(\tau^{-1}) x_{\sigma\tau^{-1}} = \sum_{\eta \in G} \chi_\beta(\sigma^{-1}\eta) x_\eta = \sum_{\eta \in G} \text{Tr}\,[\beta_{\sigma^{-1}\eta}] x_\eta$$

$$= \text{Tr}\left[\sum_{\eta \in G} \beta_{\sigma^{-1}} \beta_\eta\right] x_\eta = \text{Tr}\,[\beta_{\sigma^{-1}} \widehat{x}(\beta)].$$

Consequently, summing over the irreducible representations of G, we have $I = \sum_\beta \mathcal{P}_\beta$, and

$$e'_\sigma x = e'_\sigma \sum_\beta \mathcal{P}_\beta x = \sum_\beta n_\beta \text{Tr}\,[\beta_{\sigma^{-1}} \widehat{x}(\beta)]/g,$$

from which the Fourier-inverse formula $x_\sigma = \sum_{\beta \in \widehat{G}} n_\beta \text{Tr}\,[\beta_{\sigma^{-1}} \widehat{x}(\beta)]/g$ follows. ∎

A decomposition for $x \in \mathcal{F}(G)$

Here we consider, in general, any irreducible representation β in dimension of n_β of a finite group G with g elements. Define the $g \times g$ matrix

$$(T_\beta)_{\sigma\tau} = n_\beta \text{Tr}\,[\beta_{\tau^{-1}\sigma}]/g,$$

in which the rows (σ) and columns (τ) are indexed by the elements of G arranged in a fixed, arbitrary, order. Let also

$$\mathcal{P}_\beta = n_\beta \sum_{\gamma \in G} \chi_\beta(\gamma^{-1}) \phi_\gamma / g$$

indicate be the canonical projection for the left regular representation ϕ associated with $\beta \in \widehat{G}$. It then follows that $\mathcal{P}_\beta = T_\beta$. To see this note that the single nonzero entry in row σ of the permutation matrix ϕ_γ is at column τ if and only if $\sigma\gamma = \tau$, or $\gamma = \sigma^{-1}\tau$. Consequently,

$$(\mathcal{P}_\beta)_{\sigma\tau} = n_\beta \sum_{\gamma \in G} \chi_\beta(\gamma^{-1}) \left(\phi_\gamma\right)_{\sigma\tau} / g = n_\beta \chi_\beta((\sigma^{-1}\tau)^{-1})/g = n_\beta \text{Tr}\, \beta_{\tau^{-1}\sigma}/g = (T_\beta)_{\sigma\tau}.$$

Moreover, then, for an arbitrary vector $x \in \mathcal{F}(G)$,

$$(\mathcal{P}'_\beta x)_\sigma = \sum_\tau (T'_\beta)_{\sigma\tau} x_\tau = n_\beta \text{Tr} \left[\beta_{\sigma^{-1}} \sum_\tau x_\tau \beta_\tau \right] / g = n_\beta \text{Tr}\,[\beta_{\sigma^{-1}} \widehat{x}(\beta)]/g = x_\beta(\sigma), \quad \sigma \in G,$$

thus showing that the scalar indexing

$$x_\beta(\sigma) = n_\beta \text{Tr}\,[\beta_{\sigma^{-1}} \widehat{x}(\beta)]/g, \quad \sigma \in G \tag{4.19}$$

of G is a linear superposition in the projection space for $\mathcal{P}_\beta$. Since $\sum_{\beta \in \widehat{G}} \mathcal{P}'_\beta = I$,

$$x_\sigma = \sum_{\beta \in \widehat{G}} (\mathcal{P}'_\beta x)_\sigma = \sum_{\beta \in \widehat{G}} x_\beta(\sigma) = \sum_{\beta \in \widehat{G}} n_\beta \text{Tr}\,[\beta_{\sigma^{-1}} \widehat{x}(\beta)]/g, \tag{4.20}$$

which, in particular, reveals the Fourier-inverse formula. The resulting equality

$$x = \sum_{\beta \in \widehat{G}} x_\beta$$

shows that any indexing $x \in \mathcal{F}(G)$ of G decomposes as the linear superposition $\sum_{\beta \in \widehat{G}} x_\beta$ of points x_β in the (left) regular canonical projection spaces of G.

Remarks.

(1) We often refer to the canonical projections derived from regular representations as regular canonical projections.
(2) Since, in the regular case, $\mathcal{P}' = \overline{\mathcal{P}}$, the above results can be expressed equivalently in terms of the complex conjugates $\overline{\mathcal{P}}$ of $\mathcal{P}$.
(3) The choice of left or right regular has the effect of commuting $\beta_{\sigma^{-1}}$ and $\widehat{x}(\beta)$ in the expression for $x_\beta(\sigma)$. For Abelian groups there is no distinction between the two;
(4) The matrix transpose $\mathcal{P}'$ appearing in these derivations is necessary only for consistency with the adopted definition $\sum_\tau x_\tau \beta_\tau$ of the Fourier transform $\widehat{x}(\beta)$ and choosing between the left and right regular action.

A decomposition for $x \in \mathcal{F}(S_3)$

Following with the notation on page 104, let $x' = (r_1, r_2, r_3, t_1, t_2, t_3)$ indicate a point in $\mathcal{F}(S_3)$. Again here, the labels and the data indexed by those labels are indicated with the same notation. First we evaluate the Fourier transform

$$\widehat{x}(\beta) = r_1 \begin{bmatrix} 1 & 0 \\ 0 & 1 \end{bmatrix} + r_2 \begin{bmatrix} -1 & -1 \\ 1 & 0 \end{bmatrix} + r_3 \begin{bmatrix} 0 & 1 \\ -1 & -1 \end{bmatrix} + t_1 \begin{bmatrix} -1 & -1 \\ 0 & 1 \end{bmatrix} + t_2 \begin{bmatrix} 1 & 0 \\ -1 & -1 \end{bmatrix} + t_3 \begin{bmatrix} 0 & 1 \\ 1 & 0 \end{bmatrix}$$

of x at the two-dimensional irreducible representation β of S_3, introduced earlier on page 64. It gives

$$\widehat{x}(\beta) = \begin{bmatrix} r_1 - r_2 - t_1 + t_2 & -r_2 + r_3 - t_1 + t_3 \\ r_2 - r_3 - t_2 + t_3 & r_1 - r_3 + t_1 - t_2 \end{bmatrix} = \begin{bmatrix} v_1 & v_2 \\ w_1 & w_2 \end{bmatrix},$$

where its components are exactly the bases $\mathcal{B}_v = \{v_1, v_2\}$ and $\mathcal{B}_w = \{w_1, w_2\}$ that appear in the decomposition of the irreducible subspace $\mathcal{V}_\beta$, on page 105. Evaluation of the regular indexing gives

$$x_\beta = 1/3\ (2r_1 - r_3 - r_2,\ -r_1 + 2r_2 - r_3,\ -r_2 + 2r_3 - r_1,\ 2t_1 - t_3 - t_2,\ 2t_2 - t_1 - t_3,\ -t_2 + 2t_3 - t_1),$$

which, when compared with $\mathcal{P}_\beta$ on page 110, shows that $x_\beta = \mathcal{P}_\beta x$. Therefore, x_β is indeed in the projection space for $\mathcal{P}_\beta$. These derivations, of course, coincide with those for $x \in \mathcal{F}(D^3)$.

A decomposition for $x \in \mathcal{F}(D_4)$

Let $x' = (r_1, r_2, r_3, r_4, t_1, t_2, t_3, t_4)$ indicate a point in $\mathcal{F}(D_4)$. The Fourier transform of x at the two-dimensional irreducible representation

$$\beta_{k,1} = \begin{bmatrix} 1 & 0 \\ 0 & 1 \end{bmatrix}, \begin{bmatrix} 0 & -1 \\ 1 & 0 \end{bmatrix}, \begin{bmatrix} -1 & 0 \\ 0 & -1 \end{bmatrix}, \begin{bmatrix} 0 & 1 \\ -1 & 0 \end{bmatrix}, \quad k = 1, 2, 3, 4,$$

$$\beta_{k,-1} = \begin{bmatrix} 1 & 0 \\ 0 & -1 \end{bmatrix}, \begin{bmatrix} 0 & 1 \\ 1 & 0 \end{bmatrix}, \begin{bmatrix} -1 & 0 \\ 0 & 1 \end{bmatrix}, \begin{bmatrix} 0 & -1 \\ -1 & 0 \end{bmatrix}, \quad k = 1, 2, 3, 4$$

of D_4 evaluates as

$$\begin{aligned} \widehat{x}(\beta) &= \sum_{k=1}^{4} (r_k \beta_{k,1} + t_k \beta_{k,-1}) \\ &= \begin{bmatrix} r_1 - r_3 + t_1 - t_3 & -r_2 + r_4 + t_2 - t_4 \\ r_2 - r_4 + t_2 - t_4 & r_1 - r_3 - t_1 + t_3 \end{bmatrix} = \begin{bmatrix} v_1 & v_2 \\ w_1 & w_2 \end{bmatrix}, \end{aligned}$$

where its components are exactly the bases $\mathcal{B}_v = \{v_1, v_2\}$ and $\mathcal{B}_w = \{w_1, w_2\}$ that appear in the decomposition of the irreducible subspace $\mathcal{V}_\beta$ of D_4, on page 108, 109. The regular indexing gives, for rotations,

$$x_\beta = \{r_1 - r_3,\ r_2 - r_4,\ -r_1 + r_3,\ -r_2 + r_4\}$$

and, for reversals,

$$x_\beta = \{t_1 - t_3,\ t_2 - t_4,\ -t_1 + t_3,\ -t_2 + t_4\}.$$

A Poisson summation formula

The regular indexing can be used to derive a class of Poisson summation formulas, of special practical importance in symmetry studies of data indexed by cosets. See, for details, Van de Ven and Di Bucchianico (2006) and Van de Ven (2007).

Specifically, let, for $\eta \in \widehat{G}$, $x \in \mathcal{F}(G)$, $\tau \in G$ and H a subgroup of G,

$$y_\sigma = \sum_{\tau \in H} \chi_\eta(\tau\sigma) x_{\tau\sigma}$$

and assume that G is Abelian with g elements. Rewrite y as $y_\sigma = \sum_{\tau \in G} \chi_\eta(\tau\sigma) x_{\tau\sigma} h(\tau)$, where h is the indicator function for H. To apply (4.20) to the indexing $y \in \mathcal{F}(G)$ first evaluate the Fourier transform

$$\begin{aligned} \widehat{y}(\beta) &= \sum_{\sigma \in G} y_\sigma \chi_\beta(\sigma) = \sum_{\sigma \in G} \left[\sum_{\tau \in G} \chi_\eta(\tau\sigma) x_{\tau\sigma} h(\tau) \right] \chi_\beta(\sigma) \\ &= \sum_{\tau \in G} \chi_\eta(\tau) h(\tau) \left[\sum_{\sigma \in G} \chi_\eta(\sigma) x_{\tau\sigma} \chi_\beta(\sigma) \right]. \end{aligned}$$

of y at $\beta \in \widehat{G}$. Writing $\widehat{G}_H$ to indicate those characters $\chi \in \widehat{G}$ identically equal to 1 over H, it is not difficult to verify (Van de Ven, 2007) that $\sum_{h \in H} \chi(h) = |H|$, for $\chi \in \widehat{G}_H$ and $\sum_{h \in H} \chi(h) = 0$, otherwise. Because

$$\sum_{\sigma \in G} \chi_\eta(\sigma) x_{\tau\sigma} \chi_\beta(\sigma) = \chi_\eta(\tau^{-1}) \chi_\beta(\tau^{-1}) \widehat{x}(\eta\beta),$$

then,

$$\widehat{y}(\beta) = \left[\sum_{\tau \in G} \chi_\beta(\tau^{-1}) h(\tau) \right] \widehat{x}(\eta\beta) = \left[\sum_{\tau \in H} \chi_\beta(\tau^{-1}) \right] \widehat{x}(\eta\beta) = |H| \widehat{x}(\eta\beta), \quad \beta \in \widehat{G}_H,$$

and $\widehat{y}(\beta) = 0$ otherwise. From (4.20) it follows that

$$y_\sigma = \sum_{\beta \in \widehat{G}} |H| [\chi_\beta(\sigma^{-1}) \widehat{x}(\eta\beta)]/g, \quad \beta \in \widehat{G}_H,$$

and 0 otherwise. That is,

$$y_\sigma = \sum_{\beta \in \widehat{G}_H} |H| [\chi_\beta(\sigma^{-1}) \widehat{x}(\eta\beta)]/g,$$

which is, up to notation, the Poisson Summation formula in Van de Ven and Di Bucchianico (2006, p. 13).

Exercises

Exercise 4.1. Show that $\sum_{\tau \in S_n} \rho_\tau / n! = \mathcal{A}_n$, where ρ is the permutation representation of S_n.

Exercise 4.2. Verify, using Theorem 4.2, that the representation $\beta_{k,d}$ of D_4 shown on page 42 is irreducible. When $n = 3$, show that $\beta_{k,d}$ and β (page 64) are equivalent.

Exercise 4.3. Propose and carry out a symmetry study for the transition frequency data shown on Table (2.28) (page 56).

Exercise 4.4. Derive the canonical projections associated with action defined by (2.32) on page 57.

Exercise 4.5. Show that if $\mathcal{P}$ is a regular canonical projection then $\mathcal{P}' = \overline{\mathcal{P}}$.

Exercise 4.6. Show that if $\chi \in \widehat{G}$ then $\overline{\chi} \in \widehat{G}$.

Exercise 4.7. Following Exercises 4.5 and 4.6, show that if $\mathcal{P}_\chi$ is a regular canonical projection of a group G then $\overline{\mathcal{P}}_\chi = \mathcal{P}_{\overline{\chi}}$ so that $\overline{\mathcal{P}}$ is also a canonical projection of G. Moreover, then, $\mathcal{P}_\chi + \overline{\mathcal{P}}_\chi$ is a real symmetric canonical projection of G (see Proposition 4.6 on page 114).

Exercise 4.8. Interpret Proposition 4.7 in terms of the standard decomposition (page 112).

Exercise 4.9. Following with the canonical decomposition for the conjugacy action of S_3 derived on page 112, indicate by $x' = (r_1, r_2, r_3, t_1, t_2, t_3)$ the data indexed by S_3 and show that the following canonical invariants

$$\mathcal{I}_1 = r_1 + r_2 + r_3 + t_1 + t_2 + t_3, \quad \mathcal{I}_{\text{Sgn}} = \pm(r_2 - r_3),$$
$$\mathcal{I}_\beta = \{t_1 - t_2 + t_1 - t_3, \quad t_2 - t_1 + t_2 - t_3\}$$

in dimensions of $1, 1, 2$ respectively, can be identified. In addition, show that $x'\mathcal{P}_1 x = r_1^2 + (t_1 + t_2 + t_3)^2/3 + (t_2 + t_3)^2/2$,

$$x'\mathcal{P}_{\text{Sgn}}x = (r_2 - r_3)^2/2, \quad x'\mathcal{P}_\beta x = 2/3[t_1(t_1 - t_2) + t_2(t_2 - t_3) + t_3(t_3 - t_1)]$$

are the components in the decomposition $x'\mathcal{P}_1 x + x'\mathcal{P}_{\text{Sgn}}x + x'\mathcal{P}_\beta x$ of $x'x$ and interpret the underlying parametric hypotheses associated with each component.

Exercise 4.10. Reduction of oriented triangles. Indicate by $V = \{(a, b), (b, a), (a, c), (c, a), (b, c), (c, b)\}$ the set of the directed edges of a triangle with vertices $\{a, b, c\}$, where (x, y) indicates a directed edge $x \to y$. Equivalently

$$V = \begin{bmatrix} 0 & 1 & 1 \\ 1 & 0 & 1 \\ 1 & 1 & 0 \end{bmatrix},$$

where $V_{ij} = 1 \iff (i, j) \in V$. Let also $x' = (ab, ba, ac, ca, bc, cb)$ indicate the vector of elementary data indexed by V. Show that

$$\tau : (i, j) \in V \xrightarrow{\rho} (\tau i, \tau j) \in V, \quad \tau \in S_3, \tag{4.21}$$

gives a representation of S_3 in dimension of six. Evaluate the canonical projections and the canonical invariants on the data indexed by the oriented triangles.

Exercise 4.11. Reduction of A-loops. Given the set of labels

$$V = \{(12, 22), (22, 32), (32, 12), (11, 21), (21, 31), (31, 11)\} \equiv \{a, b, c, d, e, f\},$$

show that

$$(\tau, \sigma) : (ij, k\ell) \in V \xrightarrow{\rho} (\tau i \sigma j, \tau k \sigma \ell) \in V, \quad (\tau, \sigma) \in C_3 \times C_2 \tag{4.22}$$

there is a representation of $C_3 \times C_2$ defined by

τ	σ	a	b	c	d	e	f
1	1	a	b	c	d	e	f
r	1	b	c	a	e	f	d
r^2	1	c	a	b	f	d	e
1	t	d	e	f	a	b	c
r	t	e	f	d	b	c	a
r^2	t	f	d	e	c	a	b

, $r = (123), \; t = (12)$.

Determine the canonical projections and their six one-dimensional canonical invariants on the data indexed by the A-loops.

Exercise 4.12. Reduction of cross-loops. Let

$$V = \{(11, 22), (22, 31), (31, 12), (12, 21), (21, 32), (32, 11)\} \equiv \{a, b, c, d, e, f\}.$$

Show that

$$(\tau, \sigma) : (ij, k\ell) \in V \overset{\rho}{\to} (\tau i \sigma j, \tau k \sigma \ell) \in V, \quad (\tau, \sigma) \in C_3 \times C_2, \tag{4.23}$$

gives a representation of $C_3 \times C_2$, defined from

τ	σ	a	b	c	d	e	f
1	1	a	b	c	d	e	f
r	1	e	f	a	b	c	d
r^2	1	c	d	e	f	a	b
1	t	d	e	a	b	c	d
r	t	b	c	d	e	f	a
r^2	t	f	a	b	c	d	e

, $r = (123), \; t = (12)$.

Determine and interpret the resulting linear representation and the six one-dimensional canonical invariants on the data indexed by the cross-loops.

Exercise 4.13. Reduction of closed loops. Let

$$V = \{(11, 21), (21, 31), (31, 32), (32, 22), (22, 12), (12, 11)\} \equiv \{a, A, b, B, c, C\}$$

and identify it (uniquely) by the restriction

$$V = \{(1, 2), (2, 3), (3, 3), (3, 2), (2, 1), (1, 1)\}$$

of $(ij, k\ell)$ to (i, k). Show that

$$\tau : (i, k) \in V \overset{\rho}{\to} (i, \tau^{-1}k) \in V \tag{4.24}$$

gives a representation of $C_3 = \{1, \tau, \tau^2\}$ with generating matrix

$$\rho_\tau = \begin{bmatrix} 0 & 0 & 0 & 0 & 1 & 0 \\ 0 & 0 & 0 & 1 & 0 & 0 \\ 1 & 0 & 0 & 0 & 0 & 0 \\ 0 & 0 & 0 & 0 & 0 & 1 \\ 0 & 0 & 1 & 0 & 0 & 0 \\ 0 & 1 & 0 & 0 & 0 & 0 \end{bmatrix}.$$

In addition, show that there is an action of S_3 on V. To see this, first verify that the action in (4.24) defines a partition $\mathcal{O}_a \cup \mathcal{O}_A$ of V into orbits $\mathcal{O}_a = \{a, b, c\}$ and $\mathcal{O}_A = \{A, B, C\}$, which in turn identifies an action of C_2 on the whole set V, obtained by transpositions between the two orbits. Show that the resulting permutation representation is generated by

$$m = \begin{bmatrix} 0 & 0 & 0 & 1 & 0 & 0 \\ 0 & 0 & 0 & 0 & 1 & 0 \\ 0 & 0 & 0 & 0 & 0 & 1 \\ 1 & 0 & 0 & 0 & 0 & 0 \\ 0 & 1 & 0 & 0 & 0 & 0 \\ 0 & 0 & 1 & 0 & 0 & 0 \end{bmatrix},$$

and that it follows from the action

$$(\sigma, \tau) : (i, k) \in V \xrightarrow{\eta} (\sigma i, \tau^{-1} k) \in V, \quad \sigma \in C_2, \quad \tau \in C_3, \tag{4.25}$$

of $C_2 \times C_3$ realized as $\eta(\sigma^f, \tau^g) = r^g m^f$, $g = 1, 2, 3$, $f = 1, 2$. Its multiplication table is defined by $m^2 = 1$, $mrm = r^2$, $r^3 = 1$, that is, $G \simeq S_3 \simeq D_3$. Determine the canonical projections and their canonical invariants.

Exercise 4.14. Consider the action

$$\varphi((\sigma, \tau), (i, j)) = (\sigma i, \tau j), \quad \sigma \in G \subseteq S_{\ell_1}, \tau \in G' \subseteq S_{\ell_2}, \quad (i, j) \in L_1 \times L_2,$$

where $\ell_i = |L_i|$ and G and G' symmetric subgroups. Derive the canonical projections derived from φ, assuming $H = H' = S_2$ and show that the resulting data reduction can be assembled in matrix form as

$$\begin{bmatrix} x_{11} & x_{12} \\ x_{21} & x_{22} \end{bmatrix} = (x_{11} + x_{12} + x_{21} + x_{22}) \begin{bmatrix} 1 & 1 \\ 1 & 1 \end{bmatrix} /4 + (x_{11} - x_{12} - x_{21} + x_{22}) \begin{bmatrix} 1 & -1 \\ -1 & 1 \end{bmatrix} /4$$
$$+ (x_{11} - x_{12} + x_{21} - x_{22}) \begin{bmatrix} 1 & -1 \\ 1 & -1 \end{bmatrix} /4 + (x_{11} + x_{12} - x_{21} - x_{22}) \begin{bmatrix} 1 & 1 \\ -1 & -1 \end{bmatrix} /4,$$

and that the decomposition gives four canonical invariants in dimension of 1, the coefficients in the above decomposition. Moreover, show that the matrices

$$A = \begin{bmatrix} 1 & 1 \\ 1 & 1 \end{bmatrix}/2, \quad D = \begin{bmatrix} 1 & -1 \\ -1 & 1 \end{bmatrix}/2, \quad C = \begin{bmatrix} 1 & -1 \\ 1 & -1 \end{bmatrix}/2, \quad R = \begin{bmatrix} 1 & 1 \\ -1 & -1 \end{bmatrix}/2$$

multiply according to

*	A	D	C	R
A	A	0	C	0
D	0	D	0	R
C	0	C	0	A
R	R	0	D	0

and generate a non commutative algebra, with multiplication induced by the usual matrix multiplication. Argue that within this vector space, contingency tables can be combined, compared, scaled and that the same construction can be extended to $V = L_1 \times \ldots \times L_f$.

Exercise 4.15. The action $\sigma s \tau^{-1}$. With the same notation from Exercise 4.14, let $s \in V = L^F$ and show that $\varphi((\sigma, \tau), s) = \sigma s \tau^{-1}, \quad s \in V, \ \sigma \in G \subseteq S_\ell, \ \tau \in G' \subseteq S_f$, gives an action of $G \times G'$ on V. Show that the case $f = \ell = 2$, with $F = \{1, 2\}$ (position) and $L = \{u, y\}$ (symbols) corresponds to $V = \{uu, uy, yu, yy\}$ and

φ :

σ, τ	uu	uy	yu	yy
11	uu	uy	uy	yy
1, (12)	uu	yu	uy	yy
(uy), 1	yy	yu	uy	uu
(uy), (12)	yy	uy	yu	uu

Derive the canonical projections for $c = \ell = 2$ and show that they lead to the decomposition

$$2\begin{bmatrix} x_{11} & x_{12} \\ x_{21} & x_{22} \end{bmatrix} = (x_{11} + x_{22})I + (x_{12} + x_{21})D + (x_{11} - x_{22})H + (x_{12} - x_{21})R,$$

where

$$I = \begin{bmatrix} 1 & 0 \\ 0 & 1 \end{bmatrix}, \quad R = \begin{bmatrix} 0 & 1 \\ -1 & 0 \end{bmatrix}, \quad H = \begin{bmatrix} 1 & 0 \\ 0 & -1 \end{bmatrix}, \quad D = \begin{bmatrix} 0 & 1 \\ 1 & 0 \end{bmatrix}.$$

Note that there are three invariants now:

(1) $\{x_{11} + x_{22}, x_{12} + x_{21}\}$ in dimension of 2,
(2) $x_{11} - x_{22}$ in dimension of 1,
(3) $x_{12} - x_{21}$ in dimension of 1.

The matrix R defines a 90° central rotation, whereas H is a horizontal line reflection and D is a 45° line reflection. These matrices multiply according to

$\cdot$	1	R	H	D
1	1	R	H	D
R	R	-1	$-D$	H
H	H	D	1	R
D	D	$-H$	$-R$	1

and generate an associative algebra. In this vector space the data indexed by V can then be investigated.

Exercise 4.16. Apply Algorithm 3 (page 221) to derive the canonical projections associated with the group action given by Table (3.5) on page 66.

Exercise 4.17. Show that $\{1, (12)(34), (14)(23), (13)(24)\}$ is isomorphic with $C_2 \times C_2$ and that its character table is given by the products

χ	$(1,1)$	$(1,t)$	$(t,1)$	(t,t)
$\chi_1\chi_1$	1	1	1	1
$\chi_1\chi_{\text{Sgn}}$	1	-1	1	-1
$\chi_{\text{Sgn}}\chi_1$	1	1	-1	-1
$\chi_{\text{Sgn}}\chi_{\text{Sgn}}$	1	-1	-1	1

(4.26)

of the irreducible characters χ_1 and χ_{Sgn} of $C_2 = \{1, t\}$. Derive the regular canonical projections and corresponding invariants.

5
Examples and Techniques

5.1 Introduction

In the next sections we bring the algebraic results from previous chapters in contact to well-known methods of statistical inference, such as the many applications of the Fisher-Cochran theorems for the distribution of quadratic forms, and illustrate how those same principles apply and extend to the broader interpretation of structured data in symmetry studies.

5.2 Analysis of Variance

In the proof of Theorem 4.4 on page 111 we observed that relative to a new basis $y = Bx$ in the original data vector space $\mathcal{V}$,

$$\sum_{i=1}^{h} P_i = \text{Diag}(I_{m_1} \otimes I_{n_1}, \ldots, I_{m_h} \otimes I_{n_h}) = I_v.$$

If B is real (as in the case of regular projections; see Exercise 4.7 on page 122) so that its orthonormal version $\mathcal{B}$ is also real, and the distribution of x is normal with mean μ and covariance matrix I_v, shortly $x \sim N(\mu, I_v)$, then $\mathcal{B}x \sim N(\mathcal{B}\mu, I_v)$ and,

$$x'x = y'y = \sum_{i=1}^{h} y'\mathcal{P}_j y = \sum_{i=1}^{h} x'\mathcal{B}^{-1} P_j \mathcal{B}x = \sum_{i=1}^{h} x'\mathcal{B}' P_j \mathcal{B}x = \sum_{i=1}^{h} (\mathcal{B}x)' P_j \mathcal{B}x,$$

which is distributed as the sum $\sum_{i=1}^{h} \chi^2_{\nu_i}(\delta_i)$ of chi-square distributions $\chi^2_{\nu_i}(\delta_i)$ with $\nu_i = \text{Tr}\, \mathcal{P}_i$ degrees of freedom and noncentrality parameter $\delta_i = \mu'\mathcal{P}_i\mu$. Clearly, these components are independently distributed.

Analysis of variance for scalar data indexed by $S_3 \simeq D_3$

Following with the notation introduced on page 104 and assuming that $x \sim N(\mu, I_v)$, we obtain

(1) $x'\mathcal{P}_1 x = (r_1 + r_2 + r_3 + t_1 + t_2 + t_3)^2/6 \sim \chi_1^2(\delta_1)$;
(2) $x'\mathcal{P}_{\text{Sgn}} x = (r_1 + r_2 + r_3 - t_1 - t_2 - t_3)^2/6 \sim \chi_1^2(\delta_{\text{Sgn}})$;
(3) $x'\mathcal{P}_\beta x = [r_1(2r_1 - r_2 - r_3) + r_2(2r_2 - r_1 - r_3) + r_3(2r_3 - r_1 - r_2)$
$+ t_1(2t_1 - t_2 - t_3) + t_2(2t_2 - t_1 - t_3) + t_3(2t_3 - t_1 - t_2)]/3 \sim \chi_4^2(\delta_\beta)$.

The corresponding parametric hypotheses of interest, in terms of the expected value μ of x, are then

$$H_{\text{Sgn}} : \sum_j (\mu_{r_j} - \mu_{t_j}) = 0$$

under which $\delta_{\text{Sgn}} = 0$ and

$$H_\beta : \mu_{r_1} = \mu_{r_2} = \mu_{r_3} \quad \text{and} \quad \mu_{t_1} = \mu_{t_2} = \mu_{t_3}$$

under which $\delta_\beta = 0$. The equality in H_{Sgn} defines the hypothesis of no difference between rotations and reversals, whereas H_β tests the joint hypotheses of no variation among rotations and no variation among reversals. Other interpretations, of course, can be obtained from different contexts.

Analysis of variance for a simple triangular array

Consider the simple triangular array

$$V = \{(0,0), (1,0), (0,1), (2,0), (0,2), (1,1)\} \equiv \{\alpha, x, y, X, Y, \gamma\},$$

where S_2 acts on the coordinates of each point according to

S_2	α	x	y	X	Y	γ
1	α	x	y	X	Y	γ
τ	α	y	x	Y	X	γ

The resulting canonical decomposition

$$\mathcal{P}_1 = \text{Diag}(1, \mathcal{A}_2, \mathcal{A}_2, 1), \qquad \mathcal{P}_2 = \text{Diag}(0, \mathcal{Q}_2, \mathcal{Q}_2, 0),$$

leads to the canonical invariants $\mathcal{P}_1\mu = \{\alpha, x + y, X + Y, \gamma\}$ and $\mathcal{P}_2\mu = \{x - y, X - Y\}$ on $\mu = (\alpha, x, y, X, Y, \gamma)$ in dimensions of 4 and 2, respectively. Correspondingly,

$$\mu'\mathcal{P}_1\mu = 1/2\left[2\alpha^2 + (x+y)^2 + (X+Y)^2 + 2\gamma^2\right], \quad \mu'\mathcal{P}_2\mu = 1/2\left[(x-y)^2 + (X-Y)^2\right].$$

Suppose that $n = 3$ independent, identically distributed normal observations are obtained in each point of V so that the analysis of variance

$\mathcal{P}$	$x'\mathcal{P}x$	Tr $\mathcal{P}$
$\mathcal{P}_1 \otimes \mathcal{A}$	687.66	4
$\mathcal{P}_2 \otimes \mathcal{A}$	11.33	2
$\mathcal{P}_1 \otimes \mathcal{Q}$	24.33	8
$\mathcal{P}_2 \otimes \mathcal{Q}$	9.66	4
total	733	18

,

source	ss	df
$\mathcal{P}_1 \otimes \mathcal{A}$	687.66	4
$\mathcal{P}_2 \otimes \mathcal{A}$	11.33	2
residual	34	12
total	733	18

for the resulting data

$$x' = (\alpha_1, \alpha_2, \alpha_3, \ldots, \gamma_1, \gamma_2, \gamma_3) = (3, 4, 3, 4, 5, 5, 6, 7, 3, 9, 7, 4, 10, 9, 9, 5, 5, 9) \in \mathbb{R}^{18},$$

is then obtained from $x'[(\mathcal{P}_1 + \mathcal{P}_2) \otimes (\mathcal{A}_3 + \mathcal{Q}_3)]x$ and from combining the error terms $\mathcal{P}_1 \otimes \mathcal{Q} + \mathcal{P}_2 \otimes \mathcal{Q}$.

Under the hypothesis $H : \{x = y,\ \ X = Y\}$, then $\mu'\mathcal{P}_2\mu = 1/2\,[(x - y)^2 + (X - Y)^2] = 0$ so that

$$E(x'(\mathcal{P}_2 \otimes \mathcal{A})x) = E(x'(\mathcal{P}_1 \otimes \mathcal{Q} + \mathcal{P}_2 \otimes \mathcal{Q})x) = 0,$$

where E indicates expected value. Therefore, under the hypothesis H,

$$F = \frac{x'(\mathcal{P}_2 \otimes \mathcal{A})x/\mathrm{Tr}\,(\mathcal{P}_2 \otimes \mathcal{A})}{x'(\mathcal{P}_1 \otimes \mathcal{Q} + \mathcal{P}_2 \otimes \mathcal{Q})x/\mathrm{Tr}\,(\mathcal{P}_1 \otimes \mathcal{Q} + \mathcal{P}_2 \otimes \mathcal{Q})}$$

has a F distribution with degrees of freedom $df_1 = \mathrm{Tr}\,(\mathcal{P}_2 \otimes \mathcal{A}) = 2$ and $df_2 = \mathrm{Tr}\,(\mathcal{P}_2 \otimes \mathcal{Q}) = 12$, and can be used to assess that hypothesis. In the present example, $F = 1.99$.

5.3 Classical Structures and Their Analysis of Variance

One-way analysis of variance

The canonical reduction for the one-way analysis of variance is simply

$$I_n = \mathcal{A} + \mathcal{Q}(D_{\mathcal{A}} + D_{\mathcal{Q}}),$$

where $\mathcal{A} + \mathcal{Q}$ is the standard reduction in dimension of $n = n_1 + \cdots + n_k$, $D_{\mathcal{A}} = \mathrm{Diag}(\mathcal{A}_{n_1}, \ldots, \mathcal{A}_{n_k})$, and $D_{\mathcal{Q}} = I_n - D_{\mathcal{A}}$. To illustrate, let $n_1 = n_2 = 3$ and $n_3 = 5$, so that $D_{\mathcal{A}} = \mathrm{diag}\,(\mathcal{A}_3, \mathcal{A}_3, \mathcal{A}_5)$. It then follows that $I_{11} = \mathcal{A}_{11} + \mathcal{Q}_{11}D_{\mathcal{A}} + \mathcal{Q}_{11}D_{\mathcal{Q}} \equiv \mathcal{P}_1 + \mathcal{P}_2 + \mathcal{P}_3$ is a canonical reduction and, given the data

$$y' = (\underbrace{12, 14, 11}_{\text{group 1}}, \underbrace{10, 9, 11}_{\text{group 2}}, \underbrace{8, 12, 15, 14, 12}_{\text{group 3}}),$$

the resulting decomposition is

$\mathcal{P}$	$y'\mathcal{P}y$	$\text{Tr}\,\mathcal{P}$	$y'\mathcal{P}y/\text{Tr}\,\mathcal{P}$
$\mathcal{P}_1$(constant)	1,489.45	1	1,489.45
$\mathcal{P}_2$ (treatment)	11.07	2	5.53
$\mathcal{P}_3$ (residual)	35.46	8	4.43
I(total)	1,536.0	11	

.

Two-way analysis of variance

The reduction of a two-way ANOVA with r row levels, c column levels, and n observations in each cell is given by

$$I_{rcn} = (\mathcal{A}_r + \mathcal{Q}_r) \otimes (\mathcal{A}_c + \mathcal{Q}_c) \otimes (\mathcal{A}_n + \mathcal{Q}_n),$$

in which $\mathcal{A}_r \otimes \mathcal{A}_c \otimes \mathcal{A}_n$ is the constant, $\mathcal{Q}_r \otimes \mathcal{A}_c \otimes \mathcal{A}_n$ the row effect, $\mathcal{A}_r \otimes \mathcal{Q}_c \otimes \mathcal{A}_n$ the column effect, $\mathcal{Q}_r \otimes \mathcal{Q}_c \otimes \mathcal{A}_n$ the interaction term and $\mathcal{A}_r \otimes \mathcal{A}_c \otimes \mathcal{Q}_n + \mathcal{Q}_r \otimes \mathcal{A}_c \otimes \mathcal{Q}_n + \mathcal{A}_r \otimes \mathcal{Q}_c \otimes \mathcal{Q}_n + \mathcal{Q}_r \otimes \mathcal{Q}_c \otimes \mathcal{Q}_n$ the error term.

To illustrate, let $V = L_1 \times L_2 \equiv \{u, v\} \times \{1, 2, 3\}$ be the set of data labels

$$\begin{bmatrix} u_1 & u_2 & u_3 \\ v_1 & v_2 & v_3 \end{bmatrix},$$

where $S_2 \times S_3$ acts by permutation. The canonical reduction follows from Proposition 4.6 on page 114 with $\mathcal{P}_1 = \mathcal{A}_2 \otimes \mathcal{A}_3$, $\mathcal{P}_2 = \mathcal{A}_2 \otimes \mathcal{Q}_3$, $\mathcal{P}_3 = \mathcal{Q}_2 \otimes \mathcal{A}_3$ and $\mathcal{P}_4 = \mathcal{Q}_2 \otimes \mathcal{Q}_3$ and is summarized on Table (5.1).

$\mathcal{P}$	$\text{Tr}\,\mathcal{P}$	invariant subspaces	interpretation
$\mathcal{P}_1$	1	$u_1 + u_2 + u_3 + v_1 + v_2 + v_3$	baseline average
$\mathcal{P}_2$	2	$2u_1 - u_2 - u_3 + 2v_1 - v_2 - v_3, \quad -u_1 + 2u_2 - u_3 - v_1 + 2v_2 - v_3$	column effect
$\mathcal{P}_3$	1	$u_1 + u_2 + u_3 - v_1 - v_2 - v_3$	row effect
$\mathcal{P}_4$	2	$2u_1 - u_2 - u_3 - 2v_1 + v_2 + v_3, \quad -u_1 + 2u_2 - u_3 + v_1 - 2v_2 + v_3$	remainder ϵ

(5.1)

Suppose that $n = 3$ independent and identically distributed are obtained at each point of the initial structure. The new underlying data structure is then $V \times \{1, 2, 3\}$ so that the corresponding data space $\mathcal{V}$ has dimension $\ell_1 \times \ell_2 \times n$. The ensuing reduction is now obtained by an additional factoring with the standard reduction $I_3 = \mathcal{A}_3 + \mathcal{Q}_3$ for the sampling part. From Proposition 4.6, $I = \mathcal{P}_1 \otimes \mathcal{A} + \cdots \mathcal{P}_4 \otimes \mathcal{A} + \mathcal{P}_1 \otimes \mathcal{Q} + \cdots \mathcal{P}_4 \otimes \mathcal{Q}$ is the canonical reduction of interest.

Reducing the standard 2^p factorial data

The canonical reduction for the 2^p factorial data is obtained as the p-fold tensor of the standard reduction for S_2, that is,

$$I_{2^p} = \underbrace{(\mathcal{A} + \mathcal{Q}) \otimes \cdots \otimes (\mathcal{A} + \mathcal{Q})}_{p \text{ times}}.$$

The resulting projections act on observations indexed by the 2^p combinations of high (1)-low (0) levels describing the labels in V. To illustrate, consider the case in which $n = 2$ observations are obtained at each of the 8 labels of a 2^3 factorial experiment, that is,

$$V = \begin{bmatrix} 0 & 0 & 0 & 0 & 1 & 1 & 1 & 1 \\ 0 & 0 & 1 & 1 & 0 & 0 & 1 & 1 \\ 0 & 1 & 0 & 1 & 0 & 1 & 0 & 1 \end{bmatrix}, \; y = \begin{bmatrix} 15 & 999 & 499 & 286 & 438 & 926 & 871 & 891 \\ 779 & 990 & 212 & 611 & 239 & 787 & 303 & 663 \end{bmatrix} \in \mathcal{V}.$$

The data reduce according to $I = uuu \otimes \mathcal{A} + \ldots ttt \otimes \mathcal{A} + uuu \otimes \mathcal{Q} + \ldots ttt \otimes \mathcal{Q}$, leading to the decomposition shown on Table (5.2)

trait	$\mathcal{P}$	$y'(\mathcal{P} \otimes \mathcal{A})y$	Tr $\mathcal{P} \otimes \mathcal{A}$
1	uuu	5, 651, 317.56	1
c	uut	488, 950.56	1
b	utu	43, 785.56	1
bc	utt	173, 264.06	1
a	tuu	33, 033.06	1
ac	tut	76.56	1
ab	ttu	143, 073.06	1
abc	ttt	7, 788.06	1
error	$I \otimes \mathcal{Q}$	$y'(I \otimes \mathcal{Q})y$	Tr $I \otimes \mathcal{Q}$
		75, 331.31	8
Sum	I	7, 143, 939.00	16

(5.2)

Latin squares

The reduction of a Latin square experiment has the form

$$I = [\underbrace{(\mathcal{A} + \mathcal{Q})}_{\text{rows}} \otimes \underbrace{(\mathcal{A} + \mathcal{Q})}_{\text{columns}}] \underbrace{(\mathcal{A} + \mathcal{Q})}_{\text{letters}}.$$

To illustrate, consider the following experiment described in Youden (1951, p. 96), in which the data

	I	II	III	IV
1	A	B	C	D
2	C	D	A	B
3	B	C	D	A
4	D	A	B	C

→

36	38	36	30
17	18	26	17
30	39	41	34
30	45	38	33

are the melting point temperature readings of four chemical cells (1, 2, 3, 4) obtained from four thermometers (I,II,III,IV) in four different days (A,B,C,D). The numerical entries are the readings converted to degrees Centigrade. The experimental background is such that there is no reason to assume an interaction between cells and thermometers. Write the data as

$$y' = (36, 38, 36, 30, 17, 18, 26, 17, 30, 39, 41, 34, 30, 45, 38, 33)$$

and first evaluate the four projections associated with $(\mathcal{A} + \mathcal{Q}) \otimes (\mathcal{A} + \mathcal{Q})$, where $I_4 = \mathcal{A} + \mathcal{Q}$ is the standard reduction in dimension 4. As a result, SS total $= y'y = 17230$, SS constant $= y'(\mathcal{A} \otimes \mathcal{A})y = 16129$, SS thermometers $= y'(\mathcal{A} \otimes \mathcal{Q})y = 182.50$, SS cells $= y'(\mathcal{Q} \otimes \mathcal{A})y = 805$, SS days + SS residual $= y'(\mathcal{Q} \otimes \mathcal{Q})y = 113.5$. To further reduce $y'(\mathcal{Q} \otimes \mathcal{Q})y$ and determine the component due to eventual day-to-day variability, the standard reduction (indicated here by $\mathcal{A}_\bullet$ and $\mathcal{Q}_\bullet$) is applied to aggregate the data from corresponding days, obtaining

$$\text{SS days} = y'([\mathcal{Q} \otimes \mathcal{Q}]\mathcal{A}_\bullet)y = 70, \quad \text{SS residual} = y'([\mathcal{Q} \otimes \mathcal{Q}]\mathcal{Q}_\bullet)y = 43.5.$$

Table (5.3) summarizes the results. The F ratio for cells is quite significant ($268.33/7.25 = 37.01$). The F ratio for thermometers, $60.83/7.25 = 8.39$, also points to a difference among the thermometers whereas the F ratio for days is only suggestive of a probable day-to-day effect. The estimated standard deviation for a single measurement is $\sqrt{7.25} = 2.69$, which shows an improvement in the error of comparison. In fact, if the effect of days on the readings is not eliminated, the standard deviation would then be $\sqrt{113.5/9} = 3.55$.

$\mathcal{P}$	$ss = y'\mathcal{P}y$	$df = \text{Tr}\,\mathcal{P}$	ss/df
$\mathcal{A} \otimes \mathcal{A}$ (constant)	16,129.00	1	16,129.00
$\mathcal{A} \otimes \mathcal{Q}$ (thermometers)	182.50	3	60.83
$\mathcal{Q} \otimes \mathcal{A}$ (cells)	805.00	3	268.33
$(\mathcal{Q} \otimes \mathcal{Q})\mathcal{A}_\bullet$ (days)	70.00	3	23.33
$(\mathcal{Q} \otimes \mathcal{Q})\mathcal{Q}_\bullet$ (residual)	43.50	6	7.25
I (total)	17,230.00	16	

(5.3)

5.4 The Analysis of Variance for Dihedral Data

Table (5.4) shows the analysis of variance for the data $x' = (r', t')$ indexed by the dihedral group D_4, where $r' = (r_1, \ldots, r_4)$ indicates the rotations part and $t' = (t_1, \ldots, t_4)$ the reversals part of the data.

The canonical projections $\mathcal{P}_1, \ldots, \mathcal{P}_5$ for the regular representation of D_4 were derived on page 110, in correspondence with the symmetric, α, Sgn, γ and β irreducible representations.

The reduction of the data space ($\mathbb{R}^8$) and the determination of their invariants follow from the canonical projections. The resulting decomposition $x'x = x'\mathcal{P}_1 x + \cdots + x'\mathcal{P}_5 x$ of the sum of squares $x'x$ is then

source	$x'\mathcal{P}x$	df
$\mathcal{P}_1$	$\left[\sum_i (r_i + t_i)\right]^2 /8$	1
$\mathcal{P}_2$	$\left[\sum_i (r_i - t_i)\right]^2 /8$	1
$\mathcal{P}_3$	$(r_2 - r_1 + r_4 - r_3 + t_2 - t_1 + t_4 - t_3)^2 /8$	1
$\mathcal{P}_4$	$(r_2 - r_1 + r_4 - r_3 - t_2 + t_1 - t_4 + t_3)^2 /8$	1
$\mathcal{P}_5$	$[(r_1 - r_3)^2 + (r_2 - r_4)^2 + (t_1 - t_3)^2 + (t_2 - t_4)^2]/2$	4
total	$\sum_i x_i^2$	8

. (5.4)

The following interpretations are relevant: $x'\mathcal{P}_1 x$ is the overall reference constant; $x'\mathcal{P}_2 x$ compares rotations and reversals; $x'\mathcal{P}_3 x$ and $x'\mathcal{P}_4 x$ combine and contrast within-rotation and within-reversal variability, respectively, whereas $x'\mathcal{P}_5 x$ assesses the presence of point symmetry.

Sampling considerations. When a sample of size n is obtained at each one of the dihedral symmetries, the new decomposition of the sum of squares $x'x$ for the $8 \times n$ data points is then obtained from the new canonical decomposition $I = I_8 \otimes I_n = \sum_{i=1}^{5} \mathcal{P}_i \otimes (\mathcal{A} + \mathcal{Q})$, where $\mathcal{A}$ and $\mathcal{Q}$ define the standard decomposition in dimension of n.

5.5 Cyclic Reduction of Binary Sequences

Consider the set V of binary sequences (s) in length of 4 where $C_4 = \{1, (1234), (13)(24), (1432)\}$ acts according to $s\tau^{-1}$ (position symmetry). This is the group action introduced earlier on page 50. Table (5.5) shows the set V along with an arbitrary data vector x_s indexed by V.

1	y	u	y	u	u	u	y	y	u	u	y	u	y	y	y	u
2	y	u	u	y	u	u	y	u	y	u	u	y	y	y	u	y
3	y	u	u	u	y	u	u	u	y	y	y	u	y	u	y	y
4	y	u	u	u	u	y	u	y	u	y	u	y	u	y	y	y
s	1	16	15	14	12	8	13	7	10	4	11	6	9	5	3	2
x_s	y	z	p	q	r	s	a	b	c	d	e	f	k	l	m	n

(5.5)

The resulting orbits

$$\mathcal{O}_0 = \{1\}, \quad \mathcal{O}_1 = \{9, 5, 3, 2\}, \quad \mathcal{O}_{21} = \{13, 7, 10, 4\}, \quad \mathcal{O}_{22} = \{11, 6\},$$
$$\mathcal{O}_3 = \{15, 14, 12, 8\}, \quad \mathcal{O}_4 = \{16\},$$

are characterized essentially by the number of u symbols in the sequence, noting that $\mathcal{O}_2$ (two occurences of u) is now split into two smaller orbits distinguishing whether the u elements are adjacent to each other in the sequence. For example, the two points $yyuu$ and $yuyu$, originally in the same orbit under the full permutation group S_4, are not cyclically related.

To evaluate form the canonical projections, we start with the representation ρ of $\tau = (1234)$ acting on V according to

$$\begin{bmatrix} 1 & 16 & 15 & 14 & 12 & 8 & 13 & 7 & 10 & 4 & 11 & 6 & 9 & 5 & 3 & 2 \\ 1 & 16 & 14 & 12 & 8 & 15 & 10 & 13 & 4 & 7 & 6 & 11 & 2 & 9 & 5 & 3 \end{bmatrix},$$

giving a 16×16 permutation matrix that can be directly evaluated using Algorithm 1 (page 221). The representation of interest is then $\rho_{\tau^k} = \rho_\tau^k$, for $k = 1, 2, 3, 4$.

The table of irreducible characters for C_4 is

C_4	1	τ	τ^2	τ^3
χ_1	1	1	1	1
χ_2	1	i	-1	$-i$
χ_3	1	-1	1	-1
χ_4	1	$-i$	-1	i

, (5.6)

from which we derive the one-dimensional canonical projections

$$\mathcal{P}_j = \frac{1}{4} \sum_{\tau \in C_4} \overline{\chi}_j(\tau) \rho_\tau, \quad j = 1 \ldots, 4.$$

The data invariants are summarized in (5.7), along with the dimensions $\operatorname{Tr} \mathcal{P}$ of each reduced subspace of $\mathcal{V} = \mathbb{R}^{16}$. Note that the orbits $\mathcal{O}_{22} = \{11, 6\}$ and $\mathcal{O}_{21} = \{13, 7, 10, 4\}$ are now displayed one adjacent to the other. We also observe that $\mathcal{P}_1 x$ is a one-dimensional invariant averaging the data over each orbit; $\mathcal{P}_3 x$ is a one-dimensional invariant comparing the resulting (mean) measurements $p + r$ and $q + s$ when the positions of the symbols are rotated by 90° and $(\mathcal{P}_2 + \mathcal{P}_4)x$ is a two-dimensional invariant. One dimension is accounted by the comparison of p and r, whereas the other by the comparison of q and s. These comparisons result from rotating the positions in the symbols by 180°.

Similar interpretations apply to each one the other orbits so that we may summarize the invariants obtained in this experimental setting as within orbit averaging, 90° and 180° shift effects.

data	$4\mathcal{P}_1 x$	$4\mathcal{P}_3 x$
y	$4y$	0
z	$4z$	0
p	$p+q+r+s$	$p-q+r-s$
q	$p+q+r+s$	$-p+q-r+s$
r	$p+q+r+s$	$p-q+r-s$
s	$p+q+r+s$	$-p+q-r+s$
e	$2e+2f$	$2e-2f$
f	$2e+2f$	$-2e+2f$
Tr ($\mathcal{P}$)	6	4

,

data	$4\mathcal{P}_2 x$	$4\mathcal{P}_4 x$	$4(\mathcal{P}_2+\mathcal{P}_4)x$
y	0	0	0
z	0	0	0
p	$p-is-r+iq$	$p+is-r-iq$	$2p-2r$
q	$ir+q-ip-s$	$-ir+q+ip-s$	$2q-2s$
r	$-p+is+r-iq$	$-p-is+r+iq$	$-2p+2r$
s	$-ir-q+ip+s$	$ir-q-ip+s$	$-2q+2s$
e	0	0	0
f	0	0	0
Tr ($\mathcal{P}$)	3	3	6

. (5.7)

5.6 Dihedral Reduction of Binary Sequences

Given continuation to the previous example, consider now the action $s\tau^{-1}$ of the dihedral group D_4 on the space V of binary sequences in length of 4. The resulting action

$D_4 \backslash s$	1	16	15	14	12	8	13	11	7	10	6	4	9	5	3	2
1	1	16	15	14	12	8	13	11	7	10	6	4	9	5	3	2
(24)	1	16	15	8	12	14	7	11	13	4	6	10	3	5	9	2
(13)	1	16	12	14	15	8	10	11	4	13	6	7	9	2	3	5
(12)(34)	1	16	14	15	8	12	13	6	10	7	11	4	5	9	2	3
(13)(24)	1	16	12	8	15	14	4	11	10	7	6	13	3	2	9	5
(14)(23)	1	16	8	12	14	15	4	6	7	10	11	13	2	3	5	9
(1234)	1	16	14	12	8	15	10	6	13	4	11	7	2	9	5	3
(1432)	1	16	8	15	14	12	7	6	4	13	11	10	5	3	2	9

, (5.8)

shows that D_4 and C_4 generate the same set of orbits.

Table (5.9) summarizes the results, based on the canonical projections derived on page 110.

data	$4\mathcal{P}_1x$	$4\mathcal{P}_2x$	$4\mathcal{P}_3x$	$4\mathcal{P}_4x$	$4\mathcal{P}_5x$
x	$4x$	0	0	0	0
y	$4y$	0	0	0	0
p	$p+q+r+s$	0	0	$p-q+r-s$	$2p-2r$
q	$p+q+r+s$	0	0	$-p+q-r+s$	$2q-2s$
r	$p+q+r+s$	0	0	$p-q+r-s$	$-2p+2r$
s	$p+q+r+s$	0	0	$-p+q-r+s$	$-2q+2s$
e	$2e+2f$	0	0	$2e-2f$	0
f	$2e+2f$	0	0	$-2e+2f$	0
a	$a+b+c+d$	0	$a-b-c+d$	0	$2a-2d$
b	$a+b+c+d$	0	$-a+b+c-d$	0	$2b-2c$
c	$a+b+c+d$	0	$-a+b+c-d$	0	$-2b+2c$
d	$a+b+c+d$	0	$a-b-c+d$	0	$-2a+2d$
k	$k+l+m+n$	0	0	$k-l+m-n$	$2k-2m$
l	$k+l+m+n$	0	0	$-k+l-m+n$	$2l-2n$
m	$k+l+m+n$	0	0	$k-l+m-n$	$-2k+2m$
n	$k+l+m+n$	0	0	$-k+l-m+n$	$-2l+2n$
Tr ($\mathcal{P}$)	6	0	1	3	6

(5.9)

We observe that the effect of introducing the dihedral symmetries is that of isolating the effect of orbit $\mathcal{O}_{21}$ indexing the data $\{a, b, c, d\}$. In fact, the dihedral symmetries split the invariants determined by $\mathcal{P}_3$ under C_4 into the invariants determined by $\mathcal{P}_3$ and $\mathcal{P}_4$ under D_4, as summarized in the following table

D_4	$\mathcal{P}_1$	$\mathcal{P}_3+\mathcal{P}_4$		$\mathcal{P}_5$
C_4	$\mathcal{P}_1$		$\mathcal{P}_3$	$\mathcal{P}_2+\mathcal{P}_4$

Dihedral stratifications for voting preferences

The data on Table (5.10) were introduced earlier on Chapter 1 on page 12 to describe the frequencies with which the rankings of four candidates, their names indicated by $\{a, g, c, t\}$, were selected. For example, 29 voters ordered the candidates according to (a, g, c, t), or $x_{agct} = 29$. The complete election result is then an example of frequency data indexed by S_4. We will stratify the data and summarize the results on the basis of the dihedral decomposition described in Chapter 4 on pages 108, 109, and 121. This is obtained by observing that S_4 has three distinct cyclic orbits, namely those generated by (1234), (1243) and (1324). To each generator we add a reversal, thus obtaining three dihedral subgroups, respectively here $D_4 =< (1234), (13) >$, $D_4' =< (1243), (14) >$ and $D_4'' =< (1324), (12) >$.

D_4 ranking	$agct$	$gcta$	$ctag$	$tagc$	$cgat$	$tcga$	$atcg$	$gatc$
votes	29	26	49	28	22	29	50	35
D_4' ranking	$agct$	$gtac$	$tcga$	$catg$	$tgca$	$ctag$	$acgt$	$gatc$
votes	29	44	29	34	37	49	19	35
D_4'' ranking	$agct$	$ctga$	$gatc$	$tcag$	$gact$	$tcga$	$agtc$	$ctag$
votes	29	57	35	67	50	29	11	49

(5.10)

For the first dihedral stratification, the canonical invariants are the rows of the Fourier transform

$$\widehat{x}_\beta = \begin{bmatrix} agct - ctag + cgat - atcg & -gcta + tagc + tcga - gatc \\ gcta - tagc + tcga - gatc & agct - ctag - cgat + atcg \end{bmatrix}.$$

For frequency data we may express the invariants in the log scale, in which case

$$\mathcal{B}_v = \left(\log \frac{agct\ cgat}{atcg\ ctag},\ \log \frac{gcta\ tcga}{gatc\ tagc}\right), \quad \mathcal{B}_w = \left(\log \frac{gcta\ tcga}{gatc\ tagc},\ \log \frac{agct\ atcg}{cgat\ ctag}\right)$$

describe the two two-dimensional invariants. Similarly, for the second dihedral stratification,

$$\widehat{x}_\beta = \begin{bmatrix} agct - tcga + tgca - acgt & -gtac + catg + ctag - gatc \\ gtac - catg + ctag - gatc & agct - tcga - tgca + acgt \end{bmatrix},$$

and for the third stratification,

$$\widehat{x}_\beta = \begin{bmatrix} agct - gatc + gact - agtc & -ctga + tcag + tcga - ctag \\ ctga - tcag + tcga - ctag & agct - gatc - gact + agtc \end{bmatrix}.$$

5.7 Projections in the Space of Scalar-Valued Functions

The canonical projections are collective ways of summarizing the data vector $x \in \mathbb{R}^v$. In this section we study the role of canonical projections restricted to transitive actions. This is a local interpretation of any group action φ on a set V of indices for the data.

We indicate by $\mathcal{F}$ the vector space of scalar-valued functions defined in V. For each $\tau \in G$ with g elements, define $\tau^* : \mathcal{F} \to \mathcal{F}$, which takes $x \in \mathcal{F}$ into $\tau^*(x) \in \mathcal{F}$ given by $\tau^*(x)(s) = x(\varphi_\tau s)$. It is simple to verify that τ^* is linear, and that $\tau \to \tau^*$ is a homomorphism in G into Aut $(\mathcal{F})$. Moreover, τ^* is unitary with respect to the scalar product

$$(x, y)_s = \frac{1}{g} \sum_{\tau \in G} x(\varphi_\sigma s)\overline{y}(\varphi_\sigma s), \quad s \in V,$$

in $\mathcal{F}$. Applying Theorem 4.4 on page 111 to the representation τ^*, we have

Proposition 5.1. *The mapping τ^* is a unitary linear representation of G in $GL(\mathcal{F})$, and, for each irreducible character χ of G in dimension of n_χ, the corresponding canonical projection evaluates as $\mathcal{P}_\chi(x)(s) = \sum_{\tau \in G} n_\chi \overline{\chi}_{\tau^{-1}} x(\varphi_\tau s)/g$.*

Direct verification shows that the symmetric projection $\mathcal{P}_1(x)$ evaluates at $s \in V$ as

$$\mathcal{P}_1(x)(s) = \frac{1}{g} \sum_{\tau \in G} x(\varphi_\tau s) = \frac{1}{g} |G_s| \sum_{f \in \mathcal{O}_s} x(f) = \frac{1}{g} \frac{g}{|\mathcal{O}_s|} \sum_{f \in \mathcal{O}_s} x(f) = \frac{1}{|\mathcal{O}_s|} \sum_{f \in \mathcal{O}_s} x(f),$$

the average of x in the orbit $\mathcal{O}_s$ of s under action φ. We also note that Proposition 5.1 applies to the vector space $\mathcal{F}(G)$ of scalar-valued functions defined in G, with $\varphi : G \times G \to G$, and scalar product

$$(x, y) = \frac{1}{g} \sum_{\tau \in G} x_\tau \overline{y}_\tau = (x, y)_1.$$

5.8 Decompositions in the Dual Space

Elie Cartan, in his 1937 seminal book on the theory of spinors[1] makes the following remark, adapted to our current notation:

> Let two vectors x and y be referred to the same Cartesian frame of reference and let us consider the n^2 products $x_i y_i$; as a result of a rotation they obviously undergo a linear transformation T, which also possesses the property that if T and T' correspond to the rotations R and R', the transformation TT' corresponds to RR'. The n^2 quantities $x_i y_j$ therefore provide a new linear representation of the group of rotations, completely distinct from the two previous ones. (p. 22)

Recall, from Section 3.5 that if G acts on V and W giving linear representations ρ and η of G on $\mathcal{V}$ and $\mathcal{W}$, respectively, then $\rho \otimes \eta$ gives a linear representation of G on $\mathcal{V} \times \mathcal{W}$, evaluated as $(\rho \otimes \eta)_\tau = \rho_\tau \otimes \eta_\tau$. To recognize Cartan's construction, it is now sufficient to identify the entries of $x'y$ with the entries of $x \otimes y$.

As a result, the canonical decomposition for (block) matrices of the form

$$\Sigma = \begin{bmatrix} xx' & xy' \\ yx' & yy' \end{bmatrix},$$

where x and y are structured data, can be ontained. We refer to the resulting decompositions as canonical decompositions in the dual space, and to their invariants as coinvariants.

[1] The English version (Cartan, 1966) of the original text was published by MIT Press.

Planar rotations

Recall, from page 42, that the set of matrices

$$\beta_{k,d} = \frac{1}{2}\begin{bmatrix} (1+d)\omega^{k-1} & (1-d)\omega^{k-1} \\ (1-d)\omega^{-k+1} & (1+d)\omega^{-k+1} \end{bmatrix}, \quad k = 1, \ldots, n, \; d = \pm 1,$$

gives a linear representation of D_n. In particular, the planar rotations $\rho_k = X\beta_{k,1}X^{-1}$ give a representation of C_n. The transformation X is defined up to a rotation factor; that is, we may replace X by ξX for an arbitrary complex unimodular factor ξ.

When $n = 2$, the canonical decomposition determined by this representation, as the reader may want to verify, leads to $x = [z(x)X_1 + \overline{z}(x)X_2]/2$, where $z(x) = x_1 + ix_2$ and

$$X_1 = \begin{bmatrix} 1 \\ -i \end{bmatrix}, \quad X_2 = \begin{bmatrix} 1 \\ i \end{bmatrix}.$$

That is, the canonical decomposition is simply the embedding of x in the rotation space, relative to the basis X (see also Exercise 5.2). The rotation invariants are then $z(x)$ and $\overline{z}(x)$ in rotation space. The interpretation of x as an invariant in rotation space is concurrent with the view in which rotations provide a passage between components of the same vector, but referred to rotated frames of reference, e.g., Cartan (1966, p. 23).

The dual decomposition. We will derive the dual decomposition of planar rotations as representations of C_3. The structured data are indexed by $V = \{1, 2\}$. Starting with its character table

k	1	2	3
χ_1	1	1	1
χ_2	1	ω	ω^2
χ_3	1	ω^2	ω

(5.11)

the canonical projections $\mathcal{P}_j = \sum_{k=1}^{3} \overline{\chi}_j(k)[\rho_k \otimes \rho_k]/3$ are then

$$\mathcal{P}_1 = \frac{1}{2}\begin{bmatrix} 1 & 0 & 0 & 1 \\ 0 & 1 & -1 & 0 \\ 0 & -1 & 1 & 0 \\ 1 & 0 & 0 & 1 \end{bmatrix}, \quad \mathcal{P}_2 = \frac{1}{4}\begin{bmatrix} 1 & -i & -i & -1 \\ i & 1 & 1 & -i \\ i & 1 & 1 & -i \\ -1 & i & i & 1 \end{bmatrix}, \quad \mathcal{P}_3 = \frac{1}{4}\begin{bmatrix} 1 & i & i & -1 \\ -i & 1 & 1 & i \\ -i & 1 & 1 & i \\ -1 & -i & -i & 1 \end{bmatrix},$$

in dimensions of 2, 1, 1, respectively. When these projections act on the entries $f = x \otimes y$ of xy', the invariants associated with

$$\mathcal{P}_1 f = \frac{1}{2} \begin{bmatrix} x_1 y_1 + x_2 y_2 \\ x_1 y_2 - x_2 y_1 \\ -x_1 y_2 + x_2 y_1 \\ x_1 y_1 + x_2 y_2 \end{bmatrix}$$

are, as expected, the *angular* component $\alpha = x_1 y_1 + x_2 y_2$ and the *area* component $\delta = x_1 y_2 - x_2 y_1$, whereas $\mathcal{P}_2 f = (X_1 \otimes X_1)/4$ and $\mathcal{P}_3 f = (X_2 \otimes X_2)/4$ leave $w = x_1 y_1 - x_2 y_2 + i(x_1 y_2 + x_2 y_1)$ and $\overline{w}$ invariant, respectively. Therefore, we obtain the decomposition

$$xy' = \frac{1}{2} \begin{bmatrix} \alpha & \delta \\ -\delta & \alpha \end{bmatrix} + \frac{1}{2} \begin{bmatrix} \Re(w) & \Im(w) \\ \Im(w) & -\Re(w) \end{bmatrix}$$

in terms of four (dual) canonical invariants, namely, α, δ, $\Re(w)$ and $\Im(w)$. In particular, when $x = y$,

$$xx' = \frac{1}{2} \begin{bmatrix} x_1^2 + x_2^2 & 0 \\ 0 & x_1^2 + x_2^2 \end{bmatrix} + \frac{1}{2} \begin{bmatrix} x_1^2 - x_2^2 & 2x_1 x_2 \\ 2x_1 x_2 & -x_1^2 + x_2^2 \end{bmatrix}, \tag{5.12}$$

in which case $\alpha = ||x||^2$, a consequence of the fact that $\alpha = \cos \angle(x, y)||x||||y||$. Moreover, then, the list of coinvariants reduces to α, $\Re(w)$, and $\Im(w)$. This one-to-one correspondence between the dimension of xy' and the number of coinvariants reflects the fact that C_n is Abelian and all irreducible characters are one-dimensional. In addition, it can be shown that the entries of these matrices are the only invariants of the 2-dim planar rotation representations of C_n, for all $n \geq 3$. When $n = 2$, the planar rotation is a point inversion $x \to -x$ so that all components $x_1 y_1, x_1 y_2, x_2 y_1, x_2 y_2$ of xy' remain invariant. This also follows from the fact that in this case the canonical projections are $\mathcal{P}_1 = I_4$ and $\mathcal{P}_2 = 0$. It is also opportune to note that the decomposition (5.12) has the form

$$xx' = A \begin{bmatrix} 1 & 0 \\ 0 & 1 \end{bmatrix} + \begin{bmatrix} B & C \\ C & -B \end{bmatrix},$$

that coincides with the form

$$xx^* = A \begin{bmatrix} 1 & 0 \\ 0 & 1 \end{bmatrix} + \begin{bmatrix} B & D \\ D^* & C \end{bmatrix},$$

of decompositions obtained for coherence matrices xx^*, utilized in the theory of partial polarization of light. Here, A, B, and C are real positive quantities, and x^* is the Hermitian conjugate of x. See, for example, O'Neill (1963, p. 151).

Coinvariants of bilateral symmetries

Consider two scalar measurements, x and y, obtained on the two sides of an experimental unit and let S_2 act on the sides $\{1, 2\}$ by permutation. We observe the data $x' = (x_1, x_2)$, and $y' = (y_1, y_2)$. For example, x and y may indicate the intraocular pressure and the visual acuity obtained from fellow eyes.

Direct evaluation of $\rho \otimes \rho$ and the associated canonical decomposition of xy' gives

$$xy' = \frac{1}{2}\begin{bmatrix} x_1y_1 + x_2y_2 & x_1y_2 + x_2y_1 \\ x_1y_2 + x_2y_1 & x_1y_1 + x_2y_2 \end{bmatrix} + \frac{1}{2}\begin{bmatrix} x_1y_1 - x_2y_2 & x_1y_2 - x_2y_1 \\ -x_1y_2 + x_2y_1 & -x_1y_1 + x_2y_2 \end{bmatrix},$$

and in particular,

$$xx' = \frac{1}{2}\begin{bmatrix} x_1^2 + x_2^2 & 2x_1x_2 \\ 2x_1x_2 & x_1^2 + x_2^2 \end{bmatrix} + \frac{1}{2}\begin{bmatrix} x_1^2 - x_2^2 & 0 \\ 0 & -x_1^2 + x_2^2 \end{bmatrix}. \tag{5.13}$$

The two components in the canonical decomposition represent, respectively, the coinvariants of intraclass covariance and bilateral variance differentiation. To see this in the usual statistical formulation, we apply the decomposition (5.13) to

$$A = \frac{1}{n}\sum_{\alpha} z_\alpha z_\alpha', \quad \alpha = 1, \ldots, N,$$

where $z_\alpha = x_\alpha - \overline{x}$, $\overline{x} = \sum_\alpha x_\alpha / N$ and $n = N - 1$, so that then the canonical decomposition

$$A = \frac{1}{2}A_{\text{intraclass}} + \frac{1}{2}\left(s_1^2 - s_2^2\right)\begin{bmatrix} 1 & 0 \\ 0 & -1 \end{bmatrix}$$

obtains. Not surprisingly, it also says that matrix A is an intraclass matrix if and only if the second component in the decomposition vanishes, that it, when $s_1^2 = s_2^2$, as well-known. In that case, in fact, A is the usual maximum likelihood estimate of the underlying covariance structure. Algebraically, the first component in the decomposition is a matrix that commutes with all the elements in the permutation representation of S_2. We say that it *intertwines* with the permutation representation.

Coinvariants of C_4

We conclude this section with an outline of the derivation of the coinvariants of C_4 acting by permutation on $V = \{1, 2, 3, 4\}$. Here, the structured data take the form of

$$x \otimes y \equiv xy' = \begin{bmatrix} x_1y_1 & x_1y_2 & x_1y_3 & x_1y_4 \\ x_2y_1 & x_2y_2 & x_2y_3 & x_2y_4 \\ x_3y_1 & x_3y_2 & x_3y_3 & x_3y_4 \\ x_4y_1 & x_4y_2 & x_4y_3 & x_4y_4 \end{bmatrix}, \quad x, y \in \mathbb{R}^4.$$

Starting with the representation of C_4 and its character table, respectively,

$$\rho_k = \begin{bmatrix} 0 & 1 & 0 & 0 \\ 0 & 0 & 1 & 0 \\ 0 & 0 & 0 & 1 \\ 1 & 0 & 0 & 0 \end{bmatrix}^{k-1}, \quad \begin{array}{c|cccc} k & 1 & 2 & 3 & 4 \\ \hline \chi_1 & 1 & 1 & 1 & 1 \\ \chi_2 & 1 & i & -1 & -i \\ \chi_3 & 1 & -1 & 1 & -1 \\ \chi_4 & 1 & -i & -1 & i \end{array}, \quad k = 1, 2, 3, 4, \qquad (5.14)$$

the resulting canonical projections $\mathcal{P}_j = \sum_k \overline{\chi}_j(k)[\rho_k \otimes \rho_k]/4$, $j = 1, \ldots, 4$, each in dimension of 4, lead to the decomposition

$$xy' = \frac{1}{4}\begin{bmatrix} \alpha & \beta & \gamma & \delta \\ \delta & \alpha & \beta & \gamma \\ \gamma & \delta & \alpha & \beta \\ \beta & \gamma & \delta & \alpha \end{bmatrix} + \frac{1}{4}\begin{bmatrix} A & B & C & D \\ -D & -A & -B & -C \\ C & D & A & B \\ -B & -C & -D & -A \end{bmatrix} + \frac{1}{2}\begin{bmatrix} a & c & e & f \\ g & b & d & h \\ -e & -f & -a & -c \\ -d & -h & -g & -b \end{bmatrix},$$

obtained, respectively, from $\mathcal{P}_1$, $\mathcal{P}_3$, and $\mathcal{P}_{24} = \mathcal{P}_2 + \mathcal{P}_4$. We remark that $\mathcal{P}_2$ and $\mathcal{P}_4$ are complex conjugate and that $\mathcal{P}_2^*\mathcal{P}_4 = \mathcal{P}_4^*\mathcal{P}_2 = 0$. We note that the first component in the above decomposition intertwines with C_4 whereas the other two components exhibit a pattern that has essentially the symmetry of C_4.

The corresponding 16 coinvariants for xy' and 10 coinvariants for xx' are shown in Table (5.15).

entry	coinvariants of xy'	coinvariants of xx'
α	$x_1y_1 + x_2y_2 + x_3y_3 + x_4y_4$	$x_1^2 + x_2^2 + x_3^2 + x_4^2$
β	$x_1y_2 + x_2y_3 + x_3y_4 + x_4y_1$	$x_4x_1 + x_2x_1 + x_3x_2 + x_4x_3$
γ	$x_1y_3 + x_2y_4 + x_3y_1 + x_4y_2$	$2\,x_3x_1 + 2\,x_4x_2$
δ	$x_1y_4 + x_2y_1 + x_3y_2 + x_4y_3$	$x_4x_1 + x_2x_1 + x_3x_2 + x_4x_3$
A	$x_1y_1 - x_2y_2 + x_3y_3 - x_4y_4$	$x_1^2 - x_2^2 + x_3^2 - x_4^2$
B	$x_1y_2 - x_2y_3 + x_3y_4 - x_4y_1$	$-x_4x_1 + x_2x_1 - x_3x_2 + x_4x_3$
C	$x_1y_3 - x_2y_4 + x_3y_1 - x_4y_2$	$2\,x_3x_1 - 2\,x_4x_2$
D	$x_1y_4 - x_2y_1 + x_3y_2 - x_4y_3$	$x_4x_1 - x_2x_1 + x_3x_2 - x_4x_3$
a	$x_1y_1 - x_3y_3$	$x_1^2 - x_3^2$
c	$x_1y_2 - x_3y_4$	$x_2x_1 - x_4x_3$
e	$x_1y_3 - x_3y_1$	0
f	$x_1y_4 - x_3y_2$	$-x_3x_2 + x_4x_1$
g	$x_2y_1 - x_4y_3$	$x_2x_1 - x_4x_3$
b	$x_2y_2 - x_4y_4$	$x_2^2 - x_4^2$
d	$x_2y_3 - x_4y_1$	$x_3x_2 - x_4x_1$
h	$x_2y_4 - x_4y_2$	0

. (5.15)

Note that in the decomposition of xx' we have the additional 6 constraints $\beta = \delta$, $B = -D$, $e = h = 0$, $c = g$ and $f = -d$, thus bringing the total number of coinvariants to be $16 - 6 = 10$, the dimension of xx'.

Cyclic coinvariants for short nucleotide sequences. The cyclic coinvariants can be utilized to explore a nucleotide sequence. For example, the work of Doi (1991) on the evolutionary strategy of the HIV-1 virus is based on the study of frequencies of certain cyclic orbits. To illustrate, we use as a reference sequence the isolate BRU from the HIV-1, which is a 9,229 bp (base-pair) long nucleotide sequence. Specifically, we will consider the six cyclic orbits in length of 4, namely,

$$\begin{aligned} \mathcal{O}_{agct} &= \{agct, tagc, ctag, gcta\}, & \mathcal{O}_{acgt} &= \{acgt, tacg, gtac, cgta\}, \\ \mathcal{O}_{agtc} &= \{agtc, cagt, tcag, gtca\}, & \mathcal{O}_{actg} &= \{actg, gact, tgac, ctga\}, \\ \mathcal{O}_{atcg} &= \{atcg, gatc, cgat, tcga\}, & \mathcal{O}_{atgc} &= \{atgc, catg, gcat, tgac\}, \end{aligned}$$

where C_4 acts transitively according to $s\tau^{-1}$ for $s \in \mathcal{O}$ and $\tau \in C_4$. At each point $s \in \mathcal{O}$ we measure the frequency x_s with which the sequence s appeared in six adjacent subsequences in length of 1,500 bps. Following the evaluation of the 10 coinvariants of xx', as described by Table (5.15), the range ($= max - min$), mean, and signal-to-noise ratio (SNR) over the six adjacent regions were calculated, for each orbit. The results are shown in Tables (5.16), (5.17), and (5.18), respectively.

range	$\mathcal{O}_{agct}$	$\mathcal{O}_{acgt}$	$\mathcal{O}_{agtc}$	$\mathcal{O}_{actg}$	$\mathcal{O}_{atcg}$	$\mathcal{O}_{atgc}$
α	945.55	80.00	290.21	768.19	92.57	164.00
β	260.00	24.00	240.00	357.09	46.78	156.00
γ	720.61	42.00	204.00	120.00	36.00	164.00
A	1030.00	78.00	216.00	90.00	45.00	114.00
B	101.58	7.00	84.00	427.09	46.78	12.00
C	1084.60	36.00	264.00	188.00	48.00	281.58
a	581.46	71.00	176.00	421.09	47.78	96.00
c	209.15	9.00	50.00	457.09	10.00	90.78
f	141.58	8.00	146.00	50.00	56.78	68.78
b	85.78	1.00	133.00	457.09	91.78	78.78

(5.16)

mean	$\mathcal{O}_{agct}$	$\mathcal{O}_{acgt}$	$\mathcal{O}_{agtc}$	$\mathcal{O}_{actg}$	$\mathcal{O}_{atcg}$	$\mathcal{O}_{atgc}$
α	311.00	57.60	164.00	223.00	34.80	117.00
β	167.00	6.33	122.00	140.00	15.10	97.60
γ	262.00	11.70	104.00	70.00	7.33	97.60
A	161.00	56.30	−2.87	1.17	−15.80	−1.04
B	14.40	−2.00	−20.80	71.00	−9.13	2.33
C	74.10	10.70	−90.30	−20.00	−10.70	−49.40
a	109.00	−55.00	−64.10	78.80	7.46	−9.33
c	27.50	−1.17	−8.17	69.80	1.67	−8.96
f	−5.43	−3.50	−48.70	9.50	4.80	3.30
b	0.46	0.33	50.00	68.70	8.04	−0.96

(5.17)

SNR	$\mathcal{O}_{agct}$	$\mathcal{O}_{acgt}$	$\mathcal{O}_{agtc}$	$\mathcal{O}_{actg}$	$\mathcal{O}_{atcg}$	$\mathcal{O}_{atgc}$
α	0.937	4.990	2.570	0.642	1.150	4.580
β	2.720	0.545	2.440	1.210	0.640	3.900
γ	1.150	0.606	2.660	3.610	0.311	3.490
A	0.220	5.120	0.001	0.002	1.040	0.561×10^{-3}
B	0.164	0.480	0.523	0.206	0.290	0.283
C	0.048	0.674	1.080	0.121	0.364	0.259
a	0.283	5.680	1.160	0.265	0.180	0.101
c	0.146	0.144	0.291	0.195	0.200	0.110
f	0.015	0.925	1.070	0.344	0.063	0.019
b	0.205×10^{-3}	0.500	0.981	0.186	0.081	0.001

(5.18)

These data summaries are systematic extensions of the cyclic diversities $\max_{s\in\mathcal{O}} x_s / \min_{s\in\mathcal{O}} x_s$ that appear in the work of Doi (1991), defined as the ratio between the largest and the smallest local frequencies x_s, as s varies within the orbit $\mathcal{O}$. The frequency diversity can be used to probe the randomness of base substitution for the virus in certain local regions of interest. The cyclic orbits are sufficient to characterize the nucleotide sequence according to whether or not a certain orbit can be observed. The closest the frequency diversity is to 1 the more characteristic that orbit is, while conversely, the frequency diversity of a non characteristic orbit is relatively larger. Clearly, the frequency diversity is also an invariant under the action on C_4. The SNR of each coinvariant may also characterize the sequence in the sense that large values of SNR should imply small variability in the coinvariant across the (six) localized regions in the sequence. We observe, for example, that the $acgt$ to $atcg$ SNR ratio for $A = x_1{}^2 - x_2{}^2 + x_3{}^2 - x_4{}^2$ is 9,126 : 1, thus conveying the fact that these two pairs of orbits characterize the genome in remarkably distinct ways and may provide a basis for explanation at the molecular biology level.

We conclude this example with the evaluation of the coinvariants for six of the orbits in length of 4 considered in Doi's work, namely, the cyclic orbits generated by $aatt$, $acac$, $atat$, $gtgt$, $cttg$ and $ggtt$. Tables (5.19) to (5.24) show the observed coinvariants at each one of the six adjacent regions in length of $L = 1,500$ bps, whereas Tables (5.25) and (5.26) show the mean and range over the same regions. We note that for orbits $acac$, $atat$, and $gtgt$ the action of C_4 is such that $x_1 = x_3$ and $x_2 = x_4$, and consequently the coinvariants B, a, c, f, and b are all equal to 0. The SNRs for the remaining orbits are shown in Table (5.27). These ratios are all relatively small, thus implying a relatively large potential for separating the six regions of the genome.

$\mathcal{O}_{aatt}$	α	β	γ	A	B	C	a	c	f	b
1	450	336	322	190	−64	124	192	−40	152	−112
2	771	682	674	271	−36	−28	279	37	161	−88
3	291	198	274	−209	4	−428	9	29	−7	88
4	1047	992	938	77	30	−490	320	203	107	93
5	174	165	172	52	−1	−8	15	−2	13	−11
6	1	0	0	1	0	0	−1	0	0	0

(5.19)

$\mathcal{O}_{acac}$	α	β	γ	A	B	C	a	c	f	b
1	82	80	82	−18	0	−68	0	0	0	0
2	232	160	232	−168	0	−368	0	0	0	0
3	200	192	200	−56	0	−184	0	0	0	0
4	520	448	520	−264	0	−656	0	0	0	0
5	370	352	370	−114	0	−356	0	0	0	0
6	0	0	0	0	0	0	0	0	0	0

(5.20)

$\mathcal{O}_{atat}$	α	β	γ	A	B	C	a	c	f	b
1	50	48	50	−14	0	−46	0	0	0	0
2	232	160	232	168	0	136	0	0	0	0
3	314	264	314	170	0	98	0	0	0	0
4	324	324	324	0	0	−162	0	0	0	0
5	122	120	122	22	0	−28	0	0	0	0
6	4	4	4	0	0	−2	0	0	0	0

(5.21)

$\mathcal{O}_{gtgt}$	α	β	γ	A	B	C	a	c	f	b
1	52	20	52	−48	0	−98	0	0	0	0
2	4	4	4	0	0	−2	0	0	0	0
3	298	280	298	−102	0	−302	0	0	0	0
4	146	96	146	−110	0	−238	0	0	0	0
5	208	80	208	−192	0	−392	0	0	0	0
6	0	0	0	0	0	0	0	0	0	0

(5.22)

$\mathcal{O}_{cttg}$	α	β	γ	A	B	C	a	c	f	b
1	24	12	16	−16	4	−32	−4	−8	−4	−12
2	3	2	2	−1	0	−4	−1	−1	−1	0
3	158	135	148	76	−3	28	−45	−6	−21	9
4	66	33	64	56	1	52	−11	−7	4	−3
5	321	294	316	121	−2	28	−21	14	−28	28
6	68	64	60	0	−4	−30	−16	0	−16	16

(5.23)

$\mathcal{O}_{ggtt}$	α	β	γ	A	B	C	a	c	f	b
1	97	60	72	63	12	48	−48	−28	8	−15
2	60	48	40	20	−8	−8	−32	−4	−20	12
3	81	50	44	55	−6	8	−60	−10	−20	5
4	163	132	102	17	−30	−42	−72	−3	−63	55
5	319	252	266	171	−14	56	−147	−21	−63	24
6	2	1	0	0	1	0	1	1	0	1

(5.24)

mean	$\mathcal{O}_{aatt}$	$\mathcal{O}_{acac}$	$\mathcal{O}_{atat}$	$\mathcal{O}_{gtgt}$	$\mathcal{O}_{cttg}$	$\mathcal{O}_{ggtt}$
α	456.0	234.0	174.0	118.0	107.0	120.0
β	396.0	205.0	153.0	80.0	90.0	90.5
γ	397.0	234.0	174.0	118.0	101.0	87.3
A	63.7	−103.0	57.7	−75.3	39.3	54.3
B	−11.2	0.0	0.0	0.0	−0.7	−7.5
C	−138.0	−272.0	−0.7	−172.0	7.0	10.3
a	136.0	0.0	0.0	0.0	−16.3	−59.7
c	37.8	0.0	0.0	0.0	−1.3	−10.8
f	71.0	0.0	0.0	0.0	−11.0	−26.3
b	−5.0	0.0	0.0	0.0	6.3	13.7

(5.25)

range	$\mathcal{O}_{aatt}$	$\mathcal{O}_{acac}$	$\mathcal{O}_{atat}$	$\mathcal{O}_{gtgt}$	$\mathcal{O}_{cttg}$	$\mathcal{O}_{ggtt}$
α	1046	520	320	298	318	317
β	992	448	320	280	292	251
γ	938	520	320	298	314	266
A	480	264	184	192	137	171
B	94	0	0	0	8	42
C	614	656	298	392	84	98
a	321	0	0	0	44	148
c	243	0	0	0	22	29
f	168	0	0	0	32	71
b	205	0	0	0	40	70

(5.26)

SNR	$\mathcal{O}_{aatt}$	$\mathcal{O}_{cttg}$	$\mathcal{O}_{ggtt}$
α	1.630	0.985	1.420
β	1.360	0.792	1.220
γ	1.580	0.893	1.040
A	0.177	0.639	0.922
B	0.135	0.061	0.336
C	0.354	0.049	0.095
a	1.030	1.270	1.730
c	0.236	0.031	1.090
f	0.993	0.921	0.896
b	0.004	0.233	0.390

(5.27)

The examples in this section apply to a set $\{x, y, z\}$ of three or (similarly) more measurements by extending

$$x \otimes y = (x_1y_1, x_1y_2, \ldots, x_ny_n)',$$

to $x \otimes y \otimes z$ and observing that then $\rho \otimes \rho \otimes \rho$ is the linear representation of G taking the components of $x \otimes y \otimes z$ into the components of $(\rho_\tau x) \otimes (\rho_\tau y) \otimes (\rho_\tau z)$, for each $\tau \in G$. The entries of $x \otimes y \otimes \ldots$ are often called Euclidean tensors (relative to G) in Cartan's notation. The illustration on page 140 also has the interpretation that $\{x_1y_1 \pm x_2y_2, x_1y_2 \pm x_2y_1, x_1^2 \pm x_2^2, y_1^2 \pm y_2^2\}$ is the set of cyclically invariant polynomials in $\{xy', xx', yy'\}$.

5.9 Canonical Decompositions of Entropy Data

The entropy $H = -\sum_j p_j \log p_j$ of a finite set of n mutually exclusive events with corresponding probabilities $p_1, \ldots, p_n$ measures the amount of uncertainty associated with those events (Khinchin, 1957, p. 3). Its value is zero when any of the events is certain, it is positive otherwise, and attains its maximum value ($\log n$) when the events are equally like, that is, $p_1 = \ldots = p_n = 1/n$. Alternatively (Kullback, 1968, p. 7), H is the mean value of the quantities $-\log p_j$ and can be interpreted as the mean information in an observation obtained to ascertain the mutually exclusive and exhaustive (hypotheses defined by those) events. In the case of (structured) events s indexed by V, we write, accordingly, $H = -\sum_{s \in V} p_s \log p_s$.

The Standard Decomposition of Entropy

In the simplest case in which the probability distribution is $p' = (p_1, p_2)$ and S_2 acts on $V = \{1, 2\}$ by permutation, the canonical decomposition is simply $I = \mathcal{A} + \mathcal{Q}$. Writing $\ell_p' = (\log p_1, \log p_2)$, we see that

$$H = -p'\ell_p = -p'I\ell_p = -(p'\mathcal{A}\ell_p + p'\mathcal{Q}\ell_p),$$

is the canonical decomposition of the entropy, the components of which can be expressed as the log geometric mean

$$H_1 = -p'\mathcal{A}\ell_p = -\frac{1}{2}\log(p_1p_2)$$

of the components of p, and as

$$H_2 = -p'\mathcal{Q}\ell_p = -\frac{1}{2}(p_1 - p_2)\log\left(\frac{p_1}{p_2}\right).$$

However, equivalently,

$$H_2 = -\sum_{i=1}^{2} \left(p_i - \frac{1}{2}\right) \log \frac{p_i}{1/2},$$

is precisely Kullback's (1968) divergence between p and the uniform distribution $(1/2, 1/2)$, which is consistent with the interpretation of entropy as a measure of nonuniformity.

Invariant plots. Observe that both H_1 and H_2 remain invariant under S_2, in the sense that

$$(\rho_\tau p)'\mathcal{P}(\rho_\tau \ell_p) = p'\mathcal{P}\ell_p, \quad \mathcal{P} = \mathcal{A}, \mathcal{Q},$$

for all permutations ρ_τ, $\tau \in S_2$, so that they provide a natural system of coordinates for displaying and interpreting the entropy of p and any covariates jointly observed with p.

The n-component case. The standard decomposition of the entropy obtained for two-component distributions extends to n-component distributions simply by applying the standard decomposition $I = \mathcal{A} + \mathcal{Q}$ in the corresponding dimension to the entropy $H = -p'\ell_p$ of a distribution $p' = (p_1, \ldots, p_n)$, where $\ell'_p = (\log(p1), \ldots, \log(p_n))$.

The standard reduction of the entropy H in $p' = (p_1, \ldots, p_n)$ is then $H = H_1 + H_2$ with

$$H_1 = -\frac{1}{n} \log(p_1 p_2 \ldots p_n), \quad H_2 = -\frac{1}{n} \sum_{i<j} (p_i - p_j) \log \frac{p_i}{p_j}.$$

Similarly to the case $n = 2$ described earlier, $-H$ decomposes as the sum of the log geometric mean and Kullback's divergence

$$-H_2 = \sum_{i=1}^{n} \left(p_i - \frac{1}{n}\right) \log \frac{p_i}{1/n}$$

between p and the corresponding uniform distribution. This can be easily verified by direct evaluation of the RHS of the above equality.

Invariant plots in the H_1 x H_2 space

Table (5.28) shows the observed frequencies with which the words in the permutation orbit $V = \{act, cta, tac, cat, tca, atc\}$ of the DNA word act appear in nine subsequent regions along the BRU isolate of the HIV-1, introduced earlier on page 38. The table also shows the entropy (H) of each distribution and its standard decomposition: the log geometric mean ($-H_1$) and divergence ($-H_2$) relative to

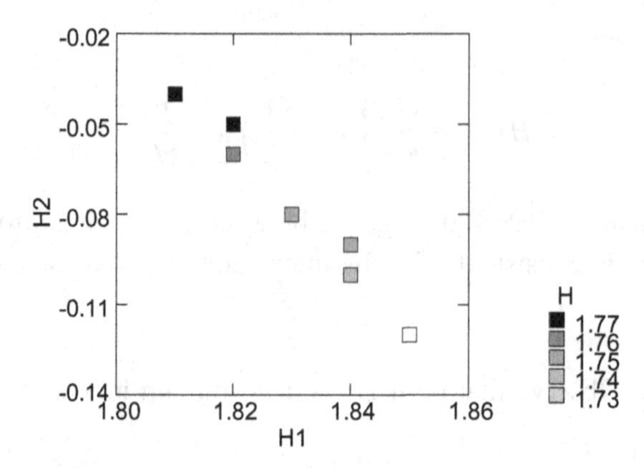

Figure 5.1: Points of constant entropy for the nine ACT orbits of Table (5.28).

the corresponding uniform distribution. In the present example S_6 acts on V by permutation of the DNA words as labels for the data.

The decompositions are shown in the invariant plot of Figure 5.1. Note, for example, the splitting of two distributions with the same entropy ($H = 1.77$) into two representations in the $H_1 \times H_2$ space, one of them with increased divergence from the corresponding uniform distribution. Differentiations of that nature are possible because the dimension of the invariant subspace associated with the projection $\mathcal{Q}$ is $\operatorname{Tr} \mathcal{Q} = m - 1 > 1$, where m is the number of components in the distribution under consideration.

s\region	1	2	3	4	5	6	7	8	9
act	8	16	16	7	17	11	12	6	14
cta	15	8	14	9	14	15	8	5	16
tac	7	17	13	15	9	11	18	5	17
cat	14	15	16	14	21	17	15	10	8
tca	11	18	10	17	11	16	14	9	13
atc	7	15	9	13	11	11	11	12	10
total	62	89	78	75	83	81	78	47	78
H	1.74	1.76	1.77	1.75	1.75	1.77	1.76	1.73	1.76
H_1	1.84	1.82	1.82	1.84	1.83	1.81	1.82	1.85	1.82
H_2	−0.10	−0.06	−0.05	−0.09	−0.08	−0.04	−0.06	−0.12	−0.06

(5.28)

It is remarked again that H_1 and H_2 are the only (nonzero) canonical components associated with the permutation action on the components of the distribution, and in that sense the space $H_1 \times H_2$ is the unique, up to equivalent representations, two-dimensional permutation-invariant space for the graphical display of entropy.

The standard decomposition of the entropy of the Sloan fonts

Table (5.29) shows the 10 Sloan fonts introduced earlier on page 3, along with their estimated entropy and standard components H_1 and H_2 of H. It gives a numerical

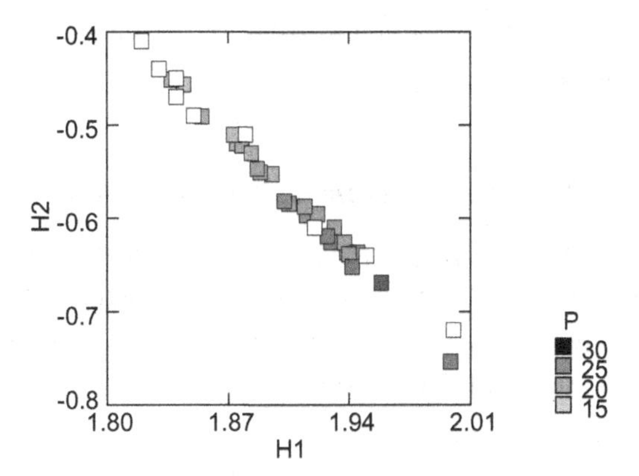

Figure 5.2: Porosity levels in the standard invariant plot for coxite data.

assessment of the role of $-H_2$ as the divergence from the uniform distribution, ranging from 0.58 when the distribution is $p' = (0.844, 0.156)$ to 0.001 when $p' = (0.516, 0.484)$.

Sloan font	Difficulty	Entropy	H_1	H_2
Z	0.844	0.433	1.010	−0.580
N	0.774	0.535	0.870	−0.337
H	0.688	0.619	0.770	−0.152
V	0.636	0.656	0.730	−0.076
R	0.622	0.663	0.725	−0.061
K	0.609	0.669	0.720	−0.048
D	0.556	0.687	0.700	−0.012
S	0.516	0.693	0.695	−0.001
O	0.470	0.692	0.695	−0.003
C	0.393	0.673	0.715	−0.040

(5.29)

Geological compositions

The geological composition data shown in Aitchison (1986, p. 354) describe the composition of albite, blandite, cornite, daubite, and endite in 25 samples of coxite, in addition to their porosity (percentage of void space). In that context, probability distributions are referred to as compositions, following Aitchison (1986, p. 26).

Figure 5.2 shows the porosity levels in the canonical components. The entropy among all compositions is within the range of $75 - 87\%$ of the max entropy ($\log 5 = 1.6$), thus being concentrated in a relatively narrow segment of the invariant space. Evaluation of the joint distribution of porosity and H_2 suggested that porosity is negatively correlated with H_2, or, equivalently, positively correlated with the divergence. In fact, the observed sample correlation coefficient based on 25 samples of these two variables is 0.78.

The regular decomposition of entropy

Indicate by $p' = (\pi_1, \pi_2, \pi_3, \theta_1, \theta_2, \theta_3)$ a probability distribution in which the components are indexed by S_3. Direct application of the canonical decomposition of the regular representation of S_3 derived on pages 106–107 shows that entropy H in p resolves into the sum of the components

$$H_1 = -p'\mathcal{P}_1\ell_p = -\frac{1}{6}\log[\pi_1\pi_2\pi_3\theta_1\theta_2\theta_3],$$

$$\begin{aligned} H_2 &= -p'\mathcal{P}_\beta\ell_p \\ &= -\frac{1}{3}\log\left[\left(\frac{\pi_1}{\pi_2}\right)^{\pi_1-\pi_2}\left(\frac{\pi_1}{\pi_3}\right)^{\pi_1-\pi_3}\left(\frac{\pi_2}{\pi_3}\right)^{\pi_2-\pi_3}\left(\frac{\theta_1}{\theta_2}\right)^{\theta_1-\theta_2}\left(\frac{\theta_1}{\theta_3}\right)^{\theta_1-\theta_3}\left(\frac{\theta_2}{\theta_3}\right)^{\theta_2-\theta_3}\right], \end{aligned}$$

and

$$H_3 = -p'\mathcal{P}_{\text{Sgn}}\ell_p = -\frac{1}{6}\log\left(\frac{\pi_1\pi_2\pi_3}{\theta_1\theta_2\theta_3}\right)^{(\pi_1+\pi_2+\pi_3-\theta_1-\theta_2-\theta_3)}.$$

Data-analytic aspects. When the distribution of the observed frequencies x is multinomial $M(x \mid p)$, the posterior distribution of p given the data x under a Dirichlet prior model $D(a)$ for p is again Dirichlet $D(x + a)$, so that the evaluation of the posterior moments for scalar functions of p often follows from numerical methods to evaluate the corresponding integrals over the simplex $S = \{p; \sum_j p_j = 1\}$. Alternatively, the posterior Dirichlet densities can be simulated from independent univariate gamma distributions (Devroye, 1986, p. 593). To illustrate, consider the decompositions of the entropy in the frequency distributions indexed by the full permutation orbit of the DNA word $\{act\}$ in Region 2, shown in Table (5.28). A similar calculation was obtained for the orbit of $\{agc\}$, also in region 2. A total of 5,000 samples of the posterior Dirichlet distributions under uniform prior were generated. Figure 5.3 shows the (90% approximate) contours for the posterior joint density of H_2 and H_1, determined by $\mathcal{P}_\beta$, for each one of the two orbits. The mean and standard deviations for the entropy components are shown in Table (5.30)

	H_{agc}	$H_{1,agc}$	$H_{2,agc}$	$H_{3,agc}$	H_{act}	$H_{1,act}$	$H_{2,act}$	$H_{3,act}$
mean	1.403	2.444	−1.029	−0.013	1.790	1.855	−0.041	−0.024
SD	0.083	0.236	0.325	0.071	0.123	0.035	0.142	0.028

(5.30)

The striking difference in the entropy of the two orbits is explained, following the interpretation of the regular component H_2 of H under the action of S_3, by the

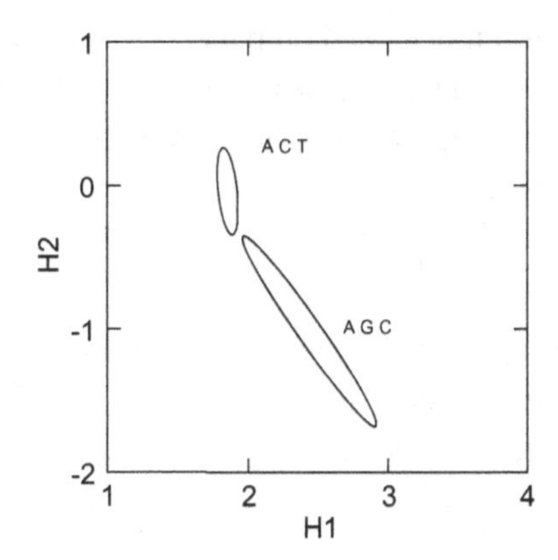

Figure 5.3: Posterior joint density contours (90%) for H_2 and H_1.

amount with which the relative freguencies of rotations and reversals differ from the uniform distribution $(1/3, 1/3, 1/3)$. Inspection of the posterior distribution of H_3 reveals that there is very little divergence between the marginal distribution of rotations and translations and the uniform distribution $(1/2, 1/2)$, which is the interpretation of the component H_3. Again, in this example, the canonical invariants derived from permutation symmetries applied to the labels for the data led to the discovery of contrasting summaries of that data. See also Viana (2006).

5.10 A Two-Way Cyclic Decomposition

When the cyclic permutations act on the c rows and ℓ columns of $V = C \times L$, the resulting $c\ell$ canonical projections matrices are given by

$$\mathcal{P}_{mn} = \frac{1}{c\ell} \sum_{k,j} \omega_c^{mk} \omega_\ell^{nj} \left(\rho_c^k \otimes \rho_\ell^j \right), \tag{5.31}$$

where $\omega_f = e^{-2\pi i/f}$ and ρ_f is a generating matrix for C_f, for $f \in \{c, \ell\}$, $k, n = 0, \ldots, c-1,\ j, m = 0, \ldots, \ell - 1$. To illustrate, consider the mining disaster frequency data, shown in Table (5.32) and discussed in Wit and McCullagh (2001). The frequencies, recorded over a 11-year period, are distributed according to days of the week and seasons of the year so that $\ell = 4$ and $c = 7$.

	Mon	Tue	Wed	Thu	Fri	Sat	Sun	total
Autumn	7	10	5	5	6	7	1	41
Winter	5	9	10	10	11	7	0	52
Spring	3	7	10	12	13	9	2	56
Summer	4	8	8	9	5	6	2	42
total	19	34	33	36	35	29	5	$N = 191$

(5.32)

Following with the notation introduced on page 148, there are 28 one-dimensional canonical projections reducing the entropy $H = -p'\ell_p$ in the joint frequency distribution. The entropy components shown on Table (5.33) were derived from the canonical decomposition

$$I_{28} = I_4 \otimes I_7 = (\mathcal{P}_1^q + \mathcal{P}_2^q + \mathcal{P}_3^q) \otimes (\mathcal{P}_1^w + \mathcal{P}_2^w + \mathcal{P}_3^w + \mathcal{P}_4^w)$$

in which the quarterly component $\mathcal{P}_2^q$ is the sum of two complex conjugate components and, similarly for three weekly components $\mathcal{P}_2^w$, $\mathcal{P}_3^w$, and $\mathcal{P}_4^w$. The component $\mathcal{P}_1^q \otimes \mathcal{P}_1^w$ is the symmetric component.

$\otimes$	$\mathcal{P}_1^w$	$\mathcal{P}_2^w$	$\mathcal{P}_3^w$	$\mathcal{P}_4^w$
$\mathcal{P}_1^q$	3.5200	−0.1600	−0.0500	−0.02000
$\mathcal{P}_2^q$	−0.0159	−0.0390	−0.0132	−0.00499
$\mathcal{P}_3^q$	0.0001	−0.0039	−0.0018	0.00004

(5.33)

The estimated overall entropy is 3.21 (the maximum entropy is $-log(1/28) = 3.33$). Indicating by H_1 the symmetric entropy component (the log geometric mean), it follows that $H_2 = N(H - H_1)$ is then Kullback's divergence against the corresponding uniform distribution. In the joint distribution $H_2 = -191(3.21 - 3.52) = 59.2$, which is clearly significant under the (null hypothesis) large-sample χ^2_{27} distribution of H_2 (Kullback, 1968, p. 113). Similarly, in the weekly marginal distribution, $H_2 = 55.4$, whereas in the quarterly distribution $H_2 = 15.3$, both also clearly significant.

5.11 Data Structures Induced by Point Groups

In this section we illustrate an application of the canonical reduction to data induced by point groups. Since these groups are well-known,[2] the data analyst has a potentially large class of structures to work with. These groups were introduced, with the present notation, earlier on page 65.

Data structures induced by C_{3v}

The symmetry transformations in $\mathbb{R}^3$ defined by the point group C_{3v} include the identity (E), a 120° rotation (C) around the z axis, a 240° rotation (C^2) around the z axis, and three reflections ($\sigma_1, \sigma_2, \sigma_3$) on dihedral vertical planes containing the z axis. That is,

$$C_{3v} = \{E, C, C^2, \sigma_1, \sigma_2, \sigma_3\}.$$

[2] The reader may refer, for example, to the Max Plank Institute for Polymer Research (Web site: www.mpip-mainz.mpg.de) for the character table of these groups.

The ammonia molecule NH_3, for example, has the symmetry of $\mathcal{C}_{3v}$. The multiplication table for $\mathcal{C}_{3v}$ is

$$\begin{array}{c|cccccc} \mathcal{C}_{3v} & E & C & C^2 & \sigma_1 & \sigma_2 & \sigma_3 \\ \hline E & E & C & C^2 & \sigma_1 & \sigma_2 & \sigma_3 \\ C & C & C^2 & E & \sigma_2 & \sigma_3 & \sigma_1 \\ C^2 & C^2 & E & C & \sigma_3 & \sigma_1 & \sigma_2 \\ \sigma_1 & \sigma_1 & \sigma_3 & \sigma_2 & E & C^2 & C \\ \sigma_2 & \sigma_2 & \sigma_1 & \sigma_3 & C & E & C^2 \\ \sigma_3 & \sigma_3 & \sigma_2 & \sigma_1 & C^2 & C & E \end{array}. \tag{5.34}$$

Note that $\mathcal{C}_{3v}$ is noncommutative and that every rotation factors as a product of two reflections (a particular case of the fact that in the Euclidean n-dimensional space every rotation factors a the product of an even number $\leq n$ of reflections, e.g., Cartan (1966, p. 10). The natural action of $\mathcal{C}_{3v}$ on the canonical basis $\{e_1, e_2, e_3\}$ of $\mathbb{R}^3$ is given by

$$r(E) = \begin{bmatrix} 1 & 0 & 0 \\ 0 & 1 & 0 \\ 0 & 0 & 1 \end{bmatrix}, \quad r(C) = 1/2 \begin{bmatrix} -1 & \sqrt{3} & 0 \\ -\sqrt{3} & -1 & 0 \\ 0 & 0 & 2 \end{bmatrix}, \quad r(C^2) = 1/2 \begin{bmatrix} -1 & -\sqrt{3} & 0 \\ \sqrt{3} & -1 & 0 \\ 0 & 0 & 2 \end{bmatrix},$$

$$r(\sigma_1) = 1/2 \begin{bmatrix} 1 & \sqrt{3} & 0 \\ \sqrt{3} & -1 & 0 \\ 0 & 0 & 2 \end{bmatrix}, \quad r(\sigma_2) = 1/2 \begin{bmatrix} 1 & -\sqrt{3} & 0 \\ -\sqrt{3} & -1 & 0 \\ 0 & 0 & 2 \end{bmatrix}, \quad r(\sigma_3) = \begin{bmatrix} -1 & 0 & 0 \\ 0 & 1 & 0 \\ 0 & 0 & 1 \end{bmatrix},$$

where the rotation on the xy plane is clockwise, σ_3 is the reflection on the yz plane, and σ_1 and σ_2 are the reflections on the two other (120°) dihedral planes containing the z axis. It extends, by linearity, to an action of $\mathcal{C}_{3v}$ on $\mathbb{R}^3$.

Observing that $\mathcal{C}_{3v} \simeq S_3$ via $(\tau, e_i) = e_{\sigma i}$, $\tau \in \mathcal{C}_{3v}$, $\sigma \in S_3$, we obtain the canonical projections associated with the natural action (r), namely,

$$\begin{aligned} \mathcal{P}_1 &= \frac{1}{6}\left[\sum_i r(C^i) + \sum_i r(\sigma_i)\right] \\ &= \text{Diag}(0, 0, 1), \quad \mathcal{P}_\beta = \frac{2}{6}[2r(E) + r(C) + r(C^2)] = \text{Diag}(1, 1, 0), \end{aligned}$$

whereas $\mathcal{P}_{\text{Sgn}} = [\sum_i r(C^i) - \sum_i r(\sigma_i)]/6 = 0$. The corresponding invariant subspaces have dimensions 1, 0, 2, respectively. Clearly, if $v' = (x, y, z) \in \mathbb{R}^3$ then $\mathcal{C}_{3v}$ induces a partition of $||v||^2$ into two orthogonal components, namely, $x^2 + y^2$ from $\mathcal{P}_3 v = (x, y, 0)$ and z^2 from $\mathcal{P}_1 v = (0, 0, z)$. The decomposition, or reduction, of

$$v \otimes v = \begin{bmatrix} x^2 & xy & xz \\ yx & y^2 & yz \\ zx & zy & z^2 \end{bmatrix}$$

follows from the canonical projections ($\mathcal{T}$) associated with the tensor representation $r \otimes r$. Here we have displayed the column vector $v \otimes v$, for convenience, as a 3×3 matrix. Specifically, $v \otimes v$ reduces as $m_1 + m_2 + m_3$, in which m_1, m_2 and m_3 are, respectively,

$$\frac{1}{2}\begin{bmatrix} x^2+y^2 & 0 & 0 \\ 0 & x^2+y^2 & 0 \\ 0 & 0 & 2z^2 \end{bmatrix}, \quad \frac{1}{2}\begin{bmatrix} 0 & xy-yx & 0 \\ -xy+yx & 0 & 0 \\ 0 & 0 & 0 \end{bmatrix}, \quad \frac{1}{2}\begin{bmatrix} x^2-y^2 & xy+yx & 2xz \\ xy+yx & -x^2+y^2 & 2yz \\ 2zx & 2zy & 0 \end{bmatrix}.$$

The associated invariant subspaces have dimensions bounded by 2, 1, 6, respectively. Note that $\mathcal{T}_2(v \otimes v) = 0$ when the algebra is commutative. The entries in $\mathcal{T}(v \otimes v)$ are the invariant quadratic polynomials[3] characterized by the point group C_{3v}. Similarly, the components for the decomposition of $v \otimes v \otimes v$ are,

$$m_1 = \frac{1}{4}\begin{bmatrix} 0 & 3x^2y-y^3 & 2x^2z+2y^2z \\ 3x^2y-y^3 & 0 & 0 \\ 2x^2z+2y^2z & 0 & 0 \\ 3x^2y-y^3 & 0 & 0 \\ 0 & -3x^2y+y^3 & 2x^2z+2y^2z \\ 0 & 2x^2z+2y^2z & 0 \\ 2x^2z+2y^2z & 0 & 0 \\ 0 & 2x^2z+2y^2z & 0 \\ 0 & 0 & 4z^3 \end{bmatrix}, \quad m_2 = \frac{1}{4}\begin{bmatrix} x^3-3xy^2 & 0 & 0 \\ 0 & -x^3+3xy^2 & 0 \\ 0 & 0 & 0 \\ 0 & -x^3+3xy^2 & 0 \\ -x^3+3xy^2 & 0 & 0 \\ 0 & 0 & 0 \\ 0 & 0 & 0 \\ 0 & 0 & 0 \\ 0 & 0 & 0 \end{bmatrix},$$

$$m_3 = \frac{1}{4}\begin{bmatrix} 3x^3+3xy^2 & x^2y+y^3 & 2x^2z-2y^2z \\ x^2y+y^3 & x^3+xy^2 & 4xyz \\ 2x^2z-2y^2z & 4xyz & 4xz^2 \\ x^2y+y^3 & x^3+xy^2 & 4xyz \\ x^3+xy^2 & 3x^2y+3y^3 & -2x^2z+2y^2z \\ 4xyz & -2x^2z+2y^2z & 4yz^2 \\ 2x^2z-2y^2z & 4xyz & 4xz^2 \\ 4xyz & -2x^2z+2y^2z & 4yz^2 \\ 4xz^2 & 4yz^2 & 0 \end{bmatrix}.$$

The corresponding invariant subspaces have dimensions bounded by 5, 4, 18, respectively. These are the C_{3v}-invariant cubic polynomials. We have $m_1 + m_2 + m_3 = v \otimes v \otimes v$.

5.12 Data Indexed by Homogeneous Polynomials

Consider the real homogeneous polynomial $aX^2 + bY^2 + wXY$ in X, Y and interpret

$$V = \{X^2, Y^2, XY\}$$

[3] The character tables, linear, quadratic and cubic functions of all point groups ara available, for example, from the Max Planck Institute for Polymer Research – www.mpip-mainz.mpg.de.

as a set of labels for the potential data $x' = (a, b, w)$. This structure is prototypic of many experimental conditions. In one example, we may interpret (a, b, w) as the clearance rates with which an organism responds to medication under drug X alone, under drug Y alone, or under a combination XY of these two drugs. In another example, the data might consist of the frequencies (a, b, w) with which the apparent position $u(x, y)$ of an object is displaced along a given X-axis, along a second Y-axis, or along any other line in the $X - Y$ plane. In eye movement studies, the X-axis might be the nasal-temporal axis and the Y axis the superior-inferior axis, centered in the apex of the cornea. In yet another interpretation, (a, b, w) may be equated to the partial derivatives $(u_{xx}, u_{yy}, 2u_{xy})$ evaluated at (0, 0) that appear in the Maclaurin expansion

$$u(x, y) = \cdots + \frac{1}{2!}[u_{xx}X^2 + u_{yy}Y^2 + 2u_{xy}XY] + \cdots \tag{5.35}$$

of a given locally smooth function $u(x, y)$ in the variables x, y. More generally (Hoffman, 1966), given a smooth function $F(x, y)$ and the Lie derivative

$$\mathcal{L} = f(x, y)\frac{\partial}{\partial x} + g(x, y)\frac{\partial}{\partial y}$$

of a given Lie group, we may interpret the operational components in $\mathcal{L}^2$ as the labels for the corresponding data $\mathcal{L}^2 F$, where $\mathcal{L}^2$ indicates the twofold application of $\mathcal{L}$.

In linear optics (Lakshminarayanan and Viana, 2005) we may consider the labels in V assembled in the form of the 2×2 matrix

$$V = \begin{bmatrix} X^2 & XY \\ YX & Y^2 \end{bmatrix},$$

and observe that the multiplication VM of V with an arbitrary $2 \times m$ matrix M generates another matrix in which each entry is then an homogenous polynomial in degree of 2. As a consequence, the operation VM has the effect of indexing the data in M with the labels in V. The matrix M may represent a linear instrument or a ray vector. If, for example, $X = \cos(\alpha)$ and $Y = \sin(\alpha)$, the resulting matrix

$$P(\alpha) = \begin{bmatrix} \cos^2(\alpha) & \sin(\alpha)\cos(\alpha) \\ \sin(\alpha)\cos(\alpha) & \sin^2(\alpha) \end{bmatrix},$$

represents a polarizer, which takes the projection of the electric field in the direction making an angle α with the x-axis. If $M = P(\alpha)$ and $V = P(\phi)$ then VMf indexes the outcoming ray Mf from the polarizer M (excited by the incoming ray f) with the labels in the analyzer V. Similar constructions apply to n-fold products $V \times \ldots$ and $V \otimes \ldots$ of V. In general, the mapping space $V = V^L$ with c^ℓ elements is in one-to-one correspondence with the homogeneous polynomial $(a_1X_1 + \cdots + a_cX_c)^\ell$.

For example, the set $V = \{uu, yy, uy, yu\}$ of binary sequences in length of 2 can be indexed by

$$(aU + bY)^2 \equiv c_{11}U^2 + c_{12}UY + c_{21}YU + c_{22}Y^2,$$

where the coefficients may be interpreted as observations indexed by the points in V. In typical applications, these are frequencies with which the corresponding words appear in a reference sequence.

Broadly, then, the labels in V are consistent with experiments in which the potential data are indexed by each one of two distinct conditions alone and by the joint presence of the two conditions. We observe (Bacry, 1967, p. 207) that V appears as a representation space for the planar rotations within the context of solving the Laplace equation in V. See also Lakhsminarayanan and Varadharajan (1997) for related application in the context of matrix optics.

5.13 Likelihood Decompositions

We conclude this chapter with a brief overview of applying canonical decompositions to the parameterization of probability models. We refer to Section 3.3 on page 66, and Exercises 4.5–4.7 on page 122.

First we make the following remark. Given a canonical projection $\mathcal{P} = \sum_\tau n\chi_{\tau^{-1}}\rho_\tau/g$ in which the representation ρ of G (with g elements) is unitary relative to the inner product $(\cdot, \cdot)$, we have

$$\begin{aligned}(x, \mathcal{P}y) &= \left(x, \sum_\tau n\chi_{\tau^{-1}}\rho_\tau y/g\right) = \sum_\tau n\chi_\tau(\rho_{\tau^{-1}}x/g, y) = \left(\sum_\tau n\chi_\tau\rho_{\tau^{-1}}x/g, y\right)\\ &= (\mathcal{P}x, y).\end{aligned} \tag{5.36}$$

Multinomial models. Let x have a multinomial distribution with underlying probabilities $p' = (p_1, \ldots, p_m)$, and write ℓ_p to indicate the vector of corresponding log probabilities (assumed to be all nonzero). It is then simple to verify that the probability distribution $P(x \mid p)$ of x can be expressed as

$$P(x \mid p) = B(x)e^{(x, \ell_p)},$$

where the inner product is given by $x'\ell_p$ and $B(x)$ is the multinomial coefficient for x given the total number N of observations. Let G act on the components of p, giving a representation ρ (isomorphically) unitary in $(\cdot, \cdot)$, and canonical projections $\mathcal{P}_\chi$ associated with the irreducible characters $\chi \in \widehat{G}$. Then $I = \sum_{\chi \in \widehat{G}} \mathcal{P}_\chi$ is the resulting canonical decomposition of I_m and, applying (5.36),

$$P(x \mid p) = B(x) \prod_{\chi \in \widehat{G}} e^{(x, \mathcal{P}_\chi \ell_p)}.$$

In the binomial case, the two components are the symmetric and the alternating ones, in which

$$(x, \mathcal{P}_1 \ell_p) = \frac{N}{2} \log p(1-p), \quad (x, \mathcal{P}_{\text{Sgn}} \ell_p) = \frac{x_1 - x_2}{2} \log \frac{p}{(1-p)},$$

respectively. Note that the symmetric component does not depend on data, given the total number $N = x_1 + x_2$ of observations, whereas the alternating component depends on the probability parameter p only through the log odds ratio. To illustrate it further, consider a multinomial distribution $P(x \mid p)$ indexed by S_3. More specifically, write $x' = (r_1, r_2, r_3, t_1, t_2, t_3)$ to indicate the frequencies indexed by the rotations r_1, r_2, r_3 and reversals t_1, t_2, t_3, and by $p' = (\pi_1, \pi_2, \pi_3, \theta_1, \theta_2, \theta_3)$ the corresponding probabilities. Direct calculations based on the canonical projections described on pages 106–107 then show that

$$P(x \mid p) = B(x) e^{[(x, \mathcal{P}_1 \ell_p) + (x, \mathcal{P}_{\text{Sgn}}) + (x, \mathcal{P}_\beta)]},$$

in which

$$(x, \mathcal{P}_1 \ell_p) = \frac{n}{6} \log[\pi_1 \pi_2 \pi_3 \theta_1 \theta_2 \theta_3],$$
$$(x, \mathcal{P}_{\text{Sgn}} \ell_p) = \frac{1}{6}(r_1 + r_2 + r_3 - t_1 - t_2 - t_3) \log \frac{\pi_1 \pi_2 \pi_3}{\theta_1 \theta_2 \theta_3},$$

and (using the fact that $\mathcal{P}_\beta \mathcal{P}_1 = 0$),

$$(x, \mathcal{P}_\beta \ell_p) = \frac{1}{3} \left[(2r_1 - r_2 - r_3) \log \frac{\pi_1}{\pi_3} + (2r_2 - r_1 - r_3) \log \frac{\pi_2}{\pi_3} \right.$$
$$\left. + (2t_1 - t_2 - t_3) \log \frac{\theta_1}{\theta_3} + (2t_2 - t_1 - t_3) \log \frac{\theta_2}{\theta_3} \right].$$

The new parameterization $\widetilde{p} = M \ell_p$ is described by

$$\widetilde{p}' = \left(\log[\pi_1 \pi_2 \pi_3 \theta_1 \theta_2 \theta_3], \log \frac{\pi_1 \pi_2 \pi_3}{\theta_1 \theta_2 \theta_3}, \log \frac{\pi_1}{\pi_3}, \log \frac{\pi_2}{\pi_3}, \log \frac{\theta_1}{\theta_3}, \log \frac{\theta_2}{\theta_3} \right)$$

and

$$M = \frac{1}{6} \begin{bmatrix} 1 & 1 & 1 & 1 & 1 & 1 \\ 1 & 1 & 1 & -1 & -1 & -1 \\ 2 & 0 & -2 & 0 & 0 & 0 \\ 0 & 2 & -2 & 0 & 0 & 0 \\ 0 & 0 & 0 & 2 & 0 & -2 \\ 0 & 0 & 0 & 0 & 2 & -2 \end{bmatrix}.$$

Similarly to the bivariate case we see that the symmetric component is now modulated by the alternating (Sgn) and the two-dimensional (β) components. If, for example, the frequencies of rotations and reversals are about the same then the distribution is essentially modulated by the within reversals, within rotations differences accounted by β.

Multivariate normal. Not surprisingly, if the distribution of x is multivariate normal with vector of means μ and covariance matrix Σ then

$$f(\rho_\tau x \mid (\mu, \Sigma)) = f(x \mid (\rho_\tau \mu, (\rho_\tau \Sigma \rho_\tau')^{-1})$$

for all $\tau \in G$ and unitary ρ, so that if Σ is centralized by G and μ is invariant under G then $f(\rho_\tau x \mid (\mu, \Sigma)) = f(x \mid (\mu, \Sigma))$ for all $\tau \in G$ and the model is invariant under G. In general, the distibution can be factorized from

$$(x-\mu, \Sigma^{-1}(x-\mu)) = (x-\mu, \Sigma^{-1}\mathcal{P}_1(x-\mu)) + \cdots + (x-\mu, \Sigma^{-1}\mathcal{P}_h(x-\mu))$$

where $I = \mathcal{P}_1 + \cdots + \mathcal{P}_h$ is a canonical decomposition of interest.

Further Reading

The theoretic aspects of the distribution of quadratic forms are covered in many texts. The reader may refer to Searle (1971), Rao (1973), Muirhead (1982), and Eaton (1983). Additional details for some of the applications described in the chapter can be found in Viana (2007).

Applications of group-theoretic principles in statistics and probability have a long history and tradition of their own, from Legendre-Gauss's least squares principle through R.A. Fisher's method of variance decomposition. According to E.J. Hannan (1965), the work of A. James (1957) appears to be among the first describing the group-theoretic nature of Fisher's argument, giving meaning to the notion of relationship algebra, or the commuting algebra of a representation of the group of symmetries of an experimental design. See also Mann (1960) and the earlier work of James (1954) on the normal multivariate analysis and the orthogonal group. At that same time, the integral of Haar, object of L. Nachbin's (1965) monograph, became a familiar tool among statisticians. An earlier report by Ledermann (1968) described the joint action of the symmetric group on the rows and columns of a data matrix and the resulting analysis of variance that follows from the decomposition of the identity operator.

Two decades later, the relevance of group invariance and group representation arguments in statistical inference would become evident in the works of Farrell (1985), Diaconis (1988), Dawid (1988), Eaton (1989), Wijsman (1990), Andersson (1990), and Bailey (1991). These earlier references are certainly required companion for the reader interested in related views of symmetry studies. The collection of contemporary work (Viana and Richards, 2001) coedited by the author clearly documents the continued interest in the applications of algebraic methods in statistics. See also Andersson and Madsen (1998) and Helland (2004). The recent work of Van de Ven and Di Bucchianico (2006) and Van de Ven (2007) is a comprehensive synthesis of the mathematical framework uderlying (fractional) factorial designs, using arguments of Abelian groups. Details about projection matrices in factorial designs are given in Gupta and Mekerjee (1989).

The reader may refer to Viana (2007b) and Viana (2008) for complementary readings related to the section on dihedral stratifications for voting preferences (page 137).

Exercises

Exercise 5.1. The data shown in Table (5.37) is described in Cox and Snell (1989, p. 6).

s	successes	trials	s	successes	trials
none	84	477	d	2	12
a	75	231	ad	7	13
b	13	63	bd	4	7
ab	35	94	abd	8	12
c	67	150	cd	3	11
ac	201	378	acd	27	45
bc	16	32	bcd	1	23
abc	102	169	abcd	23	31

(5.37)

The study assessed the individual's general knowledge about cancer, conditioning on four factors expected to account for the variation in the probability of success (defined by a high scoring). The individuals were classified into 2^4 classes, depending on the presence of exposure to (a) newspapers; (b) radio; (c) solid reading; (d) lectures.

(1) Identify these data as an example of data indexed by a group (suggestion: refer to Exercise 2.5 on page 59);
(2) Determine the multiplication table of the group and its structure;
(3) Propose a symmetry study for these data.

Exercise 5.2. Following the definitions on page 140, consider the *canonical basis* $Y_1 = X_1/\sqrt{2}$, $Y_2 = X_2/\sqrt{2}$, e.g., Bacry (1967)[p. 203]. Show that $||Y_1|| = ||Y_2|| = 0$ and that $Y_1'Y_2 = 1$.

Exercise 5.3. Signed matrices. Show that the set of all $n \times n$ matrices {Diag $(\pm 1, \ldots, \pm 1)$} with the operation of matrix multiplication defines a finite group of order 2^n. It can be identified with the space of all binary sequences in length of n.

Exercise 5.4. Table (5.38) shows the frequency counts (x_s) obtained from five regions (in length of 200) of the HIV isolate m26727. Apply Proposition 5.1 to evaluate the projections $\mathcal{P}x$ of x derived from the left action $s\tau^{-1}$ of S_4 on the binary sequences in length of 4 (s). The location number indicates the region's starting position in the genome.

s	1	16	15	14	12	8	13	11	7	10	6	4	9	5	3	2
$s(1)$	y	u	y	u	u	u	y	y	y	u	u	u	y	y	y	u
$s(2)$	y	u	u	y	u	u	y	u	u	y	y	u	y	y	u	y
$s(3)$	y	u	u	u	y	u	u	y	u	y	u	y	y	u	y	y
$s(4)$	y	u	u	u	u	y	u	u	y	u	y	y	u	y	y	y
location	x_s															
200	5	52	18	12	15	17	16	6	10	11	9	12	5	1	4	5
1000	7	30	20	20	21	20	10	6	10	8	7	8	8	7	8	8
3000	1	43	20	15	17	19	12	8	8	11	10	10	6	5	7	6
5000	6	24	18	19	21	18	11	9	12	12	12	9	6	6	9	6
6000	4	18	14	15	14	14	11	20	12	8	19	12	11	8	7	11

. (5.38)

Exercise 5.5. Consider a clinical experimental in which 60 subjects are be randomized into one of the 6 regimen sequences $ABC, \ldots, CBA$ of three medications $\{A, B, C\}$. Following each sequence, the weight loss relative to baseline is obtained from each subject. Outline a symmetry study for these data.

Exercise 5.6. For $x, y, z \in \mathbb{R}^2$, derive the permutation (S_2) coinvariants of $x \otimes y \otimes z$.

Exercise 5.7. For $x, y \in \mathbb{R}^3$, derive the permutation (S_3) coinvariants of $x \otimes y$ and obtain the matrix decompositions of xy' and xx'.

Exercise 5.8. For $x, y \in \mathbb{R}^4$, derive the permutation dihedral (D_4) coinvariants of $x \otimes y$ and obtain the matrix decompositions of xy' and xx'.

Exercise 5.9. Starting with the multiplication table

$\mathcal{D}_{2h}$	1	2	3	4	5	6	7	8
1	1	2	3	4	5	6	7	8
$2 = C_{21}$	2	1	4	3	6	5	8	7
$3 = C_{22}$	3	4	1	2	7	8	5	6
$4 = C_{23}$	4	3	2	1	8	7	6	5
$5 = \sigma_{h1}$	5	6	7	8	1	2	3	4
$6 = \sigma_{h2}$	6	5	8	7	2	1	4	3
$7 = \sigma_{h3}$	7	8	5	6	3	4	1	2
$8 = i$	8	7	6	5	4	3	2	1

for the point group $\mathcal{D}_{2h} = \{E, C_{21}, C_{22}, C_{23}, i, \sigma_{h1}, \sigma_{h2}, \sigma_{h3}\} \simeq C_2 \otimes C_2 \otimes C_2$, determine its canonical decomposition. To which factorial experiment does it correspond to?

Exercise 5.10. Pistone et al. (2001) describe the two-dimensional structure

$$\begin{matrix} \bullet & & \\ \bullet & & \\ \bullet & \bullet & \bullet \end{matrix}$$

or $V = \{(0, 2), (0, 1), (0, 0), (1, 0), (2, 0)\}$. This is called an *echelon structure*, or design. Echelon structures are suggested by the property that if a point $s = (x, y)$ is in V, then also (x', y') is in V for $0 \leq x \leq x', 0 \leq y \leq y'$.

(1) Define a group action of $S_2 = \{1, (xy)\}$ on V;
(2) calculate the resulting orbits in V;
(3) derive the corresponding linear representation ρ of S_2;
(4) evaluate the canonical projections;
(5) show that $\mathcal{P}_1$ and $\mathcal{P}_2$ are canonical projections;
(6) suppose that you measure something in V according to the instrument $x_s = m + ax + by + \alpha x^2 + \beta y^2$, $s = (x, y) \in V$, and indicate by z the resulting data indexed by V. Evaluate and interpret $\mathcal{P}_1 z$ and $\mathcal{P}_2 z$.

Exercise 5.11. In Game Theory [see, for example, Savage (1954, p. 193)], a bilinear game is defined by

$$\mathcal{L}(f; g) = \sum_{r,i} L(r; i)\, f(r)\, g(i), \qquad r \in R,\ i \in I,$$

where R and I are finite sets, f and g are probability distributions defined on R and I, respectively, and $L(r; i)$ is a real-valued function such that the quantities $\max_i L(r; i)$, $\min_r L(r; i)$, $L^* = \min_r \max_i L(r; i)$ and $L_* = \max_i \min_r L(r; i)$ exist. For bilinear games, the classical result of von Neumann says that $L^* = L_*$. Mathematically, points Savage, the solution of a bilinear game is often simplified by conditions of symmetry, introduced as follows: Assume, for convenience of notation, that R and I are disjoint and let $\sigma \in X_R$ and $\tau \in S_I$ be permutations of the sets R and I, respectively, and write $\eta = (\sigma, \tau)$ to indicate a permutation of the set $R \cup I$. We say that $\eta \in X_{R \cup I}$ leaves the game L invariant if $L(\sigma r; \tau i) = L(r; i)$.

The interpretation here is that for each permutation $\sigma \in X_R$ there is a permutation $\tau \in S_I$ such that the two permutations taken together leave the game invariant.[4] Indicate by G the group of all permutations leaving the game invariant. G is called the group of the game. Define also $\eta f = f\sigma^{-1}$ and $\eta g = g\tau^{-1}$, and say that f, respectively g, is invariant under the group of the game when $\eta f = f$, respectively $\eta g = g$, for all η in G. Show that $\mathcal{L}(\eta f; \eta g) = \mathcal{L}(f; g)$ for all η in G. Moreover, for any distributions f (with support R) and g (with support I), verify that $\bar{f} = \sum_\eta \eta f/|G|$ and $\bar{g} = \sum_\eta \eta g/|G|$ are invariant under G. This is a

[4] The reader may want to compare this interpretation with Definition 2 and Lemma 2 of Chapter 4 in Ferguson (1967).

common technique to obtain invariant distributions. The reader will recognize the same argument present in Eaton (1989), Expression (4.2).

Exercise 5.12. Indicate by $p' = (\pi_1, \pi_2, \pi_3, \theta_1, \theta_2, \theta_3)$ a probability distribution in which the components are indexed by S_3. Apply the canonical decomposition of the regular representation of S_3 to show that the entropy H in p resolves into

$$H_1 = -p'\mathcal{P}_1\ell_p = -\frac{1}{6}\log[\pi_1\pi_2\pi_3\theta_1\theta_2\theta_3],$$

$$H_2 = -p'\mathcal{P}_2\ell_p$$
$$= -\frac{1}{3}\log\left[\left(\frac{\pi_1}{\pi_2}\right)^{\pi_1-\pi_2}\left(\frac{\pi_1}{\pi_3}\right)^{\pi_1-\pi_3}\left(\frac{\pi_2}{\pi_3}\right)^{\pi_2-\pi_3}\left(\frac{\theta_1}{\theta_2}\right)^{\theta_1-\theta_2}\left(\frac{\theta_1}{\theta_3}\right)^{\theta_1-\theta_3}\left(\frac{\theta_2}{\theta_3}\right)^{\theta_2-\theta_3}\right],$$

and

$$H_3 = -p'\mathcal{P}_3\ell_p = -\frac{1}{6}\log\left(\frac{\pi_1\pi_2\pi_3}{\theta_1\theta_2\theta_3}\right)^{(\pi_1+\pi_2+\pi_3-\theta_1-\theta_2-\theta_3)},$$

where ℓ_p indicated the vector of the log probabilities (assumed to be all positive).

Exercise 5.13. Study the canonical decompositions of multinomial models indexed by C_3 and D_4.

Exercise 5.14. Action on homogeneous polynomials. Study the action of $S_{\{x,y,z\}}$ on the the set V defined by the terms $x^\alpha y^\beta z^\gamma$ of $(x+y+z)^3$. Show that it gives three orbits $\mathcal{O}_1$, $\mathcal{O}_2$, and $\mathcal{O}_3$, in dimension of 3, 6, and 1, respectively. Derive the canonical projections and invariants for the action on $\mathcal{O}_2$.

Exercise 5.15. Following Section 5.7, show that $\tau \to \tau^*$ is a homomorphism from G to Aut $(\mathcal{F})$, that τ^* is linear and unitary with respect to the scalar products

$$(x, y)_s = \frac{1}{g}\sum_{\tau\in G} x(\varphi_\sigma s)\overline{y}(\varphi_\sigma s), \quad s \in V.$$

6

Symmetry Studies of Short Symbolic Sequences

6.1 Introduction

The work of Doi (1991) on the evolutionary strategy of the HIV-1 virus defines the *frequency diversity* in each cyclic orbit $\mathcal{O}_f = \{f\tau^{-1}; \tau \in C_\ell\}$, as the ratio $\max_{s \in \mathcal{O}_f} x_s / \min_{s \in \mathcal{O}_f} x_s$ between the largest and the smallest of the observed localized frequencies x_s, as s varies within the orbit $\mathcal{O}_f$ of f. Here s and f are short DNA sequences or mappings defined in $L = \{1, 2, \ldots, \ell\}$ with values in the four nucleotides $C = \{a, g, c, t\}$. These sequences are, therefore, points in the structure $V = C^L$, where the cyclic group C_ℓ acts by cyclically moving the positions of the residues on the sequence. The frequencies x_s are calculated within a given fixed region of interest, such as conservative or hyper variable regions, associated with different interpretations of the virus' evolutionary strategies.

In this chapter we discuss a number of elementary symmetry studies in which the data are indexed by $V = C^L$. We generally refer to L as the set of positions in the sequence and to C as the set of symbols.

6.2 Symmetry Studies of Four Sequences in Length of 3

In this section we present a summary of the canonical decompositions for the set V of all mappings $s : L := \{1, 2, 3\} \mapsto C = \{a, g, c, t\}$ where subgroups of S_3 may act according to $s\tau^{-1}$ shuffling the positions of the symbols and subgroups of S_4 may act on the symbols according to σs. Table (6.1) shows the complete set V of four-sequences in length of 3 and their base-4 labels, using the convention

$(1, 2, 3, 4) = (a, g, c, t)$ (see Exercise 1.6 of Chapter 1).

$$
V = \begin{array}{|r|ccc||r|ccc||r|ccc||r|ccc|}
\hline
1 & a & a & a & 5 & a & g & a & 23 & c & g & g & 8 & t & g & a \\
22 & g & g & g & 9 & a & c & a & 24 & t & g & g & 50 & g & a & t \\
43 & c & c & c & 13 & a & t & a & 41 & a & c & c & 14 & g & t & a \\
64 & t & t & t & 18 & g & a & g & 42 & g & c & c & 20 & t & a & g \\
17 & a & a & g & 26 & g & c & g & 44 & t & c & c & 57 & a & c & t \\
33 & a & a & c & 30 & g & t & g & 61 & a & t & t & 45 & a & t & c \\
49 & a & a & t & 35 & c & a & c & 62 & g & t & t & 12 & t & c & a \\
6 & g & g & a & 39 & c & g & c & 63 & c & t & t & 51 & c & a & t \\
38 & g & g & c & 47 & c & t & c & 37 & a & g & c & 15 & c & t & a \\
54 & g & g & t & 52 & t & a & t & 34 & g & a & c & 36 & t & a & c \\
11 & c & c & a & 56 & t & g & t & 7 & c & g & a & 58 & g & c & t \\
27 & c & c & g & 60 & t & c & t & 25 & a & c & g & 46 & g & t & c \\
59 & c & c & t & 2 & g & a & a & 10 & g & c & a & 28 & t & c & g \\
16 & t & t & a & 3 & c & a & a & 19 & c & a & g & 55 & c & g & t \\
32 & t & t & g & 4 & t & a & a & 53 & a & g & t & 31 & c & t & g \\
48 & t & t & c & 21 & a & g & g & 29 & a & t & g & 40 & t & g & c \\
\hline
\end{array} \qquad (6.1)
$$

As remarked in the previous chapters, a generic strategy for data analysis is the identification of the subsets of V where the group of interest acts transitively, and proceed with the decomposition of that smaller space. Globally, the canonical projections do not identify the subspaces in the data space indexed by those selected orbits (the selection of which is arbitrary). The projections, the reader may recall from Section 4.4 on page 101, are uniquely indexed by the irreducible representations of the selected group. In the next section we summarize the overall projections.

6.3 Reductions by Position Symmetry

The action $s\tau^{-1}$ of S_3 on V is described in the Appendix to this Chapter on page 177. For example, the permutation $\tau = (123)$ acting on the sequence aac (label 33) gives $s\tau^{-1} = caa$ (label 3).

The data vector space $\mathcal{V} = \mathbb{R}^{64}$ associated with V decomposes as follows: There are $h = 3$ canonical projections $\mathbb{P}_k$ on $\mathcal{V}$, each one of the form

$$
\mathbb{P}_k = \begin{bmatrix} \mathcal{P}_k^{\lambda_1} & 0 & 0 \\ 0 & \ddots & 0 \\ 0 & 0 & \mathcal{P}_k^{\lambda_m} \end{bmatrix}, \quad k = 1, \ldots, h,
$$

where m is equal the number of integer partitions of ℓ in length of c, as introduced earlier on Chapter 1, page 52.

Specifically, in this case $\lambda_1 = 30^3$, $\lambda_2 = 210^2$, and $\lambda_3 = 1^30$ so that $m = 3$, and

$$\mathcal{P}_k^{\lambda_1} = \begin{bmatrix} Q_1^{\lambda_1} & 0 & 0 \\ 0 & \ddots & 0 \\ 0 & 0 & Q_{c_1}^{\lambda_1} \end{bmatrix}, \quad P_k^{\lambda_2} = \begin{bmatrix} Q_1^{\lambda_2} & 0 & 0 \\ 0 & \ddots & 0 \\ 0 & 0 & Q_{c_2}^{\lambda_2} \end{bmatrix}, \quad P_k^{\lambda_3} = \begin{bmatrix} Q_1^{\lambda_3} & 0 & 0 \\ 0 & \ddots & 0 \\ 0 & 0 & Q_{c_3}^{\lambda_3} \end{bmatrix}.$$

In addition, each projection:

(1) $Q_i^{\lambda_1}$ acts on a subspace of dimension $\ell_1 = 1, i = 1, \ldots, c_1 = 4$;
(2) $Q_i^{\lambda_2}$ acts on a subspace of dimension $\ell_2 = 3, i = 1, \ldots, c_2 = 12$;
(3) $Q_i^{\lambda_3}$ acts on a subspace of dimension $\ell_3 = 6, i = 1, \ldots, c_3 = 4$,

so that $\dim \mathcal{V} = 64 = 4 \times 1 + 12 \times 3 + 4 \times 6$. The first projection $\mathbb{P}_1$ is determined by

$$Q_1^{\lambda_1} = \cdots = Q_4^{\lambda_1} = 1, \quad Q_1^{\lambda_2} = \cdots = Q_{12}^{\lambda_2} = \mathcal{A}_3, \quad Q_1^{\lambda_3} = \cdots = Q_4^{\lambda_3} = \mathcal{A}_6;$$

The second projection $\mathbb{P}_2$ is determined by

$$Q_1^{\lambda_1} = \cdots = Q_4^{\lambda_1} = 0, \quad Q_1^{\lambda_2} = \cdots = Q_{12}^{\lambda_2} = 0, \quad Q_1^{\lambda_3} = \cdots = Q_6^{\lambda_3} = \mathcal{Q}_2 \otimes \mathcal{A}_3,$$

whereas $\mathbb{P}_3$ is determined by

$$Q_1^{\lambda_1} = \cdots = Q_4^{\lambda_1} = 0, \quad Q_1^{\lambda_2} = \cdots = Q_{12}^{\lambda_2} = \mathcal{Q}_3 \quad Q_1^{\lambda_3} = \cdots = Q_4^{\lambda_3} = I_2 \otimes \mathcal{Q}_3.$$

Moreover, note that $\operatorname{Tr} \mathcal{P}_1 = 20n_1$, $\operatorname{Tr} \mathcal{P}_2 = 4n_2$, and $\operatorname{Tr} \mathcal{P}_3 = 20n_3$, where n_i is the dimension of the corresponding irreducible representations. That is, $\rho \simeq 20\rho_1 \oplus 4\rho_{\text{Sgn}} \oplus 20\rho_\beta$.

Determining the classes of transitivity

The equivalent types of orbits are given by the integer partitions $\lambda = 30^3$, 210^2 and 1^30 of 3 in length of 4. The class $\lambda = 30^3$ includes four orbits of size 1, a representative of which is $\{aaa\}$; class $\lambda = 210^2$ includes 12 orbits of size 3, a representative of which is

$$\{aag, aga, gaa\},$$

identified by the labels $\{17, 5, 2\}$ respectively, whereas the class $\lambda = 1^30$ includes 4 orbits of size 6, a representative of which is

$$\{agc, gca, cag, gac, cga, acg\},$$

identified by the labels $\{37, 10, 19, 34, 7, 25\}$, respectively, in Table (6.1).

Canonical decompositions in the partition $\lambda = 210^2$

Consider the action

$s(1)$	$s(2)$	$s(3)$	label	1	(123)	(132)	(12)	(13)	(23)
a	a	g	17	17	2	5	17	2	5
a	g	a	5	5	17	2	2	5	17
g	a	a	2	2	5	17	5	17	2

$s\tau^{-1}$ of S_3 on the representative orbit $\{aag, aga, gaa\}$ abstracted from Table (6.1) and indicate by ρ the resulting representation. Table (6.2) summarizes the irreducible characters of S_3 along with the character χ_ρ of ρ, from which, applying Proposition 4.1 on page 93, we see that ρ reduces as $1 \oplus \beta$. This is not surprising, since ρ is isomorphic to the standard representation of S_3.

τ	χ_ρ	χ_1	χ_{Sgn}	χ_β
1	3	1	1	2
(123)	0	1	1	−1
(132)	0	1	1	−1
(12)	1	1	−1	0
(13)	1	1	−1	0
(23)	1	1	−1	0

(6.2)

As a consequence, the canonical decomposition in each one of these orbits is simply the standard decomposition $I = \mathcal{A} + \mathcal{Q}$.

To illustrate, consider the observed frequencies shown on Table (6.3) with which the words $\{aag, aga, gaa\}$ appear in 45 subsequent 200-bp-long regions in the BRU isolate of the HIV-1.

A canonical invariant in dimension of 2 is determined by β and, in this case of a single copy, by its associated projection, $\mathcal{Q}$. Indicating the labels and the corresponding data with the same notation, a realization of the canonical invariant for the log-frequency data is given by the pair

$$\mathcal{I}_{21} = \log \frac{aag^2}{aga\ gaa}, \quad \mathcal{I}_{22} = \log \frac{aga^2}{aag\ gaa}.$$

Table (6.4) shows, for each region, the pair $(\mathcal{I}_{21}, \mathcal{I}_{22})$, along with its invariant norm (δ), scaled to the maximum value of 1. Their joint distribution over the regions is shown in Figure 6.1.

region	*aag*	*aga*	*gaa*	region	*aag*	*aga*	*gaa*	region	*aag*	*aga*	*gaa*
1	7	9	7	16	4	4	5	31	6	7	5
2	4	9	7	17	6	10	6	32	2	1	3
3	10	10	5	18	7	6	8	33	6	11	7
4	9	10	6	19	7	10	7	34	2	5	5
5	6	7	6	20	6	6	7	35	6	4	6
6	6	10	5	21	9	7	9	36	7	11	7
7	13	6	9	22	11	9	8	37	4	3	3
8	6	16	6	23	8	7	7	38	5	8	12
9	7	8	7	24	5	11	5	39	3	3	4
10	5	5	5	25	7	7	5	40	4	12	8
11	9	13	13	26	4	8	7	41	8	8	6
12	4	5	5	27	10	8	6	42	6	6	4
13	2	10	5	28	7	9	5	43	8	7	4
14	6	6	6	29	4	6	6	44	6	11	5
15	9	14	11	30	5	5	4	45	4	3	1

(6.3)

region	$\mathcal{I}_{21}$	$\mathcal{I}_{22}$	δ	region	$\mathcal{I}_{21}$	$\mathcal{I}_{22}$	δ	region	$\mathcal{I}_{21}$	$\mathcal{I}_{22}$	δ
1	−0.252	0.507	0.029	16	−0.223	−0.223	0.014	31	0.029	0.489	0.028
2	−1.370	1.060	0.259	17	−0.511	1.020	0.118	32	0.285	−1.790	0.344
3	0.693	0.693	0.136	18	0.019	−0.443	0.021	33	−0.761	1.060	0.147
4	0.300	0.615	0.063	19	−0.357	0.713	0.057	34	−1.830	0.916	0.381
5	−0.154	0.307	0.010	20	−0.154	−0.154	0.006	35	0.405	−0.810	0.074
6	−0.329	1.210	0.155	21	0.255	−0.504	0.029	36	−0.453	0.904	0.092
7	1.140	−1.180	0.230	22	0.519	−0.083	0.029	37	0.577	−0.288	0.037
8	−0.981	1.960	0.438	23	0.270	−0.135	0.008	38	−1.350	0.067	0.202
9	−0.135	0.270	0.008	24	−0.790	1.580	0.284	39	−0.288	−0.288	0.023
10	0.000	0.000	0.000	25	0.336	0.336	0.032	40	−1.800	1.500	0.472
11	−0.736	0.372	0.061	26	−1.250	0.829	0.196	41	0.285	0.285	0.023
12	−0.446	0.223	0.022	27	0.732	0.067	0.064	42	0.405	0.405	0.046
13	−2.530	2.300	1.000	28	0.086	0.842	0.085	43	0.824	0.425	0.118
14	0.000	0.000	0.000	29	−0.810	0.405	0.074	44	−0.425	1.400	0.209
15	−0.642	0.683	0.074	30	0.223	0.223	0.014	45	1.670	0.811	0.469

(6.4)

Canonical decompositions in the partition $\lambda = 1^3 0$

Table (6.5) shows the action $s\tau^{-1}$ of S_3 on the representative orbit $\{agc, gca, cag, gac, cga, acg\}$ identified by the labels {37, 10, 19, 34, 7, 25}, respectively,

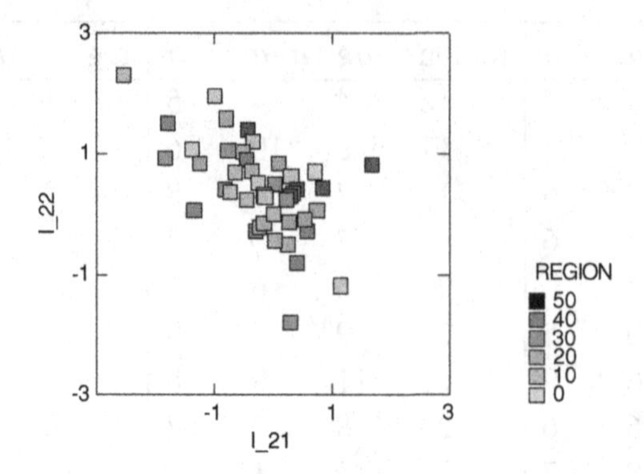

Figure 6.1: Invariant plot for the orbit $\{aag, aga, gaa\}$.

abstracted from Table (6.1).

$s(1)$	$s(2)$	$s(3)$	label	1	(123)	(132)	(12)	(13)	(23)
a	g	c	37	37	10	19	34	7	25
g	c	a	10	10	19	37	7	25	34
c	a	g	19	19	37	10	25	34	7
g	a	c	34	34	25	7	37	19	10
c	g	a	7	7	34	25	10	37	19
a	c	g	25	25	7	34	19	10	37

(6.5)

The character χ_ϕ of the resulting representation, along with the irreducible characters of S_3, are shown in Table (6.6), which clearly identifies it as the regular representation so that $\phi \simeq 1 \oplus \text{Sgn} \oplus 2\beta$.

τ	χ_ϕ	χ_1	χ_{Sgn}	χ_β
1	6	1	1	2
(123)	0	1	1	−1
(132)	0	1	1	−1
(12)	0	1	−1	0
(13)	0	1	−1	0
(23)	0	1	−1	0

(6.6)

The canonical invariants were described on pages 105 and 120. Associated with the alternating representation, we have, in the log scale,

$$\mathcal{I}_{\text{Sgn}} = \log \frac{agc \cdot gca \cdot cag}{gac \cdot cga \cdot acg},$$

comparing rotations $\{agc, gca, cag\}$ with reversals $\{gac, cga, acg\}$. On the other hand, there are two invariants, also expressed here in the log scale,

$$\mathcal{B}_v = (v_1, v_2) = \left(\log \frac{agc\ cga}{gac\ gca}, \log \frac{cag\ acg}{gac\ gca}\right),$$

$$\mathcal{B}_w = (w_1, w_2) = \left(\log \frac{gca\ acg}{cga\ cag}, \log \frac{agc\ gac}{cga\ cag}\right) \qquad (6.7)$$

determined by the bases $\mathcal{B}_v$ and $\mathcal{B}_w$ associated with the irreducible representation β.

An example with DNA frequencies – continued

Table (6.8) shows the observed frequencies with which the words in the orbit $\{agc, gca, cag, gac, cga, acg\}$ appear in 10 subsequent regions, in length of 900 bps, along the BRU isolate of the HIV-1.

region	agc	gca	cag	gac	cga	acg
1	28	19	32	15	8	4
2	26	23	35	19	2	0
3	8	10	26	13	2	1
4	18	21	31	17	1	2
5	20	23	36	8	0	1
6	20	17	20	16	2	1
7	22	22	22	12	2	1
8	13	15	22	10	1	3
9	16	23	23	13	6	5
10	28	16	25	14	5	5

(6.8)

The plot of the invariants described in (6.7) can then be obtained, summarizing the results in eight of the original regions (Regions two and five may be excluded in this analysis due to the presence of zeros in the observed frequencies of words cga and acg). The chiral aspect of the sequences along the genome is illustrated in Figure 6.2, showing the rotation to reversal ratios of frequencies gac/agc, cga/gca, and acg/cag, indicated by $rgac$, $rcga$, and $racg$ respectively. The continuous lines are smoothed contours of those ratios along the regions.

6.4 Reductions by Symbol Symmetry

The action σs of S_4 on the set V of all four-sequences in length of 3 is described in Appendix B to this chapter on page 178. For example, the permutation $\sigma = (agc)$ acting on the sequence aac (label 33) by

$$\{1, 2, 3\} \xrightarrow{s} \{a, g, c, t\} \xrightarrow{(agc)} \{a, g, c, t\}$$

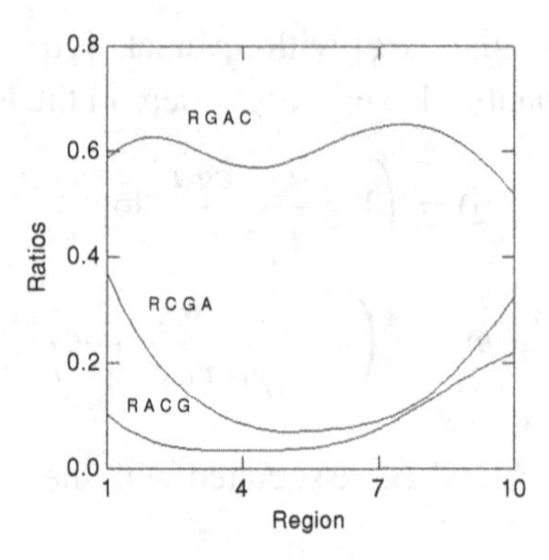

Figure 6.2: Rotation/reversal ratios for the AGC orbit.

gives $\sigma s = gga$ (label 6).

There are $h = 5$ canonical projections $\mathcal{P}_k$ of S_4 on the data space $\mathcal{V} = \mathbb{R}^{64}$, indexed by the five irreducible characters

$\chi\backslash\tau$	1	(12)	(12)(34)	(123)	(1234)
χ_1	1	1	1	1	1
χ_2	1	−1	1	1	−1
χ_3	2	0	2	−1	0
χ_4	3	1	−1	0	−1
χ_5	3	−1	−1	0	1

(6.9)

of S_4. Globally, each projection has the form

$$\mathbb{P}_k = \begin{bmatrix} \mathcal{P}_k^{\lambda_1} & 0 & 0 \\ 0 & \mathcal{P}_k^{\lambda_2} & 0 \\ 0 & 0 & \mathcal{P}_k^{\lambda_3} \end{bmatrix}, \quad k = 1, \ldots, 5,$$

with

$$\mathcal{P}_k^{\lambda_1} = \begin{bmatrix} \mathcal{Q}_1^{\lambda_1} & 0 & 0 \\ 0 & \ddots & 0 \\ 0 & 0 & \mathcal{Q}_{c_1}^{\lambda_1} \end{bmatrix}, \quad \mathcal{P}_k^{\lambda_2} = \begin{bmatrix} \mathcal{Q}_1^{\lambda_2} & 0 & 0 \\ 0 & \ddots & 0 \\ 0 & 0 & \mathcal{Q}_{c_2}^{\lambda_2} \end{bmatrix}, \quad \mathcal{P}_k^{\lambda_3} = \begin{bmatrix} \mathcal{Q}_1^{\lambda_3} & 0 & 0 \\ 0 & \ddots & 0 \\ 0 & 0 & \mathcal{Q}_{c_3}^{\lambda_3} \end{bmatrix}.$$

In addition,

(1) $\mathcal{Q}_i^{\lambda_1}$ acts on a subspace of dimension $\ell_1 = 4, i = 1, \ldots, c_1 = 1$;
(2) $\mathcal{Q}_i^{\lambda_2}$ acts on a subspace of dimension $\ell_2 = 12, i = 1, \ldots, c_2 = 3$;
(3) $\mathcal{Q}_i^{\lambda_3}$ acts on a subspace of dimension $\ell_3 = 24, i = 1, \ldots, c_3 = 1$,

so that $\dim \mathcal{V} = 64 = 1 \times 4 + 3 \times 12 + 1 \times 24$.

Moreover, we have, for the τs action,

i	Tr $\mathcal{P}_i$	n_i	Tr $\mathcal{P}_i/n_i$
1	5	1	5
2	1	1	1
3	10	2	5
4	30	3	10
5	18	3	6

so that $\rho \simeq 5\rho_1 \oplus 1\rho_2 \oplus 5\rho_3 \oplus 10\rho_4 \oplus 6\rho_5$.

6.5 Dihedral Studies

In this section we illustrate a dihedral study of frequency data suggested by the argument introduced earlier on page 137 to study the four-candidate voting preferences. In the present example, however, the frequencies of interest are those with which the words in the corresponding dihedral cosets of S_4 appear in the BRU isolate of the HIV-1 virus.

We remarked, in the voting preferences example, that S_4 has three distinct cyclic orbits, namely those generated by (1234), (1243), and (1324), so that by adding to each generator a reversal we obtained three copies of the dihedral group, specifically,

$$G_1 =< (1234), (13) >\simeq \{agct, gcta, ctag, tagc, cgat, tcga, atcg, gatc\} : \begin{array}{ccc} a_1 & \longrightarrow & g_2 \\ \uparrow & & \downarrow \\ t_4 & \longleftarrow & c_3 \end{array},$$

$$G_2 =< (1243), (14) >\simeq \{agct, gtac, tcga, catg, tgca, ctag, acgt, gatc\} : \begin{array}{ccc} a_1 & \longrightarrow & c_3 \\ \uparrow & & \downarrow \\ g_2 & \longleftarrow & t_4 \end{array},$$

and

$$G_3 =< (1324), (12) >\simeq \{agct, ctga, gatc, tcag, gact, tcga, agtc, ctag\} : \begin{array}{ccc} a_1 & \longrightarrow & t_4 \\ \uparrow & & \downarrow \\ c_3 & \longleftarrow & g_2 \end{array}.$$

The frequencies were obtained over 15 consecutive regions of the genome, each one in length of 500 bps. The dihedral analysis of variance (page 134) was complemented with the inclusion of the error term $I \otimes Q$ and understood as a large-sample approximation. The results are summarized on Tables (6.10), (6.11), and (6.12), including (on the LHS) the dihedral frequencies in each region, and the corresponding analysis of variance (on the RHS).

G_1 :

region	r_1	r_2	r_3	r_4	t_1	t_2	t_3	t_4
1	4	2	1	1	0	0	0	2
2	3	2	2	2	0	0	0	0
3	2	1	1	1	1	0	0	2
4	1	1	0	0	0	0	0	1
5	2	0	1	1	0	0	0	2
6	2	2	2	2	0	0	0	0
7	3	2	3	2	0	0	0	0
8	0	0	0	4	0	0	0	1
9	2	3	4	2	0	0	0	2
10	3	1	7	3	0	1	0	2
11	0	2	0	4	0	0	0	0
12	1	2	0	1	1	0	0	1
13	2	2	2	3	0	0	0	1
14	1	2	0	1	0	0	0	1
15	3	1	1	0	0	0	2	1

,

$\mathcal{P}$	$x'\mathcal{P}x$	df	$x'\mathcal{P}x/df$	F
$\mathcal{P}_1 \otimes \mathcal{A}$	128.100	1	128.100	
$\mathcal{P}_2 \otimes \mathcal{A}$	2.133	1	2.133	2.024
$\mathcal{P}_3 \otimes \mathcal{A}$	56.034	1	56.034	53.180
$\mathcal{P}_4 \otimes \mathcal{A}$	0.833	1	0.833	0.791
$\mathcal{P}_5 \otimes \mathcal{A}$	8.866	4	2.216	2.103
$I \otimes \mathcal{Q}$	118.010	112	1.053	
total	313.970	120		

(6.10)

G_2 :

region	r_1	r_2	r_3	r_4	t_1	t_2	t_3	t_4
1	4	1	0	5	1	1	0	2
2	3	1	0	1	4	2	0	0
3	2	1	0	0	0	1	0	2
4	1	1	0	0	2	0	0	1
5	2	2	0	2	3	1	0	2
6	2	1	0	3	1	2	1	0
7	3	4	0	4	1	3	0	0
8	0	1	0	1	0	0	0	1
9	2	1	0	1	2	4	0	2
10	3	1	1	1	1	7	0	2
11	0	2	0	2	0	0	0	0
12	1	1	0	5	3	0	1	1
13	2	1	0	2	0	2	1	1
14	1	2	0	2	3	0	0	1
15	3	0	0	1	1	1	0	1

,

$\mathcal{P}$	$x'\mathcal{P}x$	df	$x'\mathcal{P}x/df$	F
$\mathcal{P}_1 \otimes \mathcal{A}$	175.230	1	175.230	
$\mathcal{P}_2 \otimes \mathcal{A}$	0.208	1	0.208	0.151
$\mathcal{P}_3 \otimes \mathcal{A}$	1.874	1	1.874	1.364
$\mathcal{P}_4 \otimes \mathcal{A}$	10.207	1	10.207	7.429
$\mathcal{P}_5 \otimes \mathcal{A}$	43.637	4	10.909	7.940
$I \otimes \mathcal{Q}$	153.880	112	1.373	
total	385.020	120		

(6.11)

The canonical projections for the regular representation of D_4 were derived on page 110 and are written here as $\mathcal{P}_1, \ldots, \mathcal{P}_5$, respectively, the symmetric, alternating, α, γ, and β projections.

The chiral arrangement of the dihedral orbits, clearly evident in one case, can be assessed by the component of $x'x$ associated with $\mathcal{P}_3$, that compares rotations and reversals.

The resulting F-ratios reflect the marked difference between rotations and reversals present in the dihedral orbit induced by G_2 ($F_{1,112} = 53.18$).

G_3 :

region	r_1	r_2	r_3	r_4	t_1	t_2	t_3	t_4
1	4	1	2	7	0	0	0	1
2	3	1	0	3	2	0	0	2
3	2	1	2	5	0	0	0	1
4	1	1	1	5	3	0	1	0
5	2	3	2	4	3	0	1	1
6	2	2	0	0	3	0	0	2
7	3	2	0	5	3	0	3	3
8	0	1	1	1	0	0	1	0
9	2	1	2	4	3	0	1	4
10	3	1	2	2	1	1	1	7
11	0	0	0	3	2	0	2	0
12	1	1	1	3	0	0	1	0
13	2	1	1	2	0	0	1	2
14	1	4	1	2	0	0	0	0
15	3	0	1	1	0	0	1	1

,

$\mathcal{P}$	$x'\mathcal{P}x$	df	$x'\mathcal{P}x/df$	F
$\mathcal{P}_1 \otimes \mathcal{A}$	240.820	1	240.820	
$\mathcal{P}_2 \otimes \mathcal{A}$	7.499	1	7.499	4.601
$\mathcal{P}_3 \otimes \mathcal{A}$	24.300	1	24.300	14.911
$\mathcal{P}_4 \otimes \mathcal{A}$	1.633	1	1.633	1.002
$\mathcal{P}_5 \otimes \mathcal{A}$	49.205	4	12.301	7.547
$I \otimes \mathcal{Q}$	182.530	112	1.629	
total	505.990	120		

(6.12)

Further Reading

The reader interested in a didactical introduction to sequence analysis in biology may refer to Durbin et al. (1998). Aspects of entropy, long-range correlation, and local complexity in nucleotide sequences are discussed, for example, in Herzel et al. (1994), Peng et al. (1992), and Salamon and Konopka (1992), respectively.

Pattern recognition methods for microarray data analysis accounting for the geometry of the data space have been recently introduced by Vencio et al. (2007).

A statistical mechanics view of computational biology was introduced by Blossey (2006). The contributions of Bernd Sturmfels and colleagues to the development and applications of algebraic geometric and computational methodos in biology are documented in Pachter and Sturmfels (2005).

The reader may refer to Viana (2008a) for the derivation and data-analytic applications of the regular invariants of S_4.

Exercises

Exercise 6.1. Determine the fixed points and the orbit stabilizers for the action described on Table 6.5.

Exercise 6.2. Following the decomposition for $x \in \mathcal{F}(S_3)$ on page 120, construct the invariant plots for the data shown on Table (2.7), page 38. Contrast with the decomposition for $x \in \mathcal{F}(C_3)$ described in that section.

Exercise 6.3. Following Section 6.4 and the action σs of S_4 on the set V of all four sequences in length of 3 described in the Appendix to this Chapter, page 178, determine and interpret the resulting letter symmetry orbits.

Exercise 6.4. Table (1.21), introduced earlier on in Chapter 1, page 13, describes the frequencies with which the 16 purine-pyrimidine sequences in length of 4 appear in 10 subsequent 200-bp-long regions of BRU isolate of the HIV-1. Carry out a symmetry study for position and letter symmetries for those data.

Exercise 6.5. Compare the canonical invariants determined by the two-dimensional irreducible representation (β) of S_3 in its permutation ($\rho = 1 \oplus \beta$) and regular ($\phi = 1 \oplus 2\beta \oplus \text{Sgn}$) representations.

Exercise 6.6. Evaluate the letter-symmetry orbits generated by the action σs of S_3 on the set V of all ternary sequences in length of 3 given by Table (6.13).

$s(1)$	$s(2)$	$s(3)$	label	σ : 1	(12)	(13)	(23)	(123)	(132)
1	1	1	1	1	14	27	1	14	27
2	2	2	14	14	1	14	27	27	1
3	3	3	27	27	27	1	14	1	14
1	1	2	10	10	5	18	19	23	9
1	2	1	4	4	11	24	7	17	21
2	1	1	2	2	13	26	3	15	25
2	2	1	5	5	10	23	9	18	19
2	1	2	11	11	4	17	21	24	7
1	2	2	13	13	2	15	25	26	3
1	1	3	19	19	23	9	10	5	18
1	3	1	7	7	17	21	4	11	24
3	1	1	3	3	15	25	2	13	26
3	3	1	9	9	18	19	5	10	23
3	1	3	21	21	24	7	11	4	17
1	3	3	25	25	26	3	13	2	15
2	2	3	23	23	19	5	18	9	10
2	3	2	17	17	7	11	24	21	4
3	2	2	15	15	3	13	26	25	2
3	3	2	18	18	9	10	23	19	5
3	2	3	24	24	21	4	17	7	11
2	3	3	26	26	25	2	15	3	13
1	2	3	22	22	20	6	16	8	12
1	3	2	16	16	8	12	22	20	6
2	1	3	20	20	22	8	12	6	16
3	1	2	12	12	6	16	20	22	8
2	3	1	8	8	16	20	6	12	22
3	2	1	6	6	12	22	8	16	20

(6.13)

Appendix A

The action $s\tau^{-1}$ of S_3 on the set of all four-sequences in length of 3.

$\sigma \backslash s$	1	22	43	64	17	33	49	6	38	54	11	27	59	16	32	48
1	1	22	43	64	17	33	49	6	38	54	11	27	59	16	32	48
(12)	1	22	43	64	17	33	49	6	38	54	11	27	59	16	32	48
(13)	1	22	43	64	2	3	4	21	23	24	41	42	44	61	62	63
(23)	1	22	43	64	5	9	13	18	26	30	35	39	47	52	56	60
(123)	1	22	43	64	2	3	4	21	23	24	41	42	44	61	62	63
(132)	1	22	43	64	5	9	13	18	26	30	35	39	47	52	56	60
$\sigma \backslash s$	**5**	**9**	**13**	**18**	**26**	**30**	**35**	**39**	**47**	**52**	**56**	**60**	**2**	**3**	**4**	**21**
1	5	9	13	18	26	30	35	39	47	52	56	60	2	3	4	21
(12)	2	3	4	21	23	24	41	42	44	61	62	63	5	9	13	18
(13)	5	9	13	18	26	30	35	39	47	52	56	60	17	33	49	6
(23)	17	33	49	6	38	54	11	27	59	16	32	48	2	3	4	21
(123)	17	33	49	6	38	54	11	27	59	16	32	48	5	9	13	18
(132)	2	3	4	21	23	24	41	42	44	61	62	63	17	33	49	6
$\sigma \backslash s$	**23**	**24**	**41**	**42**	**44**	**61**	**62**	**63**	**37**	**34**	**7**	**25**	**10**	**19**	**53**	**29**
1	23	24	41	42	44	61	62	63	37	34	7	25	10	19	53	29
(12)	26	30	35	39	47	52	56	60	34	37	10	19	7	25	50	20
(13)	38	54	11	27	59	16	32	48	7	19	37	10	25	34	8	14
(23)	23	24	41	42	44	61	62	63	25	10	19	37	34	7	29	53
(123)	26	30	35	39	47	52	56	60	19	7	25	34	37	10	20	50
(132)	38	54	11	27	59	16	32	48	10	25	34	7	19	37	14	8
$\sigma \backslash s$	**8**	**50**	**14**	**20**	**57**	**45**	**12**	**51**	**15**	**36**	**58**	**46**	**28**	**55**	**31**	**40**
1	8	50	14	20	57	45	12	51	15	36	58	46	28	55	31	40
(12)	14	53	8	29	51	36	15	57	12	45	55	40	31	58	28	46
(13)	53	20	29	50	12	15	57	36	45	51	28	31	58	40	46	55
(23)	20	14	50	8	45	57	36	15	51	12	46	58	40	31	55	28
(123)	29	8	53	14	36	51	45	12	57	15	40	55	46	28	58	31
(132)	50	29	20	53	15	12	51	45	36	57	31	28	55	46	40	58

Appendix B

The (letter symmetry) action σs of S_4 on four sequences in length of 3, given by their labels defined on Table (6.1) on page 166, fixed points and stabilizers. The partial number of fixed points in each table is indicated by fix*.

$\tau\backslash s$	1	22	43	64	17	33	49	6	38	54	11	27	59	16	32	48	fix*
1	1	22	43	64	17	33	49	6	38	54	11	27	59	16	32	48	16
(34)	1	22	64	43	17	49	33	6	54	38	16	32	48	11	27	59	4
(23)	1	43	22	64	33	17	49	11	27	59	6	38	54	16	48	32	4
(24)	1	64	43	22	49	33	17	16	48	32	11	59	27	6	54	38	4
(12)	22	1	43	64	6	38	54	17	33	49	27	11	59	32	16	48	4
(13)	43	22	1	64	27	11	59	38	6	54	33	17	49	48	32	16	4
(14)	64	22	43	1	32	48	16	54	38	6	59	27	11	49	17	33	4
(234)	1	43	64	22	33	49	17	11	59	27	16	48	32	6	38	54	1
(243)	1	64	22	43	49	17	33	16	32	48	6	54	38	11	59	27	1
(123)	22	43	1	64	38	6	54	27	11	59	17	33	49	32	48	16	1
(124)	22	64	43	1	54	38	6	32	48	16	27	59	11	17	49	33	1
(132)	43	1	22	64	11	27	59	33	17	49	38	6	54	48	16	32	1
(134)	43	22	64	1	27	59	11	38	54	6	48	32	16	33	17	49	1
(142)	64	1	43	22	16	48	32	49	33	17	59	11	27	54	6	38	1
(143)	64	22	1	43	32	16	48	54	6	38	49	17	33	59	27	11	1
(12)(34)	22	1	64	43	6	54	38	17	49	33	32	16	48	27	11	59	0
(13)(24)	43	64	1	22	59	11	27	48	16	32	33	49	17	38	54	6	0
(14)(23)	64	43	22	1	48	32	16	59	27	11	54	38	6	49	33	17	0
(1234)	22	43	64	1	38	54	6	27	59	11	32	48	16	17	33	49	0
(1243)	22	64	1	43	54	6	38	32	16	48	17	49	33	27	59	11	0
(1324)	43	64	22	1	59	27	11	48	32	16	38	54	6	33	49	17	0
(1342)	43	1	64	22	11	59	27	33	49	17	48	16	32	38	6	54	0
(1432)	64	1	22	43	16	32	48	49	17	33	54	6	38	59	11	27	0
(1423)	64	43	1	22	48	16	32	59	11	27	49	33	17	54	38	6	0
$\lvert G_s\rvert$	6	6	6	6	2	2	2	2	2	2	2	2	2	2	2	2	

$\tau\backslash s$	5	9	13	18	26	30	35	39	47	52	56	60	2	3	4	21	fix*
1	5	9	13	18	26	30	35	39	47	52	56	60	2	3	4	21	16
(34)	5	13	9	18	30	26	52	56	60	35	39	47	2	4	3	21	4
(23)	9	5	13	35	39	47	18	26	30	52	60	56	3	2	4	41	3
(24)	13	9	5	52	60	56	35	47	39	18	30	26	4	3	2	61	3
(12)	18	26	30	5	9	13	39	35	47	56	52	60	21	23	24	2	2
(13)	39	35	47	26	18	30	9	5	13	60	56	52	42	41	44	23	2
(14)	56	60	52	30	26	18	47	39	35	13	5	9	62	63	61	24	2
(234)	9	13	5	35	47	39	52	60	56	18	26	30	3	4	2	41	0
(243)	13	5	9	52	56	60	18	30	26	35	47	39	4	2	3	61	0
(123)	26	18	30	39	35	47	5	9	13	56	60	52	23	21	24	42	0
(124)	30	26	18	56	60	52	39	47	35	5	13	9	24	23	21	62	0
(143)	35	39	47	9	5	13	26	18	30	60	52	56	41	42	44	3	0
(134)	39	47	35	26	30	18	60	56	52	9	5	13	42	44	41	23	0
(142)	52	60	56	13	9	5	47	35	39	30	18	26	61	63	62	4	0
(143)	56	52	60	30	18	26	13	5	9	47	39	35	62	61	63	24	0
(12)(34)	18	30	26	5	13	9	56	52	60	39	35	47	21	24	23	2	0
(13)(24)	47	35	39	60	52	56	9	13	5	26	30	18	44	41	42	63	0
(14)(23)	60	56	52	47	39	35	30	26	18	13	9	5	63	62	61	44	0
(1234)	26	30	18	39	47	35	56	60	52	5	9	13	23	24	21	42	0
(1243)	30	18	26	56	52	60	5	13	9	39	47	35	24	21	23	62	0
(1324)	47	39	35	60	56	52	26	30	18	9	13	5	44	42	41	63	0
(1342)	35	47	39	9	13	5	60	52	56	26	18	30	41	44	42	3	0
(1432)	52	56	60	13	5	9	30	18	26	47	35	39	61	62	63	4	0
(1423)	60	52	56	47	35	39	13	9	5	30	26	18	63	61	62	44	0
$\lvert G_s\rvert$	2	2	2	2	2	2	2	2	2	2	2	2	2	2	2	2	

$\tau\backslash s$	23	24	41	42	44	61	62	63	37	34	7	25	10	19	53	29	fix*
1	23	24	41	42	44	61	62	63	37	34	7	25	10	19	53	29	16
(34)	24	23	61	62	63	41	42	44	53	50	8	29	14	20	37	25	0
(23)	42	44	21	23	24	61	63	62	25	19	10	37	7	34	57	45	1
(24)	63	62	41	44	42	21	24	23	45	36	15	57	12	51	29	53	1
(12)	3	4	42	41	44	62	61	63	34	37	19	10	25	7	50	14	2
(13)	21	24	3	2	4	63	62	61	7	10	37	19	34	25	55	31	2
(14)	23	21	44	42	41	4	2	3	40	46	55	28	58	31	8	20	2
(234)	44	42	61	63	62	21	23	24	57	51	12	45	15	36	25	37	0
(243)	62	63	21	24	23	41	44	42	29	20	14	53	8	50	45	57	0
(123)	41	44	2	3	4	62	63	61	10	7	25	34	19	37	58	46	0
(124)	63	61	42	44	41	2	4	3	46	40	31	58	28	55	14	50	0
(143)	2	4	23	21	24	63	61	62	19	25	34	7	37	10	51	15	0
(134)	24	21	63	62	61	3	2	4	55	58	40	31	46	28	7	19	0
(142)	3	2	44	41	42	24	21	23	36	45	51	12	57	15	20	8	0
(143)	21	23	4	2	3	44	42	41	8	14	53	20	50	29	40	28	0
(12)(34)	4	3	62	61	63	42	41	44	50	53	20	14	29	8	34	10	0
(13)(24)	61	62	3	4	2	23	24	21	15	12	45	51	36	57	31	55	0
(14)(23)	42	41	24	23	21	4	3	2	28	31	58	40	55	46	12	36	0
(1234)	44	41	62	63	61	2	3	4	58	55	28	46	31	40	10	34	0
(1243)	61	63	2	4	3	42	44	41	14	8	29	50	20	53	46	58	0
(1324)	62	61	23	24	21	3	4	2	31	28	46	55	40	58	15	51	0
(1342)	4	2	63	61	62	23	21	24	51	57	36	15	45	12	19	7	0
(1432)	2	3	24	21	23	44	41	42	20	29	50	8	53	14	36	12	0
(1423)	41	42	4	3	2	24	23	21	12	15	57	36	51	45	28	40	0
$\lvert G_s\rvert$	2	2	2	2	2	2	2	2	1	1	1	1	1	1	1	1	

$\tau\backslash s$	8	50	14	20	57	45	12	51	15	36	58	46	28	55	31	40	fix*	fix
1	8	50	14	20	57	45	12	51	15	36	58	46	28	55	31	40	16	64
(34)	7	34	10	19	45	57	15	36	12	51	46	58	31	40	28	55	0	8
(23)	12	51	15	36	53	29	8	50	14	20	55	31	40	58	46	28	0	8
(24)	14	20	8	50	25	37	10	19	7	34	28	40	58	31	55	46	0	8
(12)	20	53	29	8	58	46	28	55	31	40	57	45	12	51	15	36	0	8
(13)	40	58	46	28	51	15	36	57	45	12	50	14	20	53	29	8	0	8
(14)	53	14	50	29	12	36	57	15	51	45	10	34	25	7	19	37	0	8
(234)	10	19	7	34	29	53	14	20	8	50	31	55	46	28	40	58	0	1
(243)	15	36	12	51	37	25	7	34	10	19	40	28	55	46	58	31	0	1
(123)	28	55	31	40	50	14	20	53	29	8	51	15	36	57	45	12	0	1
(124)	29	8	20	53	10	34	25	7	19	37	12	36	57	15	51	45	0	1
(143)	36	57	45	12	55	31	40	58	46	28	53	29	8	50	14	20	0	1
(134)	37	10	34	25	15	51	45	12	36	57	14	50	29	8	20	53	0	1
(142)	50	29	53	14	28	40	58	31	55	46	25	37	10	19	7	34	0	1
(143)	55	46	58	31	36	12	51	45	57	15	34	10	19	37	25	7	0	1
(12)(34)	19	37	25	7	46	58	31	40	28	55	45	57	15	36	12	51	0	0
(13)(24)	46	28	40	58	19	7	34	25	37	10	20	8	50	29	53	14	0	0
(14)(23)	57	15	51	45	8	20	53	14	50	29	7	19	37	10	34	25	0	0
(1234)	25	7	19	37	14	50	29	8	20	53	15	51	45	12	36	57	0	0
(1243)	31	40	28	55	34	10	19	37	25	7	36	12	51	45	57	15	0	0
(1324)	45	12	36	57	7	19	37	10	34	25	8	20	53	14	50	29	0	0
(1342)	34	25	37	10	31	55	46	28	40	58	29	53	14	20	8	50	0	0
(1432)	51	45	57	15	40	28	55	46	58	31	37	25	7	34	10	19	0	0
(1423)	58	31	55	46	20	8	50	29	53	14	19	7	34	25	37	10	0	0
$\lvert G_s\rvert$	1	1	1	1	1	1	1	1	1	1	1	1	1	1	1	1		120

7

Symmetry Studies of Curvature Data

7.1 Introduction

In this chapter we will discuss the application of symmetry methods to the study of curvature data, such as that obtained from the anterior surface of the (human) cornea. The objective is illustrating the effect that the symmetries, imposed by certain geometric and optic constraints, have in the underlying probability laws and parametric structures for the curvature data. Once these symmetries are identified, the corresponding canonical decompositions can be applied to reduce the data, and the invariance properties for the underlying probability laws can be proposed and assessed. Finally, a classification of the resulting types of curvature profiles can be obtained.

In contrast with Chapter 6, the structured data of interest here are, in one case, the mapping space in itself ($\mathcal{V} = C^L$), whereas in that chapter C^L represented the labels for the frequency data. Specifically, in this chapter, we assume that C is a finite set of real scalars, $L = \{1, \ldots, \ell\}$, and embed the mappings $y \in C^L$ into mappings (using the same notation)

$$y : k \mapsto y_k[\cos(k\phi), \sin(k\phi)] \in \mathbb{R}^2, \quad \phi = 2\pi/\ell, \tag{7.1}$$

so that the data y are indexed by

$$V = \{[\cos(k\phi), \sin(k\phi)] : k = 1, \ldots, \ell\} \subset \mathbb{R}^2. \tag{7.2}$$

The nature and interpretations of V, in each application, will suggest which group of symmetries is suitable for acting on V. In particular, we remark that because $V \subset \mathbb{R}^2$, any linear representation in dimension of 2 acting on V is a potential candidate for reducing the curvature data. The restrictions on that choice are discussed in the next section.

7.2 Keratometry

Keratometry is the measurement of corneal curvature, usually restricted to a small region of the anterior surface of the cornea. The measurements are obtained using a sample of reflected points of light along an annulus 3 to 4 mm. in diameter, centered about the apex of the cornea. Computerized keratometry uses the relative separation of reflected points of light along concentric rings to calculate the curvature of the measured surface. Using several scaled copies of V as a set of concentric light-reflecting rings and sampling at specific circularly equidistant intervals, a numerical model for the curvature of the measured surface can be obtained and estimated. In this case, the observed curvature data are indexed by intersections

$$V_i = \{d_i[\cos(k\phi), \sin(k\phi)] : k = 1, \ldots, \ell\}, \quad 1.5\ mm \le d_i \le 2mm, \quad i = 1, \ldots m, \tag{7.3}$$

of m rings and ℓ equally-spaced semimeridians. In ophthalmic geometric optics it is of interest to determine the angular variation between the extreme corneal curvature values along a given ring. The difference between the steep (maximum) and the flat (minimum) curvatures, as well as the angular variation between these extremes across rings, are related to the amount of regular astigmatism present in the optics of the eye. The different curvatures of these various refractive surfaces diffuse light rays and interfere with a sharp formation of the image on the retina. The concentric rings in L represent idealized locations in the anterior surface of the cornea where curvature measurements are obtained.

7.3 Astigmatic and Stigmatic Constraints

The simplest geometrical representation of a corneal curvature mapping is that derived from a spherical-cylindrical surface with the location of steep (κ_s) and flat (κ_f) curvatures oriented with a 90° angular separation. This is simply Euler Theorem of classical differential geometry. The resulting refractive profile,

$$\pi(\theta) = (\eta - \eta')[\kappa_s \cos^2(\theta - \alpha) + \kappa_f \sin^2(\theta - \alpha)], \quad 0 \le \theta \le 2\pi, \quad 0 \le \alpha \le \pi, \tag{7.4}$$

at each ring V can be expressed as $\pi(\theta) = s + c\cos^2(\theta - \alpha)$, where $s = (\eta - \eta')\kappa_f$, $c = (\eta - \eta')(\kappa_s - \kappa_f)$ and α are respectively the spherical, cylindrical, and axial (or reference angle for the $\{k_s, k_f\}$ orthogonal directions) components of the spherocylindrical corrective element, and $\{\eta', \eta\}$ are refractive indices. The notation $\{s, c, \alpha\}$ is commonly used in refractive references and mantained here in this specific context. For example, Figure 7.1 shows the refractive profiles indicated by R_{20}, with $(s, c, \alpha) = (4.25, 1.5, 20°)$, and by R_{60}, with $(s, c, \alpha) = (-2.75, 1, 60°)$.

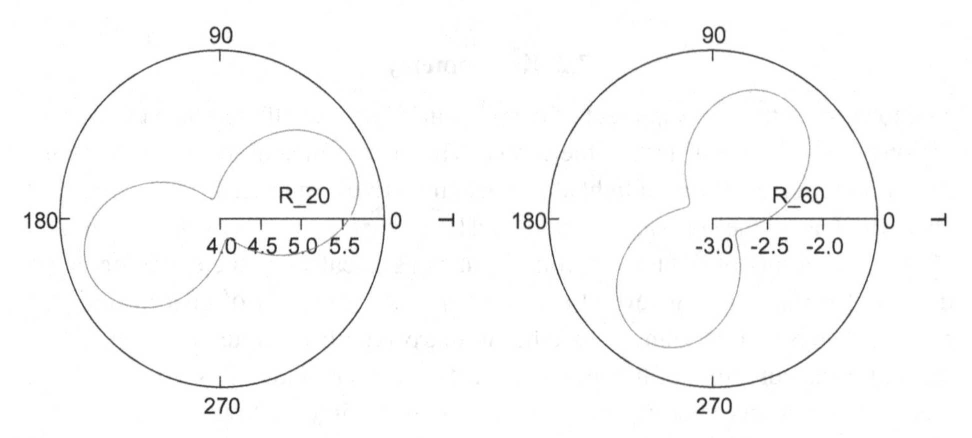

Figure 7.1: Refractive profiles R_{20}, with $(s, c, \alpha) = (4.25, 1.5, 20°)$, and R_{60}, with $(s, c, \alpha) = (-2.75, 1, 60°)$.

Probability laws and optical constraints

First note that given two components y_m and y_n of a refractive profile y (7.1), then their angular variation is given by $0 \leq |m - n|)\phi \leq \pi$.

We say that a probability law $\mathcal{L}(y)$ of y satisfies the astigmatic property when the mean angular variation between two order statistics is 90° if and only if these are the extreme order statistics. In contrast, a spherical ($c = 0$) surface would lead to a constant mean curvature and, in particular, the mean angular variation between any two ordered curvatures should be a constant. We say that a probability law $\mathcal{L}(y)$ satisfies the stigmatic property when the mean angular variation between any two order statistics is functionally independent of these order statistics.

7.4 Ranking Permutations

Implicit in the above definitions is the understanding that the mean angular variation in the astigmatic case is only approximately 90°. The limit mean angular variation, as $\ell \to \infty$, should reach the right-angle value. Because the components of y are in a totally ordered set, it is possible to order these components according to the specified total order relation ($\leq$) in C. We indicate by $\overline{y}$ the ordered version of y and write

$$\overline{\tau} = \{y \in \mathcal{V}; y\tau^{-1} = \overline{y}\} \tag{7.5}$$

to indicate the set of all mappings (or vectors) y in $\mathcal{V}$ which are ordered by the permutation τ^{-1}, that is,

$$y_{\tau^{-1}1} \leq y_{\tau^{-1}2} \leq \ldots \leq y_{\tau^{-1}\ell}.$$

In particular, $\overline{1} = \{y \in \mathcal{V}; y = \overline{y}\}$ is the set of all ordered mappings and, clearly, $\overline{\tau} = \overline{1}\tau^{-1}$, for all τ in S_ℓ. Because of the random nature of y, the permutations involved in the ordering of the components of y are also random. We refer to these random permutations as ranking permutations. Note that when C is a finite set then $\overline{\tau}$ is obviously measurable with respect to $\mathcal{L}(y)$, for all $\tau \in S_\ell$.

Partitions

If for any two distinct permutations τ and σ we have $P(\overline{\tau} \cap \overline{\sigma}) = 0$ with respect to $P \equiv \mathcal{L}(y)$, we say that $\mathcal{V} = \cup_{\tau \in S_\ell} \overline{\tau}$ is a stochastically disjoint partition of $\mathcal{V}$.

Clearly, if $y \in \overline{\tau} \cap \overline{\sigma}$ then both σ and τ are ranking permutations for y. The following proposition establishes a basic connection between the symmetries in the probability law $\mathcal{L}(y)$ of $y \in \mathcal{V}$ (describing the observable curvatures) and the probability law $\mathcal{L}(\tau)$ of the ranking permutations $\tau \in S_\ell$.

Proposition 7.1. *If $\mathcal{L}(y) = \mathcal{L}(y\tau^{-1})$ for all $\tau \in S_\ell$, and $\bigcup_\tau \overline{\tau}$ is a stochastically disjoint partition of $\mathcal{V}$, then $\mathcal{L}(\tau)$ is the uniform probability law in S_ℓ.*

Proof. Under the stated assumptions, $P = \mathcal{L}(y)$ induces the law $\pi = \mathcal{L}(\tau)$ in S_ℓ given by $\pi(\tau) = P(\overline{\tau})$, so that

$$1 = \sum_{\tau \in S_\ell} P(\overline{\tau}) = \sum_{\tau \in S_\ell} (\overline{1}\tau) = \ell! P(\overline{1}),$$

and consequently $\pi(\tau) = P(\overline{\tau}) = P(\overline{1}) = 1/\ell!$ for all $\tau \in S_\ell$, that is, $\mathcal{L}(\tau)$ is uniform in S_ℓ. ∎

Two-color topography

To illustrate, consider the simplest case in which $C = \{a, b\}$, such as with binary-colored topography mappings, and V is a single ring with two points only, say $V = \{1, 2\}$. Suppose also that $a \leq b$. The data space C^L has four points, namely, $C^L = \{aa,\ bb,\ ab,\ ba\}$, where S_2 acts according to $y\tau^{-1}$, giving three orbits

$$\mathcal{O}_{11} = \{aa\}, \quad \mathcal{O}_{12} = \{bb\}, \quad \mathcal{O}_{21} = \{ab,\ ba\}.$$

For each $\tau \in S_2 = \{1, (12)\}$, we apply the definition $\{y \in \mathcal{V};\ y_{\tau^{-1}1} \leq y_{\tau^{-1}2}\}$ of $\overline{\tau}$ to obtain

$$\overline{1} = \{aa,\ bb,\ ab\}, \quad \overline{(12)} = \{aa, bb, ba\} = \overline{1}(12).$$

Next we introduce the probability laws $P \equiv \mathcal{L}(y)$ satisfying the permutation-symmetry $\mathcal{L}(y\tau^{-1}) = \mathcal{L}(y)$ for all τ in S_ℓ. These laws (See Exercise 2.1) have the form of convex linear combinations

$$P = f_{11}w_{11} + f_{12}w_{12} + f_{21}w_{21},$$

where $w_i = 1/|\mathcal{O}_i|$ if $y \in \mathcal{O}_i$ and $w_i = 0$ otherwise, and $\sum_i f_i = 1$, $f_i \geq 0$, for $i \in \{11, 12, 21\}$. It then follows that

$$P(\overline{1}) = f_{11} + f_{12} + \frac{1}{2} f_{21} = P(\overline{(12)}),$$

and, because $f_{11} + f_{12} + f_{21} = 1$, the condition $P(\overline{1}) + P(\overline{(12)}) = 1$ is equivalent to $f_{21} = 1$, or $f_{11} = f_{12} = 0$. In this case, P induces a well-defined probability law π in S_2, given by $\pi(\tau) = P(\overline{\tau})$, which is also invariant, and hence uniform. Here we see that when $\mathcal{L}(y)$ is S_2-invariant then the law $\mathcal{L}(\tau)$ of the ranking permutations τ in S_2 is uniform if and only if $\mathcal{V} = \overline{1} \cup \overline{(12)}$ is a stochastic partition.

Three-level gray scale

Consider the case $c = \ell = 3$, and interpret the values in C as three distinct levels in a gray scale, say, and write these levels as $\{1, 2, 3\}$ to borrow their natural ordering. We observe a map $y = (y_1, y_2, y_3)$. In the resulting decomposition $\mathcal{V} = \mathcal{V}_{300} \cup \mathcal{V}_{210} \cup \mathcal{V}_{111}$ of $\mathcal{V}$, each component of $\mathcal{V}_{300} = \mathcal{O}_{11} \cup \mathcal{O}_{12} \cup \mathcal{O}_{13}$ is a single-mapping orbit, the components of $\mathcal{V}_{210} = \mathcal{O}_{21} \cup \ldots \cup \mathcal{O}_{26}$ are three-mapping orbits, and $\mathcal{V}_{111} = \mathcal{O}_{31}$ has six mappings in it.

Table (7.18) in Appendix A to this chapter summarizes the orbits and classes $\overline{\tau}$ generated by the action $y\tau^{-1}$. In each column, the boxed labels indicate the mappings defining the corresponding class $\overline{\tau}$. For example, the mapping number 3, $y = (3, 1, 1)$, is ordered by the permutations in the set $\{(13), (132)\}$, giving the mapping number 19, $y = (1, 1, 3)$. Their product, in general, however, is not an ordering permutation of the given mapping.

Note that each one of the 10 orbits has exactly 1 element from $\overline{1}$, a fact that characterizes $\overline{1}$ (and $\overline{\tau}$ in general) as cross sections in $\mathcal{V}$. More precisely, a subset $\Gamma \subset \mathcal{V}$ is a cross section in $\mathcal{V}$ if, for each $y \in \mathcal{V}$, $\Gamma \cap \mathcal{O}(y)$ consists of exactly 1 point (see, for example, Eaton (1989, p. 58) on conditions under which there is a stochastic representation of the form $\mathcal{L}(y) = \mathcal{L}(x\tau)$ for the law of y, where x is a random variable defined in $\overline{1}$ and independent of τ uniformly distributed in S_ℓ).

The invariant laws in $\mathcal{V}$ are convex combinations $P = \sum_i f_i w_i$, where $w_i = 1/|\mathcal{O}_i|$ for $y \in \mathcal{O}_i$ $w_i = 0$ otherwise, and $\sum_i f_i = 1$, $f_i \geq 0$, for $i = 1, \ldots, 10$. More precisely,

$$P = \sum_{i=1}^{3} f_{1i} w_{1i} + \sum_{j=1}^{6} f_{2j} w_{2j} + f_{31} w_{31},$$

where $w_{1i} = 1$ inside each orbit of type λ_{300} and $w_{1i} = 0$ elsewhere; $w_{2j} = 1/3$ inside each orbit of type λ_{210} and zero elsewhere; $w_{31} = 1/6$ inside the single orbit of type λ_{111} and zero elsewhere, and

$$\sum_{i=1}^{3} f_{1i} + \sum_{j=1}^{6} f_{2j} + f_{31} = 1.$$

As a consequence, we obtain

$$P(\overline{\tau}) = \sum_{i=1}^{3} f_{1i} + \frac{1}{3}\sum_{j=1}^{6} f_{2j} + \frac{1}{6} f_{31}, \quad \text{for all } \tau \in S_3. \tag{7.6}$$

Similarly to the previous example, the condition $\sum_{\tau \in S_3} P(\overline{\tau}) = 1$ is obtained (and hence the law of the ranking permutations is uniform in S_3) if and only if $\mathcal{V} = \cup_{\tau \in S_3} \overline{\tau}$ is a stochastic partition.

Expression (7.6) reflects the fact that the space $\mathcal{V}$ decomposes as the sum of three orbits of size 1, corresponding to frame $3^1 0^2$, six orbits of size 3 corresponding to frame $2^1 1^1 0^1$ and one orbit of size 6 corresponding to frame 1^3, so that $|\mathcal{V}| = 27 = 3 \times 1 + 6 \times 3 + 1 \times 6$. Combinatorial results discussed earlier on page 52 show that, in general, there are n_λ orbits of size m_λ corresponding to frame λ, with $|\mathcal{V}| = c^\ell = \sum_\lambda m_\lambda n_\lambda$, where

$$m_\lambda = \frac{\ell!}{(a_1!)^{m_1}(a_2!)^{m_2}\ldots(a_k!)^{m_k}}, \quad n_\lambda = \frac{c!}{m_1! m_2! \ldots m_k!},$$

so that (7.6) extends to

$$P(\overline{\tau}) = \sum_\lambda \frac{1}{m_\lambda} \sum_{j=1}^{n_\lambda} f_{j_\lambda},$$

where λ varies over the (m) different frames $\lambda = a_1^{m_1} \ldots a_k^{m_k}$, with $m_1 a_1 + \cdots + m_k a_k = \ell$ and $m_1 + \cdots + m_k = c$.

Consequently we see that given a permutation invariant probability law $\mathcal{L}(y)$, then the law $\mathcal{L}(\tau)$ of the ranking permutations τ in S_ℓ is uniform if and only if $\mathcal{V} = \cup_{\tau \in S_\ell} \overline{\tau}$ is a stochastic partition.

7.5 Classification of Astigmatic Mappings

The argument introduced in the previous section can be extended to determine the symmetries that are consistent with a given contraint, such as those imposed by the astigmatic properties of maximum and minimum. Similarly to the simple ordering case, the subsets $\overline{\tau} \subset \mathcal{V}$ of those mappings turned astigmatic by the permutation $\tau \in S_4$ could then be determined, leading to a table similar to (7.18).

Here, we take an alternative approach, illustrated as follows: Consider the case in which y has $\ell = 8$ components and $C = \{a, b, c\}$ is ordered according to $a \leq b \leq c$, so that $\mathcal{V}$ has $3^8 = 6{,}561$ curvature mappings. A mapping in $\mathcal{V}$ consistent

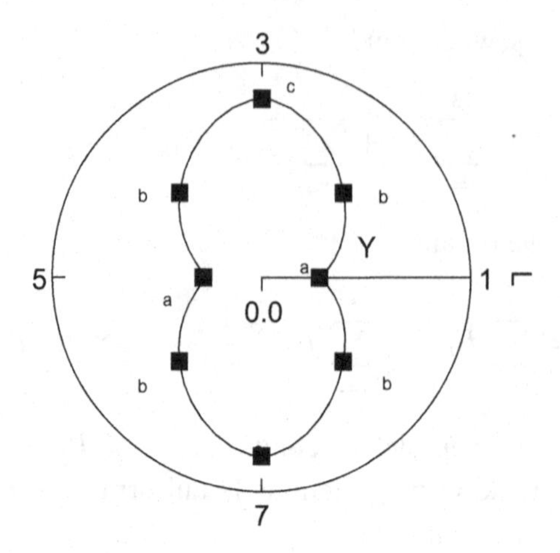

Figure 7.2: An astigmatic curvature mapping with the symmetry of K_4.

with the astigmatic property, given by Figure 7.2, is defined by the constraints

$$y_1 = y_5 = a; \quad y_2 = y_4 = y_6 = y_8 = b, \quad y_3 = y_7 = c, \quad a < b < c. \tag{7.7}$$

It belongs to the partition $\lambda = 422$ in an orbit of size 420 (see Exercise 7.10), whereas the other two orbits are defined by $a^4b^2c^2$ and $c^4a^2b^2$. With the notation introduced in (7.3), the domain of these mappings is a single ring ($m = 1$) with its components 45° apart ($\ell = 8$), specifically,

$$y : k \mapsto y_k[\cos(k\phi), \sin(k\phi)] \in \mathbb{R}^2, \quad \phi = 2\pi/8,$$
$$V = \{[\cos(k\phi), \sin(k\phi)] : k = 1, \ldots, 8\} \subset \mathbb{R}^2.$$

The astigmatic restrictions imposed on $y \in \mathcal{V}$, in addition to those suggested by the circular shape of V, bring down the number of subgroups of S_8 that consistently act on $\mathcal{V}$. Clearly, the resulting data summaries, classification, and inference are also determined by those experimental choices of symmetries. This aspect of choice is true to the nature of a broad symmetry study like this, and this is why it is emphasized here.

Specifically, the astigmatic restrictions given by (7.7) on y determine the automorphism group

$$\text{Aut}(Y) =< 1, (15)(24)(68), (28)(37)(46), (15)(26)(37)(48) > \equiv < 1, v, h, o >$$

acting on V, which is simply the Abelian group $K_4 = \{1, v, h, o\} \simeq C_2 \times C_2$ introduced earlier on in Chapter 1. Table (7.8) describes the action $y\tau^{-1}$ of K_4 on V, where we rearranged the indices of y in the order of $\{1, 5, 3, 7, 2, 4, 6, 8\}$ to highlight the three permutation orbits.

$$
\begin{array}{c|cccccccc}
\tau & 1 & 5 & 3 & 7 & 2 & 4 & 6 & 8 \\
\hline
1 & 1 & 5 & 3 & 7 & 2 & 4 & 6 & 8 \\
v & 5 & 1 & 3 & 7 & 4 & 2 & 8 & 6 \\
h & 1 & 5 & 7 & 3 & 8 & 6 & 4 & 2 \\
o & 5 & 1 & 7 & 3 & 6 & 8 & 2 & 4
\end{array}
\tag{7.8}
$$

We remark that the symmetries in K_4 are necessary to sustain the astigmatic restrictions defined by this choice of y.

It then follows that $\mathcal{P}_1 = \text{Diag}(\mathcal{A}_2, \mathcal{A}_2, \mathcal{A}_2 \otimes \mathcal{A}_2)$, $\mathcal{P}_2 = \text{Diag}(0, \mathcal{Q}_2, \mathcal{Q}_2 \otimes \mathcal{A}_2)$, $\mathcal{P}_3 = \text{Diag}(\mathcal{Q}_2, 0, \mathcal{Q}_2 \otimes \mathcal{Q}_2)$, and $\mathcal{P}_4 = \text{Diag}(0, 0, \mathcal{A}_2 \otimes \mathcal{Q}_2)$ are the resulting canonical projections, and, writing $y' = (a_1, a_2, c_1, c_2, b_1, b_2, b_3, b_4)$ to indicate a generic point in $\mathcal{V}$, that

i	$y'\mathcal{P}_i y$
1	$(a_1 + a_2)^2/2 + (c_1 + c_2)^2/2 + (b_1 + b_2 + b_3 + b_4)^2/4$
2	$(c_1 - c_2)^2/2 + (b_1 + b_2 - b_3 - b_4)^2/4$
3	$(a_1 - a_2)^2/2 + (b_1 + b_4 - b_3 - b_2)^2/4$
4	$(b_1 + b_3 - b_4 - b_2)^2/4$
total	$y'y$

is the corresponding decomposition of $y'y$. The hypotheses (H) that y satisfies the astigmatic property and (H') that $y\tau^{-1} = y$ for all $\tau \in K_4$ are related by

$$H : y \text{ satisfies the astigmatic property} \implies H' : y\tau^{-1} = y \text{ for all } \tau \in K_4 \implies y'\mathcal{P}_2 y = y'\mathcal{P}_3 y = y'\mathcal{P}_4 y = 0,$$

and provide a basis for classification and testing the astigmatic mappings in $\mathcal{V}$.

The distribution of the $y'\mathcal{P}y$ components

To study the distribution of the components $y'\mathcal{P}y$, it is convenient, for computational purposes, to work with the base 3 representation of each mapping. Therefore, we set $C = \{0, 1, 2\}$. Next, to each integer number between 0 and 6,560 (recall that $|\mathcal{V}| = 6{,}561$), the simple Maple© procedure

```
T:=proc(n) g:=(convert(n,base,3)):J:=[0,0,0,0,0,0,0,0]:
 for i from 1 to nops(g) do J[i]:=op(i,g) end do:
 matrix(8,1,J): end:
```

converts the mapping's identifying label into a point $y \in C^8$ by inserting the necessary trailing zeros to the fixed dimension arrays. For example,

$$T(27)' = \begin{bmatrix} 0 & 0 & 0 & 1 & 0 & 0 & 0 & 0 \end{bmatrix}, \quad T(3936)' = \begin{bmatrix} 0 & 1 & 2 & 1 & 0 & 1 & 2 & 1 \end{bmatrix}.$$

The original mappings, which have images in $\{1, 2, 3\}$, are then obtained after adding 1 unit to each component of the expansion. Note that $T(3936)$ corresponds to the astigmatic mapping shown in Figure 7.2, in the original basis $\{1, 2, \ldots, 8\}$. Finally, the components $y'\mathcal{P}_i y$ are evaluated for each (rearranged) y and each $i = 1, 2, 3, 4$.

Table (7.10) shows the joint frequency distribution $\mathcal{L}_{23}$ of $y'\mathcal{P}_2 y$ and $y'\mathcal{P}_3 y$, whereas Table (7.11) shows the joint frequency distribution $\mathcal{L}_{24}$ of $y'\mathcal{P}_2 y$ and $y'\mathcal{P}_4 y$. Note that the two marginal frequency distributions $\mathcal{L}_2$ and $\mathcal{L}_3$ of $\mathcal{L}_{23}$ coincide. All values of $y'\mathcal{P}y$ were evaluated as nonnegative integers by multiplying all projections by 4.

Computer sorting of the 6,561 mappings by increasing values of $y'\mathcal{Q}_2 y$, $y'\mathcal{Q}_3 y$ and $y'\mathcal{Q}_4 y$ identified the 26 mappings in which all three components vanish, that is, the mappings with the symmetry of K_4. These mappings, indicated by $s_0, \ldots s_{25}$ are described in Appendix B. The graphic display of mappings $s_1, \ldots, s_{24}$ is shown in Appendix C to this chapter.

The astigmatic mapping that we started with, on Figure 7.2, is displayed as mapping s_{15}.

The likelihood of an astigmatic mapping

If a mapping is selected at random according to a uniform probability in $\mathcal{V}$ and only the values of the three components $y'\mathcal{P}_2 y$, $y'\mathcal{P}_3 y$, and $y'\mathcal{P}_4 y$ are known, then H' would be rejected for all but the 26 mappings in which all three components are zero. One would eventually see the hypothesis H', and hence H, being rejected with probability $1 - 26/6,561 > 0.99$.

If, otherwise, H' is accepted, we know that there is at least one astigmatic mapping with the symmetry of K_4. Are there any other ones among the remaining 25 mappings?

Equivalently, what are the mappings in $\mathcal{V}$ that are consistent with the astigmatic contours given by (7.4), as equations $u \cos^2(k\phi - \alpha) + v \sin^2(k\phi - \alpha)], \phi = 2\pi/8$, $k = 0, \ldots, 7$, $0 \leq \alpha \leq \pi$ in u, v?. The solutions, for $alpha = 0$, are listed in Table (7.9), along with their derived spherical $s = u$ and cylindrical $c = v - u$ coefficients.

mapping	u	v	spherical (s)	cylindrical (c)	α
0	1	1	1	0	0
11	3	1	3	−2	0
13	2	2	2	0	0
15	1	3	1	2	0
25	3	3	3	0	0

(7.9)

In addition to mapping 15, we identified mapping 11: [2 1 0 1 2 1 0 1], which is simply mapping 15, [0 1 2 1 0 1 2 1], rotated by 90°, and the three spherical ones: mappings 0, 13, and 25. Note that the spherical mappings 0 and 25 are similar to mapping 13 and are not displayed in Appendix B. The mappings

$$1312 : \left[2\,3\,2\,1\,2\,3\,2\,1\right] \quad 5248 : \left[2\,1\,2\,3\,2\,1\,2\,3\right]$$

are the two astigmatic companions to mappings 11 and 13 at the directions $\alpha = 45^\circ$ and $\alpha = 135^\circ$. These mappings are identified in Appendix B, along with the isomorphic K_4-symmetric mappigs at those directions.

The odds on correctly accepting the astigmatic hypothesis H having the hypothesis H' of K_4 symmetry as the criterion are then limited to only $5 : 21$ at each direction. This reduced power, we must recall, is mostly a consequence of the counting (uniform) probability distribution assumed for the mapping space.

$\mathcal{L}_{23}$	0	1	2	3	4	6	8	9	11	12	16	17	18	24
0	80	0	108	0	72	96	54	0	0	48	18	0	24	12
1	0	216	0	288	0	0	0	216	96	0	0	48	0	0
2	108	0	144	0	96	128	72	0	0	64	24	0	32	16
3	0	288	0	384	0	0	0	288	128	0	0	64	0	0
4	72	0	96	0	108	144	48	0	0	72	0	0	0	0
6	96	0	128	0	144	192	64	0	0	96	0	0	0	0
8	54	0	72	0	48	64	36	0	0	32	12	0	16	8
9	0	216	0	288	0	0	0	192	64	0	0	32	0	0
11	0	96	0	128	0	0	0	64	0	0	0	0	0	0
12	48	0	64	0	72	96	32	0	0	48	0	0	0	0
16	18	0	24	0	0	0	12	0	0	0	0	0	0	0
17	0	48	0	64	0	0	0	32	0	0	0	0	0	0
18	24	0	32	0	0	0	16	0	0	0	0	0	0	0
24	12	0	16	0	0	0	8	0	0	0	0	0	0	0

(7.10)

$\mathcal{L}_{24}$	0	1	2	3	4	6	8	9	11	12	16	17	18	24
0	242	0	324	0	216	288	162	0	0	144	54	0	72	36
1	0	648	0	864	0	0	0	648	288	0	0	144	0	0
4	216	0	288	0	324	432	144	0	0	216	0	0	0	0
9	0	216	0	288	0	0	0	144	0	0	0	0	0	0
16	54	0	72	0	0	0	36	0	0	0	0	0	0	0

(7.11)

7.6 Astigmatic Covariance Structures

The results discussed in the previous section have the implication that in the search for astigmatic laws we need to restrict the symmetries of interest to proper subgroups of the full symmetric group.

This same principle applies to the determination of mean and covariance structures that are necessary for meeting the astigmatic restrictions. To illustrate the construction of these parametric structures, we apply the techniques introduced in Section 3.5 on page 70 with a generic point in $y \in \mathbb{R}^8$, a (real) symmetric 8×8 matrix Λ, and the linear representation ρ of K_4 on $\mathbb{R}^8$ given by action (7.8). Direct evaluation of

$$\Sigma = \sum_{\tau \in K_4} \rho_\tau \Lambda \rho_{\tau^{-1}}$$

gives the centralized version

$$\Sigma = \frac{1}{4}\left[\begin{array}{cc|cc|cccc} A & B & C & C & D & E & E & D \\ B & A & C & C & E & D & D & E \\ \hline C & C & F & G & H & H & I & I \\ C & C & G & F & I & I & H & H \\ \hline D & E & H & I & J & K & L & M \\ E & D & H & I & K & J & M & L \\ E & D & I & H & L & M & J & K \\ D & E & I & H & M & L & K & J \end{array}\right],$$

A	$2\,\Lambda_{1,1} + 2\,\Lambda_{5,5}$
B	$4\,\Lambda_{1,5}$
C	$\Lambda_{1,3} + \Lambda_{3,5} + \Lambda_{1,7} + \Lambda_{5,7}$
D	$\Lambda_{1,2} + \Lambda_{4,5} + \Lambda_{1,8} + \Lambda_{5,6}$
E	$\Lambda_{1,4} + \Lambda_{2,5} + \Lambda_{1,6} + \Lambda_{5,8}$
F	$2\,\Lambda_{3,3} + 2\,\Lambda_{7,7}$
G	$4\,\Lambda_{3,7}$
H	$\Lambda_{2,3} + \Lambda_{3,4} + \Lambda_{7,8} + \Lambda_{6,7}$
I	$\Lambda_{2,7} + \Lambda_{4,7} + \Lambda_{3,8} + \Lambda_{3,6}$
J	$\Lambda_{2,2} + \Lambda_{4,4} + \Lambda_{8,8} + \Lambda_{6,6}$
K	$2\,\Lambda_{2,4} + 2\,\Lambda_{6,8}$
L	$2\,\Lambda_{2,6} + 2\,\Lambda_{4,8}$
M	$2\,\Lambda_{2,8} + 2\,\Lambda_{4,6}$

of Λ.

7.7 Dihedral Fourier Analysis

The dihedral indexing introduced on page 43 has its justification from the Fourier-inverse formula described by Theorem 4.5, on page 118. Generic derivations of the Fourier transform for data indexed by D_3 and D_4 appeared on page 120. Similarly, in optics, the one-to-one correspondence between a coherency matrix J and the Stokes coefficients $S_0, \ldots, S_3$ that appear in its resolution $J = S_0\sigma_0 + S_1\sigma_1 + S_2\sigma_2 + S_3\sigma_3$ in terms of the Pauli matrices $\sigma_0, \ldots, \sigma_3$, is intrinsic to the definition

$$S_j = \frac{1}{2}\text{Tr}\ J\sigma_j \tag{7.12}$$

of the Stokes coefficients (O'Neill, 1963, p. 152). The uniquiness follows from the fact that $\mathrm{Tr}\, \sigma_i \sigma_j = 2\delta_{ij}$, so that the resolution

$$M = \begin{bmatrix} a & b \\ c & d \end{bmatrix} = \sum_{j=0}^{3} S_j \sigma_j = S_0 \begin{bmatrix} 1 & 0 \\ 0 & 1 \end{bmatrix} + S_1 \begin{bmatrix} 0 & 1 \\ 1 & 0 \end{bmatrix} + S_2 \begin{bmatrix} 0 & -i \\ i & 0 \end{bmatrix} + S_3 \begin{bmatrix} 1 & 0 \\ 0 & -1 \end{bmatrix}$$

of an arbitrary real or complex matrix M is obtained with $S_0 = (a + d)/2$, $S_1 = (b + c)/2$, $S_2 = i(b - c)/2$ and $S_3 = (a - d)/2$. See also, for example, Silverman (1995, p. 140) and Streater and Wightman (1964, p. 12).

We conclude this chapter with an application of the dihedral Fourier analysis to curvature data, showing its effectiveness in reproducing well-known quantities commonly used in linear optics. These techniques will be given continuation in the next chapter, in which we study the data-analytic aspects of handedness.

Rotations and reversals in canonical space

The dihedral groups in the plane were introduced earlier on page 42. The underlying planar rotations in $\mathbb{R}^2$ are in the counterclockwise direction, and the reversals are defined as the horizontal reflection followed by a rotation. It is remarked that the planar horizontal reflection, in the case of a rigid body, requires its realization as the 180° rotation in $\mathbb{R}^3$ along the x-axis (thus changing the polarity of the z axis). In canonical space, rotations are given by $z \mapsto wz$ and reversals by $z \mapsto w\bar{z}$. In particular, when $n = 4$, the correspondence with the Pauli matrices is given by $\beta_{k,d} = (i\sigma_3)^k \sigma_1^{(1-d)/2}$. The complete set of two-dimensional irreducible representations of D_n is obtained by defining, in (2.13), $\omega = e^{2\pi i \ell/n}$ for $\ell = 1, \ldots, m$, where $m = (n-1)/2$ for n odd and $m = n/2 - 1$ for n even. In the present application we concentrate in the case $\ell = 1$.

Adding to the language of optical instruments, $\beta_{k,d}$ are compensators introducing *mirror-image* relative phase differences, in the sense that if $z = e^{i\theta}$, then $\beta_{k,d}$ induces a phase difference of $2(k\phi + d\theta)$ between the two components of z. This two-sided view of a phase shift is a realization in canonical two-space of the polarity induced in the z axis under the 180° rotation along the x axis described above. Also note that the sign of index d in the operator $\beta_{k,d}$ is passed to its operand θ.

The dihedral Fourier coefficients

Given any 2×2 real matrix $M = (m_{ij})$, let $\mathcal{M}$ indicate its conjugate XMX^{-1} in canonical space. Specifically,

$$\mathcal{M} = \begin{bmatrix} f_+ & f_- \\ \overline{f}_- & \overline{f}_+ \end{bmatrix}, \quad \text{with } f_d = \frac{1}{2}[m_{11} + dm_{22} + i(m_{21} - dm_{12})], \quad d = \pm 1.$$

Let $x_{k,d} = \text{Tr}\,[\beta_{-k,d}\mathcal{M}]/n$ be the dihedral indexing introduced earlier on page 43, so that conversely, from Theorem 4.5, $\mathcal{M} = \sum_{k,d} x_{k,d}\beta_{k,d}$. That is,

$$\mathcal{M} = \sum_{k,d} x_{k,d}\beta_{k,d} \iff x_{k,d} = \frac{1}{n}\text{Tr}\,[\beta_{-k,d}\mathcal{M}], \quad k = 1, \ldots, n, \quad d = \pm 1.$$

Moreover, $x_{k,d} = 2\Re f_d \omega^{-k}/n = 2[\cos(k\phi)\Re f_d + \sin(k\phi)\Im f_d]/n$, where $\Re$ and $\Im$ indicate the real and complex parts of a complex number, respectively. In summary, then,

$$\mathcal{M} = \sum_{k,d} x_{k,d}\beta_{k,d} \iff x_{k,d} = \frac{2}{n}[\cos(k\phi)\Re f_d + \sin(k\phi)\Im f_d],$$
$$d = \pm 1, \; k = 1, \ldots, n. \tag{7.13}$$

Phase correlations. Direct evaluation shows that, for $d = \pm 1, \overline{x_d} = \sum_k x_{k,d}/n = 0$ and

$$< x_d, \gamma^\ell x_d > = \frac{1}{n}\sum_k x_{k,d}x_{k-\ell,d} = 2[(\Re f_d)^2 + (\Im f_d)^2]\cos(\ell\phi),$$
$$\ell = 0, \ldots, n-1. \tag{7.14}$$

In particular, $||\gamma^\ell x_d||^2 = < x_d, x_d > = 2[(\Re f_d)^2 + (\Im f_d)^2]$ for $\ell = 0, \ldots n-1$, so that

$$\frac{< x_d, \gamma^\ell x_d >}{< x_d, x_d >< \gamma^\ell x_d, \gamma^\ell x_d >} = \cos(\ell\phi), \quad d = \pm 1,$$

is the phase correlation between x_d and its lag ℓ companion vector $\gamma^\ell x_d$. Writing $F_d' = (\Re f_d, \Im f_d)$, the (lag 0) correlation between rotations and reversals is

$$\frac{< F_+, F_- >}{||F_+||\,||F_-||}.$$

For $\ell > 0$, the lagged correlations between rotations and reversals can be obtained with similar derivations. If the matrix M is complex, the decomposition of $\mathcal{M}$ is obtained from adjoining the decompositions of the real and imaginary matrix parts of M.

Dihedral Fourier analysis of refractive profiles

The refractive power matrix of a spherocylindrical lens is given by

$$F = \begin{bmatrix} s + c\sin^2(\alpha) & -c\sin(2\alpha)/2 \\ -c\sin(2\alpha)/2 & s + c\cos^2(\alpha) \end{bmatrix} = \begin{bmatrix} S - C_+ & -C_x \\ -C_x & S + S_+ \end{bmatrix},$$

where the scalars (s, c, α) indicate, respectively, the sphere, cylinder, and axis. The RHS notation is from Campbell (1997) and Campbell (1994), in which $S =$

$s + c/2$, $C_+ = (c/2) + \cos(2\alpha)$, $C_x = (c/2)\sin(2\alpha)$. The observable data are the scalars (s, c, α), respectively the sphere, cylinder, and axis.

The matrix F is expressed as

$$XFX^{-1} = \begin{bmatrix} M & C_0 + iC_{45} \\ C_0 - iC_{45} & M \end{bmatrix}$$

in canonical space, thus identifying the coefficients $M = s + c/2$, $C_0 = -c\cos(2\alpha)/2$ and $C_{45} = -c\sin(2\alpha)/2$ appearing in W.E. Humphrey's principle of astigmatic decomposition (Humphrey, 1976). See also Thibos et al. (1994). It then follows that

$$F = \frac{2}{n}\sum_k \{M\cos(k\phi)\beta_{k,+} + [C_0\cos(k\phi) + C_{45}\sin(k\phi)]\beta_{k,-}\} \tag{7.15}$$

is the resolution of F. Its rotation-reversal signature is defined by the ensemble of n points

$$k \mapsto \quad (M\cos(k\phi), C_0\cos(k\phi) + C_{45}\sin(k\phi)) \subset \mathbb{R}^2, \quad k = 1, \ldots n, \tag{7.16}$$

parameterized by Humphrey's coefficients. Consequently, in analogy with the results described on page 191, we have shown that the classic components

$$\{s + c\sin(\alpha)^2, \ -c\sin(2\alpha)/2\} = \{S - C_+, \ -C_X\}$$

of linear optics are precisely the canonical invariants relative to the data indexed by the dihedral rotations and reversals, that is, the coefficients of $\beta_{k,d}$ in (7.15).

Dihedral Fourier analysis-related applications

Rotators. For the (counterclockwise) rotator $R(\theta)$, it follows that $f_+ = e^{i\theta}$ and $f_- = 0$, and hence $x_{k,+} = 2\cos(\theta - k\phi)/n$, $x_{k,-} = 0$. Consequently, in canonical space,

$$R(\theta) = \frac{2}{n}\sum_k \cos(\theta - k\phi)\beta_{k,+}.$$

Since rotators in canonical space have the form of β_+, only those components appear in the decomposition.

Polarizers. For the polarizer $P(\alpha)$, the results are: $f_+ = 1/2$, $f_- = e^{2i\alpha}/2$, $x_{k,+} = \cos(k\phi)/n$, $x_{k,-} = \cos(2\alpha - k\phi)/n$ so that $P(\alpha)$ resolves in canonical space as

$$P(\alpha) = \frac{1}{n}\sum_k [\cos(k\phi)\beta_{k,+} + \cos(2\alpha - k\phi)\beta_{k,-}].$$

Its rotation-reversal signature is defined by the ensemble of n points $p_k(\alpha)' = (\cos(k\phi), \cos(2\alpha - k\phi))$ in $\mathbb{R}^2$. If u_k is observed with uncertain α, its value may then be numerically estimated as the value that minimizes, for example, $\sum_k ||p_k(\alpha) - u_k||^2$.

A Hamiltonian. Consider the Hamiltonian

$$\mathcal{H} = \begin{bmatrix} a_1^2 & 2a_1a_2e^{i\delta} \\ 2a_1a_2e^{-i\delta} & a_2^2 \end{bmatrix} = \begin{bmatrix} a_1^2 & 0 \\ 0 & a_2^2 \end{bmatrix} + 2a_1a_2 \begin{bmatrix} 0 & e^{i\delta} \\ e^{-i\delta} & 0 \end{bmatrix}$$

of a monochromatic plane wave represented as the superposition of linearly polarized basis states with real amplitudes a_1, a_2, and a relative phase δ. Transforming the first component to canonical space gives $f_+ = (a_1^2 + a_2^2)/2$ and $f_- = (a_1^2 - a_2^2)/2$. The second component gives $f_+ = 0$ and $f_- = 2a_1a_2e^{i\delta}$. Combining the like coefficients gives

$$x_{k,+} = \frac{1}{n}(a_1^2 + a_2^2)\cos(k\phi), \quad x_{k,-} = \frac{1}{n}(a_1^2 - a_2^2)\cos(k\phi) + 2a_1a_2\cos(k\phi - \delta),$$

in which $a_1^2 + a_2^2$ is the total light intensity, $2a_1a_2\cos(\delta)$ is the angular orientation of the elliptical motion traced out by the electric vector of the light wave, $2a_1a_2\sin(\delta)$ is the handedness or sense of circulation of the electric vector, and $a_1^2 - a_2^2$ is the eccentricity of the elliptical motion (Silverman, 1995, p. 181). The orientation and the handedness appear out of the term $\cos(k\phi - \delta)$ and the components $x_1 = a_1^2 - a_2^2$, $x_2 = i(a_1^2 + a_2^2)$, $x_3 = -2a_1a_2$ define an isotropic vector (Cartan, 1966, p. 51).

Further Reading

The linear and ophthalmic optics aspects of this chapter are detailed in Viana (2003), Lakshminarayanan and Viana (2005), and Viana and Lakshminarayanan (2006).

Patterned covariance matrices arise from a variety of contexts. Wilks (1946), in one of the early papers with patterned structure, considered a set of measurements on k equivalent psychological tests and proposed a covariance matrix with equal diagonal elements and equal off-diagonal elements. Votaw (1948) extended this model to a set of blocks in which each block had a pattern. Olkin (1973) studied a multivariate version in which each element was a matrix, and the blocks were patterned. The circular covariance matrix, carrying the cyclic symmetry of C_n, has a long history. There, typical measurements arise from a physical, spatial, or temporal model, as for example, measurements on the petals of a flower. For a discussion of this model, see Olkin and Press (1969) and Olkin (1973).

Some patterns arising from certain group operations will not have linear structure, and Diaconis (1989) provides a discussion of how such patterns can arise.

An algebraic synthesis of the possible classes of patterned matrices was obtained by Andersson (1975). See, also of related interest, the aspects of group symmetry covariance models described in Perlman (1987) and Andersson (1992).

The argument of testing for certain patterns of symmetry by averaging over a class of permutation matrices is applied, for example, by Gao and Marden (2001). In particular, the reader may consult Diaconis (1990) for specific aspects of group invariance applied to the characterization of patterned matrices.

The reader interested in the algebraic treatment of dioptric power and optical aberrations may want to consult Lakshminarayanan et al. (1998). For an earlier algebraic formulation of developmental vision, see Hoffman (1966).

Exercises

Exercise 7.1. Refer to Table (7.18) and extend the definition of $\overline{\tau}$ given by (7.5) on page 182 to construct an indexing $\{x_\tau\}_{\tau \in G}$ of data by the underlying group G. This can be done by by summarizing the data (originally indexed by V) within each set $\overline{\tau}$. For example, $\tau \mapsto \sum_{s \in \overline{\tau}} x_s$. Give examples.

Exercise 7.2. The uniform mean angular variation. Following with the definitions introduced in this chapter, assume that $\cup_{\tau \in S_\ell} \overline{\tau}$ is a stochastically disjoint partition of $\mathcal{V}$ and let τ be the ranking permutation in S_ℓ ordering the components of y, so that $\tau 1$ is the location of the flat (minimum) curvature and $\tau \ell$ the location of the steep (maximum) curvature. The angular displacement between the extreme curvatures may be defined as follows: fix a reference point $m \in \{1, 2, \ldots, \ell\}$ and an orientation $\sigma = (12 \ldots \ell)$ in S_ℓ for counting steps along V. If, in the direction determined by σ, it takes n steps to move from the point of flat curvature to the reference point and N steps from the point of steep curvature to the reference point, we then write $\sigma^n \tau 1 = m$ and $\sigma^N \tau \ell = m$. Consequently $n - N = \tau 1 - \tau \ell$, in the direction given by σ and $n - M = \tau \ell - \tau 1$, in the direction defined by σ^{-1}. This justifies the definition of the angular variation $\alpha(\tau)$ between the extreme values of y as $\alpha_{\ell 1}(\tau) = |\tau \ell - \tau 1| 2\pi/\ell$. Show that the factor $|\tau \ell - \tau 1|$ can be expressed as $d' \rho_\tau r$, where $d' = (-1, 0, \ldots, 0, 1)$, $r' = (1, 2, \ldots, \ell)$ and ρ is the permutation representation of S_ℓ, and that, consequently, $|\tau \ell - \tau 1|^2 = d' \rho_\tau (r r') \rho'_\tau d$.

Exercise 7.3. Following Problem 7.2, derive the (uniform) mean-squared angular variation by applying the result, from page 115, $\sum_{\tau \in S_\ell} \rho_\tau H \rho'_\tau / \ell! = v_0 \mathcal{A} + v_1 \mathcal{Q}$, where v_0 is the sum of the components of H, and $v_0 + (n-1) v_1 = \operatorname{Tr} H$ and, in this case, $H = r r'$.

Exercise 7.4. Average linear ranks. See also Section 4.5 on page 116. Show that the mean linear rank $R(y)$ of y, under a uniform law in S_ℓ, is given by $R(y) = \sum_{\overline{y}} \rho'_\tau r / |\overline{y}|$, where $r' = (1, 2, \ldots, \ell)$ and ρ is the permutation representation of

S_ℓ. For $y = (1, 1, 2)$, show that $\overline{y} = \{1, (12)\}$ and consequently

$$\frac{1}{2}[\rho(1)' + \rho((12))'] = \frac{1}{2}\begin{bmatrix} 1 & 1 & 0 \\ 1 & 1 & 0 \\ 0 & 0 & 2 \end{bmatrix}\begin{bmatrix} 1 \\ 2 \\ 3 \end{bmatrix} = \begin{bmatrix} 1.5 \\ 1.5 \\ 3 \end{bmatrix}.$$

For $y = (1, 2, 1)$, show that $\overline{y} = \{(23), (132)\}$, and

$$\frac{1}{2}[\rho((23))' + \rho((132))'] = \frac{1}{2}\begin{bmatrix} 1 & 1 & 0 \\ 0 & 0 & 2 \\ 1 & 1 & 0 \end{bmatrix}\begin{bmatrix} 1 \\ 2 \\ 3 \end{bmatrix} = \begin{bmatrix} 1.5 \\ 3 \\ 1.5 \end{bmatrix}.$$

Exercise 7.5. Curvature profiles in phase (or canonical) space. Show that (7.4) can be written as $w'_{\alpha,\theta} D_{\nu,\kappa} w_{\alpha,\theta}$, with

$$D_{\nu,\kappa} = (\nu - \nu')\begin{bmatrix} \kappa_s & 0 \\ 0 & \kappa_f \end{bmatrix}, \quad w'_{\alpha,\theta} = (\cos(\theta - \alpha), \sin(\theta - \alpha)),$$

so that the refractive profile transforms to phase space according to

$$X w'_{\alpha,\theta} = (\omega_{\theta-\alpha}, \omega_{-\theta+\alpha}) \equiv \omega_{\alpha,\theta}, \quad X = \begin{bmatrix} 1 & i \\ 1 & -i \end{bmatrix}, \quad \omega_\phi = e^{i\phi},$$

and

$$X D_{\nu,\kappa} X^{-1} = \frac{\nu - \nu'}{2}\begin{bmatrix} \kappa_s + \kappa_f & \kappa_s - \kappa_f \\ \kappa_s - \kappa_f & \kappa_s + \kappa_f \end{bmatrix} = \frac{1}{2}\begin{bmatrix} 2s + c & c \\ c & 2s + c \end{bmatrix} \equiv \Delta_{s,c},$$

where s and c are the spherical and cylindrical parameters. Consequently, in phase space, the refraction profile is given by $\pi_{\theta:s,c,\alpha} = \omega'_{\alpha,\theta}\Delta_{s,c}\omega_{\alpha,\theta}$. Conclude then that $\pi_{\theta:s,c,\alpha} = (2s + c)\cos(2(\theta - \alpha)) + c$.

Exercise 7.6. Joint profiles. Given two profiles $\pi_{\theta:s,c,\alpha_1}$ and $\pi_{\theta:s,c,\alpha_2}$ with common spherical and cylindrical components and varying orientations α_1, α_2, argue that the two profiles can be jointly described by

$$[\omega_{\alpha_1,\theta} \otimes \omega_{\alpha_2,\theta}]'(\Delta_{s,c} \otimes \Delta_{s,c})[\omega_{\alpha_1,\theta} \otimes \omega_{\alpha_2,\theta}] = \pi_{\theta:s,c,\alpha_1}\pi_{\theta:s,c,\alpha_2},$$

and that, similarly, two or more profiles can be jointly be described by their product $\prod_j \pi_{\theta:s,c,\alpha_j}$.

Exercise 7.7. The coefficients in the joint profile. Examine the coefficients appearing in a joint profile $\pi_{\theta:s,c,\alpha_1}\pi_{\theta:s,c,\alpha_2}$ with common spherical and cylindrical components and interpret the coefficients in c^2, s^2, and sc.

Exercise 7.8. Dihedral decompositions. In phase space, the dihedral (D_n) decomposition of $\Delta_{s,c}$ is given by

$$\Delta_{s,c} = \sum_{k,d} x_{k,d}\beta_{-k,d}, \quad k = 0, \ldots, n-1, \; d = \pm 1,$$

where $\beta_{k,d}$ are opposite direction compensators of phase shift $2\omega_k$, with $\omega_k \equiv 2k\pi/n$, $x_{k,1} = (2s + c)\cos(\omega_k)$ and $x_{k,-1} = c\sin(\omega_k)$. Following Exercise 7.5, obtain the dihedral decomposition of the single profile and show that the coefficients $x_{k,j}$ and $\Delta_{s,c}$ are in one-to-one correspondence via the dihedral Fourier formula from Theorem 4.5. How does the argument extend to joint profiles?

Exercise 7.9. Review Exercises 7.5–7.8 by separating out the direction parameter $alpha$ from the argument parameter θ. Show that then (7.4) can be written as $w'_\theta D_{\nu,\kappa,\alpha} w_\theta$, with

$$D_{\nu,\kappa,\alpha} = (\nu - \nu') \begin{bmatrix} \kappa_s \cos^2(\alpha) + \kappa_f \sin^2(\alpha) & \frac{1}{2}(\kappa_s - \kappa_f)\sin(2\alpha) \\ \frac{1}{2}(\kappa_s - \kappa_f)\sin(2\alpha) & \kappa_f \cos^2(\alpha) + \kappa_s \sin^2(\alpha) \end{bmatrix},$$

$$w'_\theta = (\cos(\theta), \sin(\theta)).$$

From (7.13), evaluate and interpret the dihedral Fourier coefficients of $D_{\nu,\kappa,\alpha}$.

Exercise 7.10. Show that the set V of all mappings $y : \{1, 2, \ldots, 8\} \mapsto \{a, b, c\}$ decomposes under the action $y\tau^{-1}$ of S_8 according to $|V| = \sum_\lambda m_\lambda n_\lambda$, where the components m_λ and n_λ are given by

λ	m_λ	n_λ	$m_\lambda n_\lambda$
800	1	3	3
710	8	6	48
620	28	6	168
611	56	3	168
530	56	6	336
521	168	6	1,008
440	70	3	210
431	280	6	1,680
422	420	3	1,260
332	560	3	1,680
total			6,561

(7.17)

Appendix A

$\mathcal{O}$	y_1	y_2	y_3	label	τ : 1	(12)	(13)	(23)	(123)	(132)
11	1	1	1	1	$\boxed{1}$	$\boxed{1}$	$\boxed{1}$	$\boxed{1}$	$\boxed{1}$	$\boxed{1}$
12	2	2	2	14	$\boxed{14}$	$\boxed{14}$	$\boxed{14}$	$\boxed{14}$	$\boxed{14}$	$\boxed{14}$
13	3	3	3	27	$\boxed{27}$	$\boxed{27}$	$\boxed{27}$	$\boxed{27}$	$\boxed{27}$	$\boxed{27}$
21	1	1	2	10	$\boxed{10}$	$\boxed{10}$	2	4	2	4
21	1	2	1	4	4	2	4	$\boxed{10}$	$\boxed{10}$	2
21	2	1	1	2	2	4	$\boxed{10}$	2	4	$\boxed{10}$
22	2	2	1	5	5	5	$\boxed{13}$	11	$\boxed{13}$	11
22	2	1	2	11	11	$\boxed{13}$	11	5	5	$\boxed{13}$
22	1	2	2	13	$\boxed{13}$	11	5	$\boxed{13}$	11	5
23	1	1	3	19	$\boxed{19}$	$\boxed{19}$	3	7	3	7
23	1	3	1	7	7	3	7	$\boxed{19}$	$\boxed{19}$	3
23	3	1	1	3	3	7	$\boxed{19}$	3	7	$\boxed{19}$
24	3	3	1	9	9	9	$\boxed{25}$	21	$\boxed{25}$	21
24	3	1	3	21	21	$\boxed{25}$	21	9	9	$\boxed{25}$
24	1	3	3	25	$\boxed{25}$	21	9	$\boxed{25}$	21	9
25	2	2	3	23	$\boxed{23}$	$\boxed{23}$	15	17	15	17
25	2	3	2	17	17	15	17	$\boxed{23}$	$\boxed{23}$	15
25	3	2	2	15	15	17	$\boxed{23}$	15	17	$\boxed{23}$
26	3	3	2	18	18	18	$\boxed{26}$	24	$\boxed{26}$	24
26	3	2	3	24	24	$\boxed{26}$	24	18	18	$\boxed{26}$
26	2	3	3	26	$\boxed{26}$	24	18	$\boxed{26}$	24	18
31	1	2	3	22	$\boxed{22}$	20	6	16	12	8
31	1	3	2	16	16	12	8	$\boxed{22}$	20	6
31	2	1	3	20	20	$\boxed{22}$	12	8	6	16
31	3	1	2	12	12	16	20	6	8	$\boxed{22}$
31	2	3	1	8	8	6	16	20	$\boxed{22}$	12
31	3	2	1	6	6	8	$\boxed{22}$	12	16	20

(7.18)

Appendix B

The 26 K_4-symmetric mappings, their distinct (at lines indicated by $*$) isomorphic companions, and astigmatic companions (at the line indicated with †)

j	mapping ID	s_j								$\alpha = 45^\circ$	$\alpha = 135^\circ$
0	0	0	0	0	0	0	0	0	0	0	0
1	82	1	0	0	0	1	0	0	0	*2214	246
2	164	2	0	0	0	2	0	0	0	*4428	492
3	738	0	0	1	0	0	0	1	0	246	2214
4	820	1	0	1	0	1	0	1	0	2460	2460
5	902	2	0	1	0	2	0	1	0	*4674	2706
6	1476	0	0	2	0	0	0	2	0	492	4428
7	1558	1	0	2	0	1	0	2	0	*2706	4674
8	1640	2	0	2	0	2	0	2	0	4920	4920
9	2460	0	1	0	1	0	1	0	1	820	820
10	2542	1	1	0	1	1	1	0	1	*3034	1066
11	2624	2	1	0	1	2	1	0	1	* †5248	1312
12	3198	0	1	1	1	0	1	1	1	1066	3034
13	3280	1	1	1	1	1	1	1	1	3280	3280
14	3362	2	1	1	1	2	1	1	1	*5494	3526
15	3936	0	1	2	1	0	1	2	1	1312	5248
16	4018	1	1	2	1	1	1	2	1	*3526	5494
17	4100	2	1	2	1	2	1	2	1	5740	5740
18	4920	0	2	0	2	0	2	0	2	1640	1640
19	5002	1	2	0	2	1	2	0	2	*3854	1886
20	5084	2	2	0	2	2	2	0	2	*6068	2132
21	5658	0	2	1	2	0	2	1	2	*1886	3854
22	5740	1	2	1	2	1	2	1	2	4100	4100
23	5822	2	2	1	2	2	2	1	2	*6314	4346
24	6396	0	2	2	2	0	2	2	2	*2132	6068
25	6478	1	2	2	2	1	2	2	2	*4346	6314

Appendix C

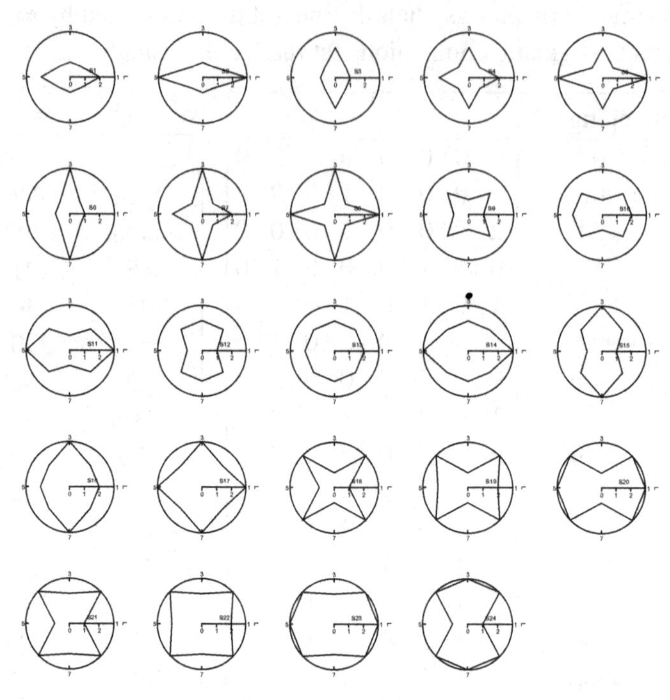

Figure 7.3: K_4-symmetric mappings.

8

Symmetry Studies of Planar Chirality

8.1 Introduction

Chirality is a term coined by Lord Kevin[1] meant to designate the quality of "any geometrical figure, or group of points of which a plane mirror image, ideally realized, cannot be brought to coincide with itself." The handedness in molecules was first identified by Pasteur[2] and since then the investigation of its presence, effects, and quantification in the natural sciences has received continued interest.

For example, molecular chirality is a necessary and sufficient condition for a substance to exibit optical activity, a property discovered by Huygens[3] in 1690 when studying a crystal of calcite. Optically active substances are capable of rotating the plane of polarization of a ray of plane polarized light. The presence of the symmetry of an improper rotation (an axial rotational followed by reflection on a plane orthogonal to the axis of rotation) in a molecule is a sufficient condition for its nonchirality as its mirror image created by the plane of reflection, after a rotation, is superimposable. In particular, molecules with the symmetry of a planar reflection or with point symmetry are achiral and hence optically inactive.

Many molecules such as amino acids and sugars are chiral, which in turn can cause DNA molecules to be chiral. The differential effect of the two pairs becomes observable only in the presence of a collective of chiral molecules, or by probing the pair with circularly polarized light, which is in itself a chiral mechanism. Planar chirality adds the constraint that the (planar and bounded) objects cannot be lifted from the plane, and consequently, the allowed symmetry operations are restricted to planar transformations.

In this chapter we apply the methods of symmetry studies to statistically probe the handedness of simple planar images. Our probing mechanisms, not surprisingly, are the two-dimensional irreducible linear representations of the dihedral groups acting on subsets of the Euclidean plane where the images are defined. These

[1] Scottish physicist William Thomson (Lord Kelvin), 1824–1907.
[2] French chemist Louis Pasteur, 1822–1895.
[3] Dutch mathematician Christiaan Huygens, 1629–1695.

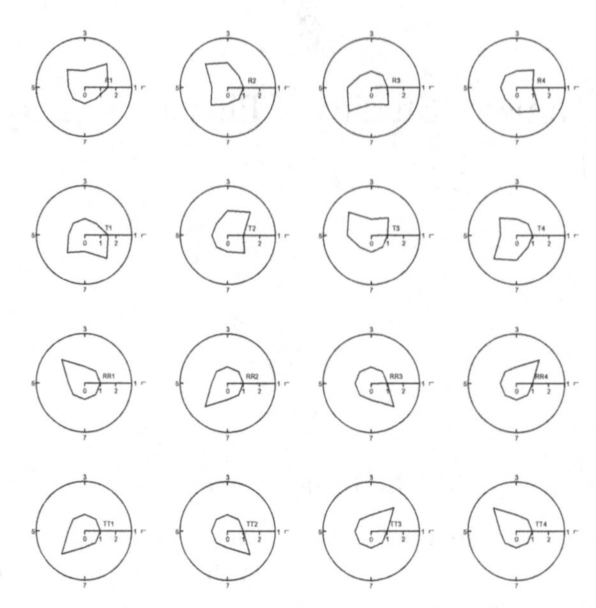

Figure 8.1: Rotations and reversals for mappings $s_7 = [2\ 3\ 1\ 1\ 1\ 1\ 1\ 1]$ (top two rows) and $s_{27} = [1\ 1\ 1\ 2\ 1\ 1\ 1\ 1]$ (bottom two rows).

actions have the effect of inspecting an image by looking it up from both sides. This is precisely the role of dihedral rotations and reversals.

The role of rotations and reversals can be appreciated from their action on the oriented edges of the squares illustrated on page 41, or by their action on the graphs (as subsets of the Euclidean plane) shown in Figure 8.1.

The contours display the four fourfold rotations (first row) followed by the reversals (second row) acting on the mapping

$$s_7 = \begin{bmatrix} 2 & 3 & 1 & 1 & 1 & 1 & 1 & 1 \end{bmatrix},$$

whereas the bottom part shows the corresponding actions on the mapping

$$s_{27} = \begin{bmatrix} 1 & 1 & 1 & 2 & 1 & 1 & 1 & 1 \end{bmatrix}.$$

These two mappings, identified here with the same labels from Chapter 7, can be contrasted by their handedness: the reader may want to try a paper model of the planar contours determined by the two mappings, rotating the models (say counterclockwise around its center) and annotating what you see first from one side, then repeating the process and inspecting the motion from the other side. Mapping s_{27} will not register any difference: it is said to be achiral; in contrast, the rotating of the model determined by mapping s_7 will reveal a sharp leading edge when seen from one side and a sharp trailing edge from the other side. This shape is then said to be chiral. This is also apparent in the rotations and reversals rows of the top and bottom displays.

Handedness in canonical space

For a nonzero $z \in \mathbb{C}$, a rotation $\beta_{k,1}$ maps $(z, \bar{z})$ to $(\omega^k z, \omega^{-k}\bar{z})$, whereas a reversal $\beta_{k,-1}$ maps $(z, \bar{z})$ to $(\omega^k \bar{z}, \omega^{-k} z)$.

As one facing a mirror with both arms wide open will recognize, if the right hand moves up ($\omega^k z$) and the left hand moves down ($\omega^k \bar{z}$) then the mirror image's left hand moves down ($\omega^{-k}\bar{z}$) and the right hand moves up ($\omega^{-k} z$). By aligning a coordinate system along a person's up-down, left-right, front-back directions, the rotations and reversals are planar representations describing (all products of) n-fold axial rotations around the front-back and around the up-down axes. The fundamental observation here is the fact that it takes both rotations and reversals to jointly describe the handedness implicit in the act of moving (by planar rotation) the right hand up and the left hand down. The dihedral motion is illustrated by the sequence of n points $g : k \to (\omega^k z, \omega^k \bar{z})$, whereas its mirror motion is the corresponding sequence $\overline{g} : k \to (\omega^{-k}\bar{z}, \omega^{-k} z)$, $k = 1, \ldots, n$.

In the theory of light, these two enantiomeric (mirror-image non superimposable) motions would describe, in the continuous limit, its left- and right-circular polarization forms.

Handedness and Euclidean space

Indicating by $r_{k,d}$ the embedding of planar rotations and reversals in Euclidean space, the handedness of these two classes of transformations is seen from the fact that $(r_{k,d}e_1 \times r_{k,d}e_2) \cdot e_3 = d$, independently of k. The parity induced in the axis along e_3 demonstrates the two distinctly oriented systems: a *right-handed* coordinate system ($d = 1$) and a *left-handed* coordinate system ($d = -1$) (Silverman, 1995, p. 186). These two interpretations are the basis of what is argued here: that rotations and reversals are jointly necessary to assess the eventual handedness in an object. Essentially, one should be able to inspect the object from *both sides*. And that is precisely what the dihedral symmetries accomplish.

8.2 Characterizing Rotations and Reversals

Recall from page 192 that given a real 2×2 matrix $M = (m_{ij})$ and its conjugate $\mathcal{M} = XMX^{-1}$ in canonical space, that

$$\mathcal{M} = \begin{bmatrix} f_+ & f_- \\ \overline{f}_- & \overline{f}_+ \end{bmatrix} = \sum_{k,d} x_{k,d}\beta_{k,d} \iff x_{k,d} = \frac{2}{n}[\cos(k\phi)\Re\, f_d + \sin(k\phi)\Im\, f_d],$$

$$d = \pm 1, \quad k = 0, \ldots, n-1,$$

where $x_{k,d} = \text{Tr}\,[\beta_{-k,d}\mathcal{M}]/n$ are the data indexed by the rotations and reversals.

We also observe that f_- and f_- are the canonical invariants for the data $x_{k,d}$, in phase space, corresponding to the invariants $\mathcal{B}_v$ and $\mathcal{B}_w$ derived on page 121.

If the matrix M is complex, the decomposition of $\mathcal{M}$ is obtained by adjoining the decompositions of the real and imaginary matrix parts of M in canonical space. In particular, the data coefficients $x_{k,d}$ of any $\beta_{k^*,-d}$ for a particular k^* will be zero. More specifically,

$$\beta_{k^*,d} = \frac{2}{n} \sum_k \cos(k\phi - k^*\phi)\beta_{k,d}.$$

Direct calculation shows that

$$\mathcal{A}_d = \left\{ \begin{bmatrix} a_1 & a_2 \\ -da_2 & da_1 \end{bmatrix}; \quad a = (a_1, a_2) \in \mathbb{R}^2 \right\}, \quad d = \pm 1,$$

are the classes of matrices with vanishing reversal coefficients ($d = 1$, pure rotations) and vanishing rotation coefficients ($d = -1$, pure reversals). Elements $A_d \in \mathcal{A}_d$ in these classes resolve in canonical space as

$$A_d = \frac{n}{2} \sum_k [a_1 \cos(k\phi) + a_2 \sin(k\phi)]\beta_{k,d}, \quad d = \pm 1,$$

and satisfy $\det A_d = d||a||^2$. Writing,

$$I = \begin{bmatrix} 1 & 0 \\ 0 & 1 \end{bmatrix}, \quad R = \begin{bmatrix} 0 & 1 \\ -1 & 0 \end{bmatrix}, \quad H = \begin{bmatrix} 1 & 0 \\ 0 & -1 \end{bmatrix}, \quad D = \begin{bmatrix} 0 & 1 \\ 1 & 0 \end{bmatrix},$$

it follows that null reversals or pure rotations are characterized by linear superpositions $A_+ = a_1 I + a_2 R$ of the identity and a 90° rotation R, whereas the matrices with null rotations, or pure reversals, are linear superpositions $A_- = a_1 H + a_2 D$ of a horizontal line reflection H and a 45° line reflection D. Clearly, $A_+ \cap A_- = \{0\}$. The matrices $\{I, R, D, H\}$ multiply according to the table

$\cdot$	1	R	H	D
1	1	R	H	D
R	R	-1	$-D$	H
H	H	D	1	R
D	D	$-H$	$-R$	1

and generate an associative algebra. In canonical space, writing $a = a_1 + ia_2$,

$$A_+ = \begin{bmatrix} a & 0 \\ 0 & \overline{a} \end{bmatrix}, \quad A_- = \begin{bmatrix} 0 & a \\ \overline{a} & 0 \end{bmatrix}$$

so that, not surprisingly, a pure rotation A_+ acts as $z \mapsto az$ whereas a pure reversal A_- acts as $z \mapsto a\overline{z}$ in canonical space. The elementary operators act as $H(z) = \overline{z}$, $D(z) = i\overline{z}$ and $R(z) = -iz$. Also in canonical space, $H = \sigma_1 = -i\mathbf{k}$, $R = -i\sigma_3 = -\mathbf{i}$, $D = -\sigma_2 = i\mathbf{j}$, are the relations to the Pauli matrices $\{\sigma_1, \sigma_2, \sigma_3\}$ and the unit quaternions $\{1, \mathbf{i}, \mathbf{j}, \mathbf{k}\}$.

8.3 Canonical Classification of Handedness in Elementary Images

This and the next section conclude our introduction to the analysis of structured data with the study and statistical classification of simple images as data indexed by simple planar arrays. To illustrate, consider the binary image shown on Figure 8.2 in which the levels of gray are indexed by a planar array $V = L \times L$.

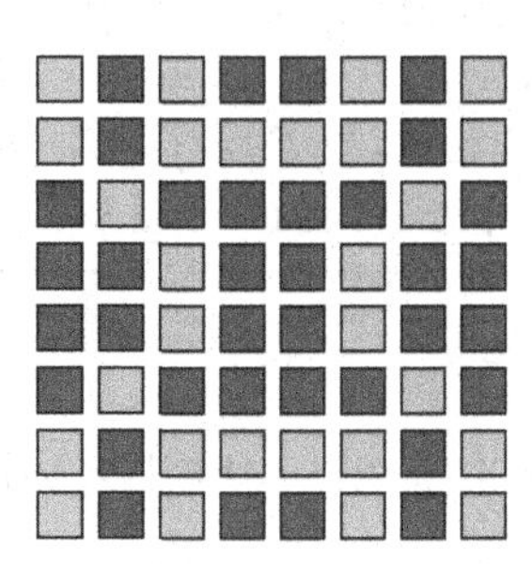

Figure 8.2: A simple planar image.

We will refer to these data $y : V \mapsto \mathbb{R}$ simply as *images*. The symmetries acting on the images are those of any linear representation ρ, in dimension of 2, of a group G leaving the array V invariant. This, in turn, gives an action

$$\tau \in G \mapsto y\rho_{\tau^{-1}} \in \mathcal{F}(V)$$

of G on $\mathcal{F}(V)$. We are interested in statistically studying the symmetries of a given image y with reference to the canonical decomposition of the sum of squares $x'x$ for the data

$$x_\tau = ||y - y\rho_{\tau^{-1}}||^2, \quad \tau \in D_4 \tag{8.1}$$

indexed by dihedral planar rotations and reversals acting on V. For each rotation or reversal, x_τ is the squared distance between the original image and the transformed image.

The notation $\gamma_i = x'\mathcal{P}_i x$, $i = 1, \ldots, 5$ for the components of the canonical reduction of $x'x$ (page 134) is used to facilitate the present narrative, so that, for example, $\sum_i \gamma_i$ is the analysis of variance for each single-sample image. With the same objective, we indicate here by $1, r, r^2, r^3$ the four-fold rotations, by h and v the horizontal and vertical line reflections, and by d and D the reflections on the 45° and 135° lines, respectively. This notation is in correspondence

$$(x_{1,1}, \ldots, x_{4,1}, x_{1,-1}, \ldots, x_{4,-1}) \simeq (r_1, \ldots, r_4, t_1, \ldots t_4) \simeq (1, r, r^2, r^3, h, d, v, D)$$

with the generic notations for dihedral data introduced earlier on in the previous chapters. A point reflection with center in the origin (0, 0) is indicated by o.

The following classification of the images is then obtained:

Aut(V)	null components
D_4	$\gamma_1, \gamma_2, \gamma_3, \gamma_4, \gamma_5$
C_4	$\gamma_3, \gamma_4, \gamma_5$
$\{1, h\}$ or $\{1, v\}$	γ_2, γ_4
$\{1, d\}$ or $\{1, D\}$	γ_2, γ_3
$\{1, o\} \simeq C_2$	γ_5
$\{1, v, h, o\} \simeq C_2 \times C_2$	$\gamma_2, \gamma_4, \gamma_5$
$\{1, d, D, o\} \simeq C_2 \times C_2$	$\gamma_2, \gamma_3, \gamma_5$

(8.2)

There are two conditions, namely,

$$0 \notin \{\gamma_2, \gamma_5\}, \quad 0 \notin \{\gamma_3, \gamma_4\},$$

each one of which is sufficient for Aut(V) $\subseteq \{\{1\}, \{1, o\}\}$, hence sufficient for determining the image to be chiral.

Each one of the images illustrated below will be followed by its dihedral indexing and the corresponding analysis of variance,

$$\begin{bmatrix} r & r^2 & r^3 & h & d & v & D \\ x_r & x_{r^2} & x_{r^3} & x_h & x_d & x_v & x_D \end{bmatrix}, \quad \begin{bmatrix} 1 & \gamma_1' \\ 2 & \gamma_2' \\ 3 & \gamma_3' \\ 4 & \gamma_4' \\ 5 & \gamma_5' \end{bmatrix},$$

where (for relative comparisons) $\gamma_i' = \gamma_i / \sum_j \gamma_j$. In addition, in selected examples, the dihedral correlation indexing

$$x_\tau = \frac{< y, y\rho_{\tau^{-1}} >}{||y||^2}, \tag{8.3}$$

is also presented. The images are classified according to their canonical decomposition components, $\gamma_1, \ldots, \gamma_5$.

D_4 symmetry

$\gamma_1 = \ldots \gamma_5 = 0.$

0	1	1	0
1	2	2	1
1	2	2	1
0	1	1	0

$$\begin{bmatrix} r & r^2 & r^3 & h & d & v & D \\ 0 & 0 & 0 & 0 & 0 & 0 & 0 \end{bmatrix} \quad \begin{bmatrix} 1 & 0 \\ 2 & 0 \\ 3 & 0 \\ 4 & 0 \\ 5 & 0 \end{bmatrix}$$

C_4 symmetry

$\gamma_3 = \gamma_4 = \gamma_5 = 0.$

$$\boxed{\begin{matrix} 0 & 1 & 2 & 0 \\ 2 & 2 & 2 & 1 \\ 1 & 2 & 2 & 2 \\ 0 & 2 & 1 & 0 \end{matrix}} \begin{bmatrix} r & r^2 & r^3 & h & d & v & D \\ 0 & 0 & 0 & 8 & 8 & 8 & 8 \end{bmatrix} \begin{bmatrix} 1 & 0.5 \\ 2 & 0.5 \\ 3 & 0 \\ 4 & 0 \\ 5 & 0 \end{bmatrix}.$$

Line ($x = 0$) symmetry

$\gamma_2 = \gamma_4 = 0.$

$$\boxed{\begin{matrix} 0 & 10 & 10 & 0 \\ 1 & 2 & 2 & 1 \\ 1 & 2 & 2 & 1 \\ 0 & 1 & 1 & 0 \end{matrix}} \begin{bmatrix} r & r^2 & r^3 & h & d & v & D \\ 324 & 324 & 324 & 324 & 324 & 0 & 324 \end{bmatrix} \begin{bmatrix} 1 & 0.74 \\ 2 & 0 \\ 3 & 0.08 \\ 4 & 0 \\ 5 & 0.16 \end{bmatrix}$$

Line ($y = 0$) symmetry

$\gamma_2 = \gamma_4 = 0.$

$$\boxed{\begin{matrix} 0 & 1 & 1 & 0 \\ 10 & 2 & 2 & 1 \\ 10 & 2 & 2 & 1 \\ 0 & 1 & 1 & 0 \end{matrix}} \begin{bmatrix} r & r^2 & r^3 & h & d & v & D \\ 324 & 324 & 324 & 0 & 324 & 324 & 324 \end{bmatrix} \begin{bmatrix} 1 & 0.74 \\ 2 & 0 \\ 3 & 0.08 \\ 4 & 0 \\ 5 & 0.16 \end{bmatrix}$$

Line ($y = x$) symmetry

$\gamma_2 = \gamma_3 = 0.$

$$\boxed{\begin{matrix} 0 & 1 & 10 & 0 \\ 1 & 2 & 2 & 10 \\ 1 & 2 & 2 & 1 \\ 0 & 1 & 1 & 0 \end{matrix}} \begin{bmatrix} r & r^2 & r^3 & h & d & v & D \\ 324 & 324 & 324 & 324 & 0 & 324 & 324 \end{bmatrix} \begin{bmatrix} 1 & 0.74 \\ 2 & 0 \\ 3 & 0 \\ 4 & 0.08 \\ 5 & 0.16 \end{bmatrix}$$

Line ($y = -x$) symmetry

$\gamma_2 = \gamma_3 = 0.$

$$\boxed{\begin{array}{cccc} 0 & 10 & 1 & 0 \\ 10 & 2 & 2 & 1 \\ 1 & 2 & 2 & 1 \\ 0 & 1 & 1 & 0 \end{array}} \begin{bmatrix} r & r^2 & r^3 & h & d & v & D \\ 324 & 324 & 324 & 324 & 324 & 324 & 0 \end{bmatrix} \begin{bmatrix} 1 & 0.74 \\ 2 & 0 \\ 3 & 0 \\ 4 & 0.08 \\ 5 & 0.16 \end{bmatrix}$$

C_2 (point) symmetry

$\gamma_5 = 0$ and $0 \notin \{\gamma_3, \gamma_4\}$. This image is chiral.

$$\boxed{\begin{array}{cccc} 0 & 1 & 0 & 1 \\ 2 & 1 & 1 & 2 \\ 2 & 1 & 1 & 2 \\ 1 & 0 & 1 & 0 \end{array}} \begin{bmatrix} r & 24.0 \\ r^2 & 0.0 \\ r^3 & 24.0 \\ h & 8.0 \\ d & 20.0 \\ v & 8.0 \\ D & 20.0 \end{bmatrix} \begin{bmatrix} 1 & 0.652 \\ 2 & 0.003 \\ 3 & 0.313 \\ 4 & 0.034 \\ 5 & 0 \end{bmatrix}$$

A binary (chiral) spiral image

$$\boxed{\begin{array}{cccccc} 1 & 1 & 1 & 1 & 1 & 1 \\ 1 & 2 & 2 & 2 & 2 & 2 \\ 1 & 2 & 1 & 1 & 1 & 2 \\ 1 & 2 & 1 & 2 & 1 & 2 \\ 1 & 2 & 2 & 2 & 1 & 2 \\ 1 & 1 & 1 & 1 & 1 & 2 \end{array}} \begin{bmatrix} r & 18.0 \\ r^2 & 18.0 \\ r^3 & 18.0 \\ h & 6.0 \\ d & 18.0 \\ v & 18.0 \\ D & 12.0 \end{bmatrix} \begin{bmatrix} 1 & 0.81 \\ 2 & 0 \\ 3 & 0.04 \\ 4 & 0.01 \\ 5 & 0.14 \end{bmatrix}$$

C_4 symmetry

$$\boxed{\begin{array}{cccccc} 0 & 2 & 2 & 1 & 1 & 0 \\ 1 & 0 & 2 & 1 & 0 & 2 \\ 1 & 1 & 0 & 0 & 2 & 2 \\ 2 & 2 & 0 & 0 & 1 & 1 \\ 2 & 0 & 1 & 2 & 0 & 1 \\ 0 & 1 & 1 & 2 & 2 & 0 \end{array}} \begin{bmatrix} r & 0.0 \\ r^2 & 0.0 \\ r^3 & 0.0 \\ h & 24.0 \\ d & 24.0 \\ v & 24.0 \\ D & 24.0 \end{bmatrix} \begin{bmatrix} 1 & 0.5 \\ 2 & 0.5 \\ 3 & 0 \\ 4 & 0 \\ 5 & 0 \end{bmatrix}$$

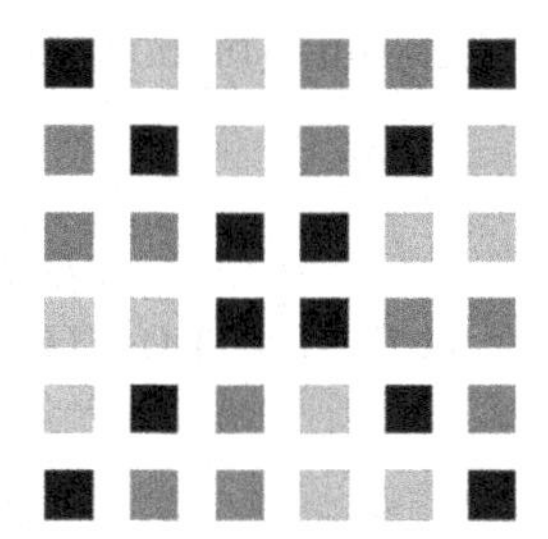

Figure 8.3: C_4-symmetry.

Figure 8.3 Shows an image with the symmetry of C_4.

Analytically generated images

$\ell(x, y) = 2x^2 + y^2$. Here $\text{Aut}(L) = \{1, v, h, o\}$.

$$\boxed{\begin{matrix} 12 & 6 & 4 & 6 & 12 \\ 9 & 3 & 1 & 3 & 9 \\ 8 & 2 & 0 & 2 & 8 \\ 9 & 3 & 1 & 3 & 9 \\ 12 & 6 & 4 & 6 & 12 \end{matrix}}, \quad \begin{bmatrix} r & r^2 & r^3 & h & d & v & D \\ 140 & 0 & 140 & 0 & 140 & 0 & 140 \end{bmatrix} \begin{bmatrix} 1 & 0.5 \\ 2 & 0 \\ 3 & 0.5 \\ 4 & 0 \\ 5 & 0 \end{bmatrix}$$

$\ell(x, y) = xy$. Here $\text{Aut}(L) = \{1, d, D, o\}$.

$$\boxed{\begin{matrix} -4 & -2 & 0 & 2 & 4 \\ -2 & -1 & 0 & 1 & 2 \\ 0 & 0 & 0 & 0 & 0 \\ 2 & 1 & 0 & -1 & -2 \\ 4 & 2 & 0 & -2 & -4 \end{matrix}}, \quad \begin{bmatrix} r & r^2 & r^3 & h & d & v & D \\ 400 & 0 & 400 & 400 & 0 & 400 & 0 \end{bmatrix} \begin{bmatrix} 1 & 0.5 \\ 2 & 0 \\ 3 & 0 \\ 4 & 0.5 \\ 5 & 0 \end{bmatrix}$$

$\ell(x, y) = x^2 y$. Here, $\text{Aut}(L) = \{1, v\}$ and it is observed that

$$\boxed{\begin{matrix} 8 & 2 & 0 & 2 & 8 \\ 4 & 1 & 0 & 1 & 4 \\ 0 & 0 & 0 & 0 & 0 \\ -4 & -1 & 0 & -1 & -4 \\ -8 & -2 & 0 & -2 & -8 \end{matrix}} \begin{bmatrix} r & r^2 & r^3 & h & d & v & D \\ 680 & 1360 & 680 & 1360 & 680 & 0 & 680 \end{bmatrix} \begin{bmatrix} 1 & 0.66 \\ 2 & 0 \\ 3 & 0 \\ 4 & 0 \\ 5 & 0.33 \end{bmatrix},$$

so that D_4, C_4, C_2, and K_4 are ruled out. The correlation indexing

$$\begin{bmatrix} r & r^2 & r^3 & h & d & v & D \\ 0.0 & -1.0 & 0.0 & -1.0 & 0.0 & 1.0 & 0.0 \end{bmatrix} \begin{bmatrix} 1 & 0.04 \\ 2 & 0.04 \\ 3 & 0.04 \\ 4 & 0.04 \\ 5 & 0.83 \end{bmatrix},$$

complements the classification, pointing to the correct automorphisms.

$\boldsymbol{\ell(x, y) = x + y}$. Similarly to the previous array, here $\mathrm{Aut}(L) = \{1, d\}$,

$$\boxed{\begin{array}{rrrrr} 0 & 1 & 2 & 3 & 4 \\ -1 & 0 & 1 & 2 & 3 \\ -2 & -1 & 0 & 1 & 2 \\ -3 & -2 & -1 & 0 & 1 \\ -4 & -3 & -2 & -1 & 0 \end{array}} \begin{bmatrix} r & r^2 & r^3 & h & d & v & D \\ 200 & 400 & 200 & 200 & 0 & 200 & 400 \end{bmatrix} \begin{bmatrix} 1 & 0.66 \\ 2 & 0 \\ 3 & 0 \\ 4 & 0 \\ 5 & 0.33 \end{bmatrix}$$

whereas the correlation indexing leads to

$$\begin{bmatrix} r & r^2 & r^3 & h & d & v & D \\ 0.0 & -1.0 & 0.0 & 0.0 & 1.0 & 0.0 & -1.0 \end{bmatrix} \begin{bmatrix} 1 & 0.04 \\ 2 & 0.04 \\ 3 & 0.04 \\ 4 & 0.04 \\ 5 & 0.83 \end{bmatrix}$$

$\boldsymbol{\ell(x, y) = 4\cos^2(x\pi/4) + 2\sin^2(x\pi/4)}$.

$$\boxed{\begin{array}{ccccc} 2 & 4 & 6 & 4 & 2 \\ 1 & 3 & 5 & 3 & 1 \\ 0 & 2 & 4 & 2 & 0 \\ 1 & 3 & 5 & 3 & 1 \\ 2 & 4 & 6 & 4 & 2 \end{array}} \begin{bmatrix} r & r^2 & r^3 & h & d & v & D \\ 252 & 0 & 252 & 0 & 252 & 0 & 252 \end{bmatrix} \begin{bmatrix} 1 & 0.5 \\ 2 & 0 \\ 3 & 0.5 \\ 4 & 0 \\ 5 & 0 \end{bmatrix}$$

Here, $\mathrm{Aut}(L) = \{1, h, v, o\} \simeq C_2 \times C_2$. The same function evaluated on a 7×7 image:

$$\boxed{\begin{array}{ccccccc} 3 & 1 & 3 & 5 & 3 & 1 & 3 \\ 4 & 2 & 4 & 6 & 4 & 2 & 4 \\ 3 & 1 & 3 & 5 & 3 & 1 & 3 \\ 2 & 0 & 2 & 4 & 2 & 0 & 2 \\ 3 & 1 & 3 & 5 & 3 & 1 & 3 \\ 4 & 2 & 4 & 6 & 4 & 2 & 4 \\ 3 & 1 & 3 & 5 & 3 & 1 & 3 \end{array}} \begin{bmatrix} r & r^2 & r^3 & h & d & v & D \\ 360 & 0 & 360 & 0 & 360 & 0 & 360 \end{bmatrix} \begin{bmatrix} 1 & 0.5 \\ 2 & 0 \\ 3 & 0.5 \\ 4 & 0 \\ 5 & 0 \end{bmatrix}$$

$\ell(x, y) = x^2 - y^2$. Similarly,

$$\begin{array}{|ccccccc|} 0 & -5 & -8 & -9 & -8 & -5 & 0 \\ 5 & 0 & -3 & -4 & -3 & 0 & 5 \\ 8 & 3 & 0 & -1 & 0 & 3 & 8 \\ 9 & 4 & 1 & 0 & 1 & 4 & 9 \\ 8 & 3 & 0 & -1 & 0 & 3 & 8 \\ 5 & 0 & -3 & -4 & -3 & 0 & 5 \\ 0 & -5 & -8 & -9 & -8 & -5 & 0 \end{array}, \quad \begin{bmatrix} r & r^2 & r^3 & h & d & v & D \\ 4704 & 0 & 4704 & 0 & 4704 & 0 & 4704 \end{bmatrix} \begin{bmatrix} 1 & 0.5 \\ 2 & 0 \\ 3 & 0.5 \\ 4 & 0 \\ 5 & 0 \end{bmatrix}.$$

8.4 Handedness of Short Symbolic Sequences

In this section we will statistically classify the handedness of elements in the mapping space of all 6,561 sequences in length of 8 taking values in $\{1, 2, 3\}$ introduced in Chapter 7, following the same steps developed in the previous section. That is, each profile is rotated, front and back, and compared with the original one. The sum of squares of the resulting data indexed by these rotations and reversals is then reduced using the canonical decomposition.

Classification of the profiles with K_4-symmetry

These are the 26 profiles identified in Chapter 7. The results are summarized in Appendix A to this chapter (page 217). It is simple to verify, from the resulting data $x_1, \ldots, x_D$ that $\gamma_1 = x'\mathcal{P}_1 x$ and $\gamma_3 = x'\mathcal{P}_3 x$ are the only eventually nonzero components in the decomposition of $x'x$, so that the classification is then primarily based on γ_3. With Table (8.2) in mind, note that all mappings have $\gamma_2 = \gamma_4 = \gamma_5 = 0$, which is the necessary condition for K_4 symmetry. Specifically, the canonical decomposition identifies three classes of mappings: Those with

(1) $\gamma_3 = 0$ corresponding to Dihedral mappings, as $\gamma_1 = \ldots, = \gamma_5 = 0$;
(2) $\gamma_3 = 32$, which, correctly, rule out all but the K_4 symmetries, correctly so; and
(3) $\gamma_3 = 512$, which do so as well, only more decisively.

The graphical display of these three classes of mappings is shown in Appendix B, Figures 8.6–8.8.

Sampling the mapping space

It is not difficult to verify, analytically, that $\gamma_2 = 0$ for all mappings in that space, so that we discuss γ_3, γ_4, and γ_5.

Table (8.4) shows the joint distribution of the canonical components γ_3 and γ_4 based on a random sample of size 100 from the mapping space (with 6,561 points).

$\gamma_3 \backslash \gamma_4$	0	2	32	162	512	Total
0	6	10	4	2	0	22
2	6	21	4	4	0	35
32	5	7	7	1	0	20
162	6	5	3	2	0	16
512	1	3	1	1	1	7
Total	24	46	19	10	1	100

(8.4)

Here is the detailed classification:

(1) There was a single mapping (ID 4592), displayed below, with $\gamma_5 = 0$ and $0 \notin \{3, 4\}$ so that it is chiral with (in this case) a center of symmetry.

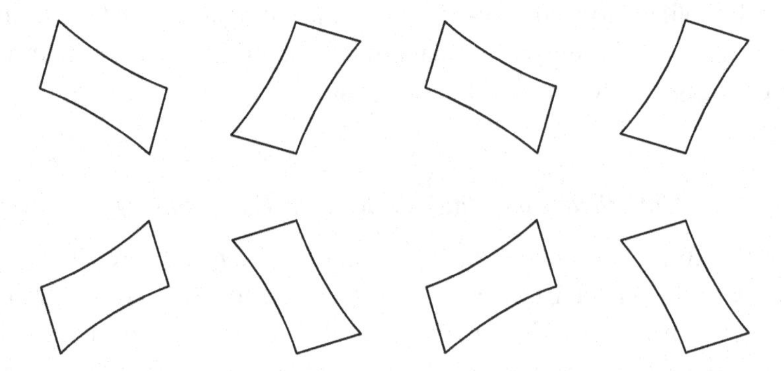

(2) There were 6 mappings, illustrated below, with $\gamma_3 = \gamma_4 = 0$ and $\gamma_5 \neq 0$. The condition on γ_5 rules out all but the simple axial reflections. Among these, indeed, one mapping with the symmetry of $\{1, d\}$ and one mapping with the symmetry of $\{1, D\}$ were identified. Both are achiral. The remaining four are chiral.

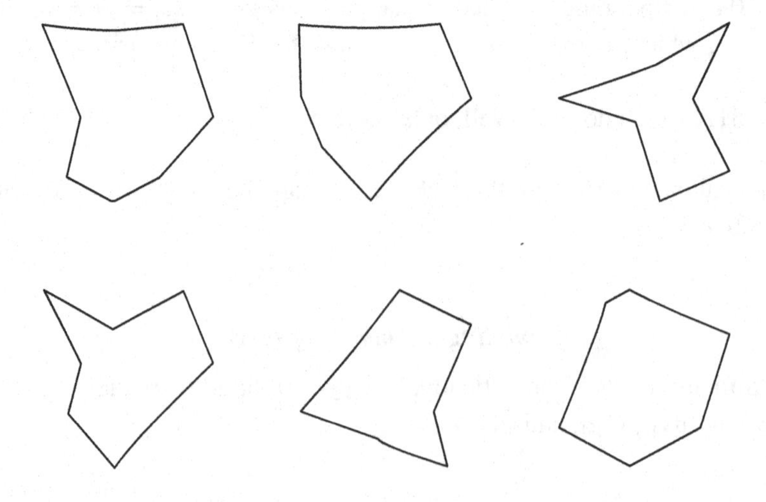

(3) A total of 16 mappings satisfy the condition $\gamma_3 = 0$ with $0 \notin \{\gamma_4, \gamma_5\}$, thus ruling out all but the chiral mappings and those with diagonal symmetries. One mapping has diagonal symmetries and 15 are chiral. The mappings are shown in Appendix C, Figure 8.9.
(4) A total of 18 mappings satisfy the condition $\gamma_4 = 0$ with $0 \notin \{\gamma_3, \gamma_5\}$, thus ruling out all but the chiral mappings and those with horizontal or vertical symmetries; of those, 9 are chiral. The mappings are shown in the Appendix C, Figure 8.10.
(5) Finally, there were 47 mappings with $\gamma_3 = \gamma_4 = \gamma_5 = 0$, ruling out all symmetries but the trivial. All mappings are chiral.

In total, then, a total of $1 + 4 + 15 + 9 + 47 = 76$ chiral mappings were identified, approximately $3/4$ of the sample.

A particular interpretation of component γ_5. Table (8.2) makes clear the role of γ_5 in the decomposition of the sum of squares for the dihedral data. It shows that large values of γ_5 should correspond to large deviations of point symmetry, or C_2 symmetry, which is quantized by x_{r^2}. This effect is shown in Figure 8.4. In addition, as pointed out above, large values of γ_5 rule out all but the symmetries of those groups of simple reversals, thus increasing the likelihood of pointing to chiral mappings.

The component γ_5 is also related to the canonical invariants derived earlier on page 121. Figure 8.5 shows the lines of constant γ_5 in the invariant space determined by the basis $\mathcal{B}_v$.

Further Reading

The difficult task of quantifying molecular chirality and relating it to its macroscopic manifestations is described in Harris et al. (1999). Recent advances in optical activity produced by a (thin film) planar grating consisting of elementary chiral elements is reported in Papakostas et al. (2003). See also Prosvirnin and Zheludev (2005). An extensive review of the many applications of chirality is documented in Petitjean (2003). The special issue (V. 16 No. 4, 2005) of *Symmetry: Culture and Science* covers a wide range of fields where chirality has been investigated, such as thermodynamics, broken symmetry and chiroptics. See also Viana and Lakshminarayanan (2005).

Exercises

Exercise 8.1. Carry out a symmetry study for an experimental data indexed by the directed edges

$$\{(a, b), (b, c), (c, a), (b, a), (a, c), (c, b)\}$$

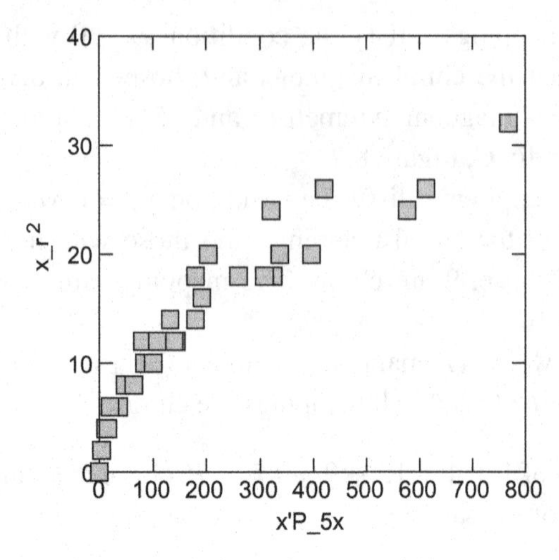

Figure 8.4: Joint distribution of $x'\mathcal{P}_5 x$ and x_{r^2}.

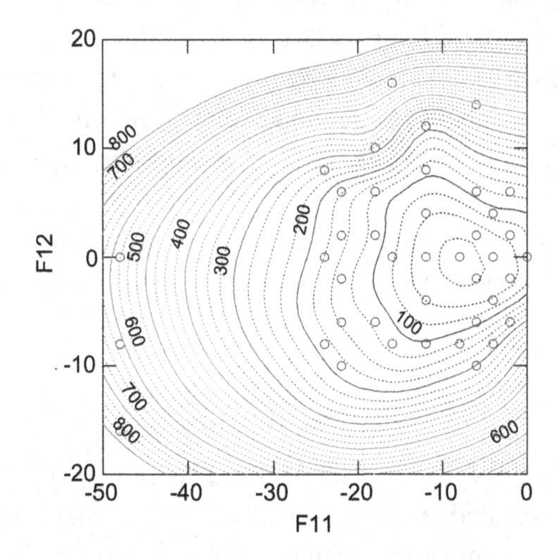

Figure 8.5: Contours of γ_5 on the dihedral Fourier invariants.

of the triangle with vertices in the set $\{a, b, c\}$, e.g., data on scalar responses indexed by a two-step order of presentation of three conditions. Identify and quantify eventual handedness effects.

Exercise 8.2. Following with Exercise 8.1, derive from Table (1.16) on page 12 the frequencies of pairwise preferences among the candidates in the set $\{a, g, c\}$ and carry out a symmetry study on these data. Then compare the results with the corresponding study for the other three sets of candidates.

Exercise 8.3. If a probability distribution is indexed by V where a group G acts upon, giving a representation ρ, then $x_\tau = J(s : s\rho_{\tau^{-1}})$, $\tau \in G$, is the divergence indexing based on Kullback's divergence measure $J(f : g) = \sum(f_i - g_i)\log f_i/g_i$ (Kullback, 1968). Apply the divergence indexing to study the voting data (1.16) on page 12.

Exercise 8.4. Following the definition (8.1) and notation on page 205, show that for an arbitrary point in $\mathbb{R}^8$ we have $x_r = x_{r^3}$.

Exercise 8.5. Show that the matrices M and $\mathcal{M}$ in Section 8.2 with $\det M = 1$ form two matrix groups, known as $SL(2, \mathbb{R})$ and $SU_{1,1}$, respectively. These are conjugate subgroups of $GL(2, \mathbb{C})$ in the sense that $XSL(2, \mathbb{R})X^{-1} = SU_{1,1}$, e.g., Carter et al. (1995, p. 57).

Exercise 8.6. Let $Q(z; k) = \epsilon^* \beta_{k,d}\epsilon/2$, where $\epsilon' = (z, \bar{z})$ in $\mathbb{C}^2$, with $z = x + iy$ and * indicates the conjugate transpose of a vector or matrix. Show that

$$Q(z;k) = \begin{cases} |z|^2 \cos k\phi & \text{if } d = 1 \text{ (rotations)}, \\ (x^2 - y^2)\cos k\phi + 2xy \sin k\phi & \text{if } d = -1 \text{ (reversals)}, \end{cases}$$

and that the set $\{Q(z; k), k = 0, \ldots, n-1\}$ is invariant under unitary rotations $z \mapsto wz$ and reversals $z \mapsto w\bar{z}$.

Exercise 8.7. With the notation of page 204, show that the commutator relations

	1	R	H	D
1	0	0	0	0
R	0	0	$-2D$	$2H$
H	0	$2D$	0	$2R$
D	0	$-2H$	$-2R$	0

for the algebra $\mathbb{A}$ generated by $\{1, R, H, D\}$ lead to the commutator relations

$$[A_d, B_f] = \begin{cases} 0 & d = f = 1, \\ (a_1d_2 - a_2b_1)R & d = f = -1, \\ a_2b_2H - a_2b_1D & d = 1,\ f = -1, \\ -a_2b_2H + a_1b_2D & d = -1,\ f = 1, \end{cases}$$

for the underlying associative algebra generated by true rotations and reversals.

Exercise 8.8. Starting with the action

$[.,.]$	e_1	e_2	e_3	e_4
R	$-e_2 - e_3$	$e_1 - e_4$	$e_1 - e_4$	$e_2 + e_3$
H	0	$2e_2$	$-2e_3$	0
D	$e_3 - e_2$	$e_4 - e_1$	$e_1 - e_4$	$e_2 - e_3$

of the elementary operators on the canonical basis $\{e_1, \ldots, e_4\}$ for $\mathbb{R}^{2\times 2}$ and indicating the resulting representation by ρ, show that $[\rho_R, \rho_H] = \rho_{[R,H]}$. These (Lie algebra) representations may be useful in the analysis of data indexed by $\mathbb{A}$.

Appendix A

j	mapping ID	s_j								$x'\mathcal{P}_3 x$	x_1	x_r	x_{r^2}	x_{r^3}	x_h	x_d	x_v	x_D
0	0	0	0	0	0	0	0	0	0	0	0	0	0	0	0	0	0	0
1	82	1	0	0	0	1	0	0	0	32	0	4	0	4	0	4	0	4
2	164	2	0	0	0	2	0	0	0	512	0	16	0	16	0	16	0	16
3	738	0	0	1	0	0	0	1	0	32	0	4	0	4	0	4	0	4
4	820	1	0	1	0	1	0	1	0	0	0	0	0	0	0	0	0	0
5	902	2	0	1	0	2	0	1	0	32	0	4	0	4	0	4	0	4
6	1476	0	0	2	0	0	0	2	0	512	0	16	0	16	0	16	0	16
7	1558	1	0	2	0	1	0	2	0	32	0	4	0	4	0	4	0	4
8	1640	2	0	2	0	2	0	2	0	0	0	0	0	0	0	0	0	0
9	2460	0	1	0	1	0	1	0	1	0	0	0	0	0	0	0	0	0
10	2542	1	1	0	1	1	1	0	1	32	0	4	0	4	0	4	0	4
11	2624	2	1	0	1	2	1	0	1	512	0	16	0	16	0	16	0	16
12	3198	0	1	1	1	0	1	1	1	32	0	4	0	4	0	4	0	4
13	3280	1	1	1	1	1	1	1	1	0	0	0	0	0	0	0	0	0
14	3362	2	1	1	1	2	1	1	1	32	0	4	0	4	0	4	0	4
15	3936	0	1	2	1	0	1	2	1	512	0	16	0	16	0	16	0	16
16	4018	1	1	2	1	1	1	2	1	32	0	4	0	4	0	4	0	4
17	4100	2	1	2	1	2	1	2	1	0	0	0	0	0	0	0	0	0
18	4920	0	2	0	2	0	2	0	2	0	0	0	0	0	0	0	0	0
19	5002	1	2	0	2	1	2	0	2	32	0	4	0	4	0	4	0	4
20	5084	2	2	0	2	2	2	0	2	512	0	16	0	16	0	16	0	16
21	5658	0	2	1	2	0	2	1	2	32	0	4	0	4	0	4	0	4
22	5740	1	2	1	2	1	2	1	2	0	0	0	0	0	0	0	0	0
23	5822	2	2	1	2	2	2	1	2	32	0	4	0	4	0	4	0	4
24	6396	0	2	2	2	0	2	2	2	512	0	16	0	16	0	16	0	16
25	6478	1	2	2	2	1	2	2	2	32	0	4	0	4	0	4	0	4

Appendix B

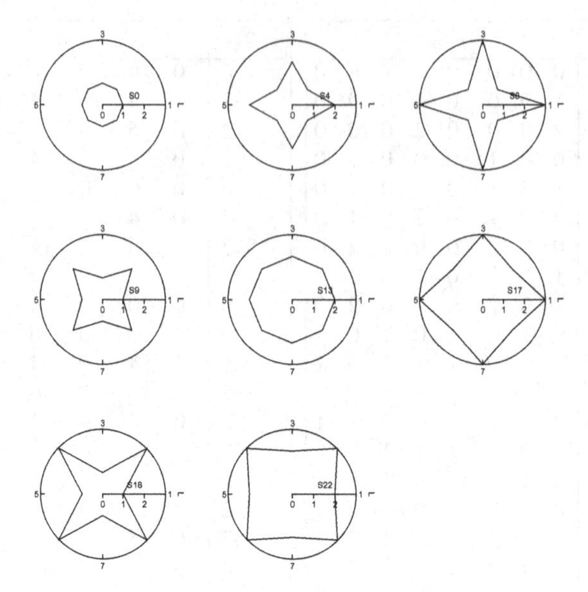

Figure 8.6: K_4 mappings with dihedral symmetry.

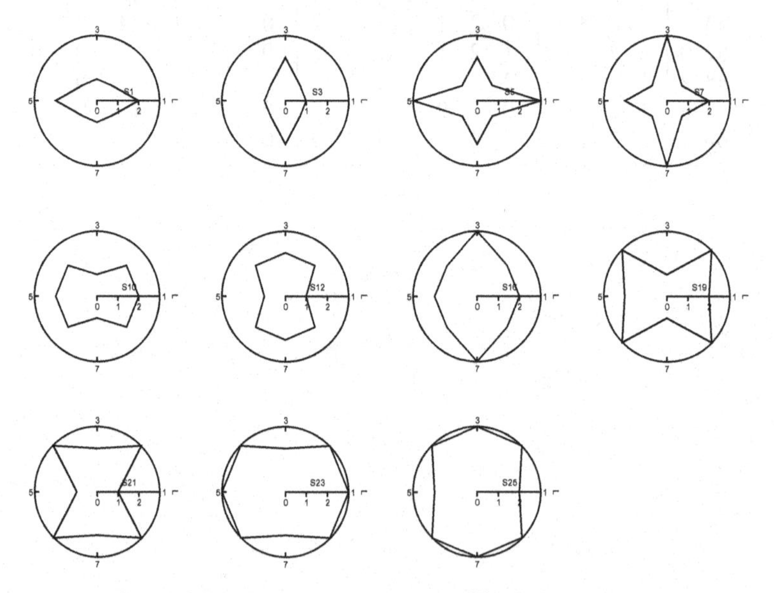

Figure 8.7: K_4 mappings without dihedral symmetry.

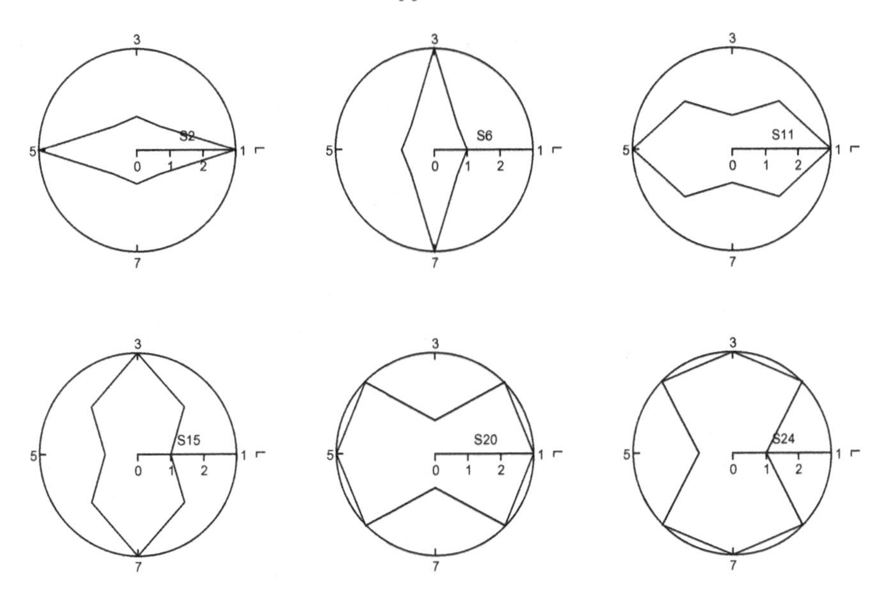

Figure 8.8: K_4 mappings without dihedral symmetry. More stringent deviations.

Appendix C

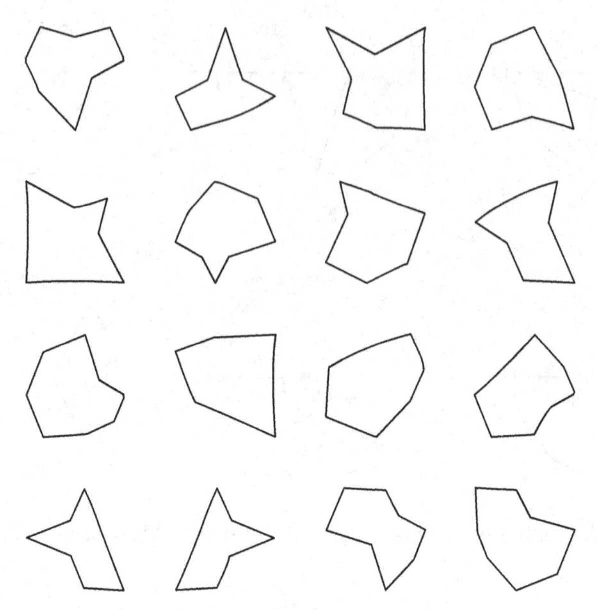

Figure 8.9: The 16 mappings satisfying the conditions $\gamma_3 = 0, \gamma_4 \neq 0, \gamma_5 \neq 0$.

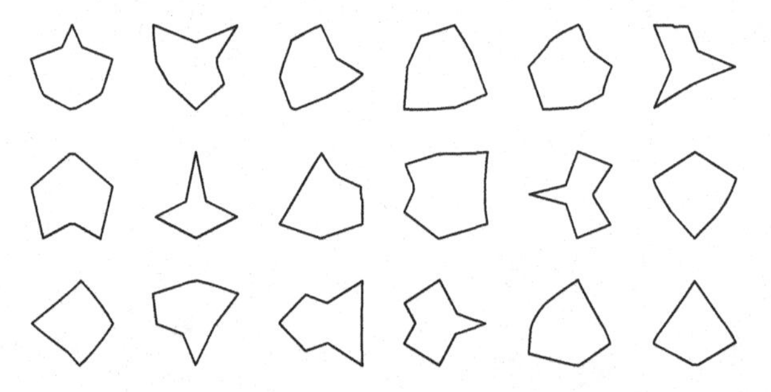

Figure 8.10: The 18 mappings satisfying the conditions $\gamma_4 = 0, \gamma_3 \neq o, \gamma_5 \neq 0$.

Appendix A: Computing Algorithms

Algorithm 1 (Permutation matrices). This procedure generates the permutation matrix for a given permutation of three objects labeled as $\{1, 2, 3\}$. To extend it to four or more objetcs, modify the input indices and the matrix dimension accordingly.

```
> restart:
> p3 := proc(a1,a2,a3) local f,m:
> f:=[a1,a2,a3]:
> m:=(i,j)->1-min(abs(f[i]-j),1):
> Matrix(3,3,m):
> end:
```

Algorithm 2 (Multiplication table for S_3).

```
> restart:
> with(group):
> t:=[[], [[1,2,3 ]], [[1,3,2 ]], [[1,2]], [[1,3]], [[2,3]] ];
> ut:=[1,2,3,4,5,6];
> m:=(i,j)->mulperms(op(i,t),op(j,t));
> M:=Matrix(6,6,m):
> CS3:=subs( seq( op(i,t)=op(i,ut), i=1..nops(ut) ), evalm(M) );
```

Algorithm 3 (Canonical invariants for action (3.5) on page 66). The charater table of $C_{2h} \simeq C_2 \times C_2$ is shown on page 98. Notation: $\{u, c, o, s\} \equiv \{E, C_2, i, \sigma_h\}$.

```
> restart:
> with(LinearAlgebra):
> p8 := proc( a1,a2,a3,a4,a5,a6,a7,a8 )
>   local f,ff,m:
>   f:=[a1,a2,a3,a4,a5,a6,a7,a8]:
>   ff:=[7,8,1,2,3,4,5,6]:  ## re-labeling
>   m:=(i,j)->1-min(abs(f[i]-ff[j]),1):
>   Matrix(8,8,m):
> end proc:
```

```
> c:=p8(1,2,7,8,5,6,3,4); o:=p8(2,1,8,7,6,5,4,3);
> s:=p8(8,7,2,1,4,3,6,5); u:=p8(7,8,1,2,3,4,5,6);
> Q1:=(u+c+o+s)/4; Q2:=(u-c+o-s)/4;
> Q3:=(u+c-o-s)/4; Q4:=(u-c-o+s)/4;
> x:=<abb,abB,aBb,aBB,Abb,AbB,ABb,ABB>;
> MatrixVectorMultiply(Q1,x);
> MatrixVectorMultiply(Q2,x);
> MatrixVectorMultiply(Q3,x);
> MatrixVectorMultiply(Q4,x);
```

Algorithm 4 (Regular projections for S_3).

```
> restart:
> with(group):
> t:=[[],[[1,2]],[[1,3]],[[2,3]],[[1,2,3]],[[1,3,2]]];
> ut:=[1,2,3,4,5,6];
> m:=(i,j)->mulperms(op(i,t),op(j,t));
> M:=Matrix(6,6,m):
> CS3:=subs( seq( op(i,t)=op(i,ut), i=1..nops(ut) ), evalm(M) );
> delta:=(i,j)->floor(2^(-abs(i-j)));
> f:=(i,j,k)->floor(2^(-abs(CS3[k,i]-CS3[1,j]))):

> c1:=<1,1,1,1,1,1>: ## the character table
> c2:=<2,0,0,0,-1,-1>:
> c3:=<1, -1,-1,-1,1,1>:

> P1:=Matrix(6,6,(i,j)->add(c1[k]*f(i,j,k),k=1..6))/6;
    ## the canonical projections
> P2:=Matrix(6,6,(i,j)->add(c2[k]*f(i,j,k),k=1..6))*2/6;
> P3:=Matrix(6,6,(i,j)->add(c3[k]*f(i,j,k),k=1..6))/6;
```

Algorithm 5 (Regular projections for S_4).

```
> restart:
> with(group):
> t:=[[],[[3,4]],[[2,3 ]],[[2,4 ]],[[1,2 ]],[[1,3 ]],[[1,4]],
> [[2,3,4 ]],[[2,4,3 ]],[[1,2,3 ]],[[ 1,2,4]],[[1,3,2 ]],
> [[1,3,4 ]],[[1,4,2 ]],[[1,4,3 ]],[[1,2],[3,4]],
> [[1,3],[2,4 ]],[[1,4],[2,3 ]],[[1,2,3,4 ]],
> [[1,2,4,3 ]],[[1,3,2,4 ]],[[1,3,4,2 ]],[[1,4,3,2 ]],
> [[1,4,2,3 ]]];
> ut:=[1,2,3,4,5,6,7,8,9,10,11,12,13,14,15,16,17,18,19,20,
> 21,22,23,24];
> m:=(i,j)->mulperms(op(i,t),op(j,t));
> M:=Matrix(24,24,m):
> CS4:=subs( seq( op(i,t)=op(i,ut), i=1..nops(ut) ), evalm(M) ):
> delta:=(i,j)->floor(2^(-abs(i-j)));
> f:=(i,j,k)->floor(2^(-abs(CS4[k,i]-CS4[1,j]))):
```

```
> c1:=<1,1,1,1,1,1,1,1,1,1,1,1,1,1,1,1,1,1,1,1,1,1,1,1>:
     ## the character table
> c2:=<3,1,1,1,1,1,1,0,0,0,0,0,0,0,0,-1,-1,-1,-1,-1,-1,-1,-1,-1>:
> c3:=<2, 0,0,0,0,0,0, -1,-1,-1,-1,-1,-1,-1,-1,2,2,2, 0,0,0,
     0,0,0>:
> c4:=<3,-1,-1,-1,-1,-1,-1, 0,0,0,0,0,0,0,0,-1,-1,-1, 1,1,1,
     1,1,1>:
> c5:=<1, -1,-1,-1,-1,-1,-1,1,1,1,1,1,1,1,1,1,1,1, -1,-1,-1,
     -1,-1,-1>:

> P1:=Matrix(24,24,(i,j)->add(c1[k]*f(i,j,k),k=1..24))/24;
     ## the canonical projections
> P2:=Matrix(24,24,(i,j)->add(c2[k]*f(i,j,k),k=1..24))*3/24;
> P3:=Matrix(24,24,(i,j)->add(c3[k]*f(i,j,k),k=1..24))*2/24;
> P4:=Matrix(24,24,(i,j)->add(c4[k]*f(i,j,k),k=1..24))*3/24;
> P5:=Matrix(24,24,(i,j)->add(c5[k]*f(i,j,k),k=1..24))/24;
```

Algorithm 6. This code generates the canonical projections of the regular representation of D_4.

```
> restart:
> with(group):
> t:=[[],[[1,2,3,4 ]],[[1,3],[2,4]],[[1,4,3,2 ]],[[1,4],[2,3 ]],
> mulperms([[1,4],[2,3 ]], [[1,2,3,4]]),mulperms([[1,4],
     [2,3 ]], [[1,3],[2,4]]),
> mulperms([[1,4],[2,3 ]], [[1,4,3,2]]) ];
> ut:=[1,2,3,4,5,6,7,8];
> m:=(i,j)->mulperms(op(i,t),op(j,t));
> M:=matrix(8,8,m):
> CD4:=subs( seq( op(i,t)=op(i,ut), i=1..nops(ut) ), evalm(M) ):
> delta:=(i,j)->floor(2^(-abs(i-j)));
> f:=(i,j,k)->floor(2^(-abs(CD4[k,i]-CD4[1,j]))):

> c1:=<1,1,1,1,1,1,1,1>:
> c2:=<1,1,1,1,-1,-1,-1,-1>:
> c3:=<1,-1,1,-1,1,-1,1,-1>:
> c4:=<1,-1,1,-1,-1,1,-1,1>:
> c5:=<2,0,-2,0,0,0,0,0>:

> P1:=Matrix(8,8,(i,j)->add(c1[k]*f(i,j,k),k=1..8))/8;
> P2:=Matrix(8,8,(i,j)->add(c2[k]*f(i,j,k),k=1..8))/8;
> P3:=Matrix(8,8,(i,j)->add(c3[k]*f(i,j,k),k=1..8))/8;
> P4:=Matrix(8,8,(i,j)->add(c4[k]*f(i,j,k),k=1..8))/8;
> P5:=Matrix(8,8,(i,j)->add(c5[k]*f(i,j,k),k=1..8))*2/8;
```

Appendix B: Glossary of Selected Symbols, Notations, and Terms

The following is a list of selected symbols and their definitions. Any exceptions are noted in the text.

- $\Re z$ and $\Im z$: the real and complex parts of a complex number z;
- I_m: the $m \times m$ identity matrix;
- $< v_1, v_2, v_3 >$: the subspace spanned by the vectors v_1, v_2, v_3;
- V: a finite set of labels or indices, a structure;
- $\tau, \sigma, \eta, \ldots$: group elements;
- G: a finite group (typically) with g elements;
- S_ℓ: the group of permutations on $\{1, 2, \ldots, \ell\}$;
- S_V: the group of permutations on V;
- C_ℓ: the cyclic group of order ℓ;
- $\mathcal{O}_s$: a symmetry orbit containing s;
- ρ_τ or $\rho(\tau)$: a linear representation ρ evaluated at $\tau \in G$;
- x_τ or $x(\tau)$: a scalar function x evaluated at $\tau \in G$;
- $\mathcal{A}_n = ee'/n$ and $\mathcal{Q}_n = I_n - \mathcal{A}_n$, where $e' = (1, \ldots 1)$ with n components;
- $|X|$: the number of elements in a set X;
- C^L: the set of mappings defined in L with values in C;
- Diag$(a, b, \ldots)$: a diagonal matrix with diagonal entries $a, b, \ldots$;
- $\mathcal{V}, \mathcal{W}$: linear subspaces;
- $GL(\mathcal{V})$: the general linear group of invertible linear transformations in the vector space $\mathcal{V}$;
- $\mathcal{F}(X)$: the vector space of scalar-valued functions defined on X;
- $\widehat{G}$: the set of all nonequivalent irreducible representations of G;
- $A \otimes B$: the Kronecker product of matrices A and B;
- A^* indicates the conjugte-transpose, or Hermitian transpose of matrix A;
- $\widehat{x}(\beta)$: the (group) Fourier transform of $x \in \mathcal{F}(G)$ evaluated at $\beta \in \widehat{G}$;

- $\rho \simeq m\,\beta \oplus \cdots \oplus n\,\gamma$ indicates the existence of a basis in $\mathcal{V}$ relative to which

$$\rho_\tau = \text{Diag}\,(I_m \otimes \beta_\tau, \ldots, I_n \otimes \gamma_\tau), \quad \tau \in G;$$

- Regular projections: the canonical projection for a regular representation.

Figure 8.11: Detail from the Church of Agios Eleftherios (Mikrí Mitrópoli) 12th Century, Athens, Greece.

Bibliography

Aigner, M. 1979. *Combinatorial theory*, Springer-Verlag, New York, NY.
Aitchison, J. 1986. *The statistical analysis of compositional data*, Chapman and Hall, New York, NY.
Alexander, K. R., W. Xie, and D. J. Derlacki. 1997. *Visual acuity and contrast sensitivity for individual Sloan letters*, Vision Res. **37**, no. 6, 813–819.
Andersson, S. and J. Madsen. 1998. *Symmetry and lattice conditional independence in a multivariate normal distribution*, Annals of Statistics **26**, 525–572.
Andersson, S. 1975. *Invariant normal models*, Annals of Statistics **3**, no. 1, 132–154.
______. 1992. *Normal statistical models given by group symmetry*, DMV Seminar Lecture Notes, Günzburg, Germany.
______. 1990. *The lattice structure of orthogonal linear models and orthogonal variance component models*, Scan J Statistics **17**, 287–319.
Bacry, H. 1963. *Leçons sur la theorie des groupes*, Université D'Aix-Marseille, Marseille, France.
______. 1967. *Leçons sur la theorie des groupes et les symétries des particules elémentaires*, Dunot, Paris, France.
Bailey, R. A. 1991. *Strata for randomized experiments*, Journal of the Royal Statistical Society B, no. 53, 27–78.
Blossey, R. 2006. *Computational biology*, Chapman & Hall/CRC Mathematical and Computational Biology Series, Chapman & Hall/CRC, Boca Raton, FL. A statistical mechanics perspective.
Bryan, G. H. 1920, The Mathematical Gazette **10**, no. 145, 44–45.
Campbell, C. 1994. *Ray vector fields*, Journal of the Optical Society of America **11**, 618–622.
______. 1997. *The refractive group*, Optometry and Vision Science **74**, 381–387.
Cartan, E. 1966. *The theory of spinors*, MIT Press, Cambridge, MA.
Carter, R., G. Segal, and I. Macdonald. 1995. *Lectures on Lie groups and Lie algebras*, Cambridge, New York, NY.
Cartier, P. 2001. *A mad day's work: from Grothendiek to Connes and Kontsevich – the evolution of concepts of space and symmetry*, Bulletin (New Series) of the American Mathematical Society **38**, no. 4, 389–408.
Coleman, A. J. 1997. *Groups and physics – pragmatic opinions of a senior citizen*, Notices of the American Mathematical Society **44**, no. 1, 8–17.
Cox, D. R. and E. J. Snell. 1989. *Analysis of binary data*, second edition, Chapman and Hall, New York.
Dawid, A. P. 1988. *Symmetry models and hypothesis for structured data layouts*, Journal of the Royal Statistical Society B, no. 50, 1–34.

Devroye, L. 1986. *Non-uniform random variate generation*, Springer, New York, NY.

Diaconis, P. 1989. *A Generalization of Spectral Analysis with Applications to Ranked Data*, Ann. Statist. **17**, no. 3, 949–979.

______. 1990. *Patterned matrices*, Proceedings of Symposia in Applied Mathematics **40**, 37–57.

______. 1988. *Group representation in probability and statistics*, IMS, Hayward, California.

Doi, H. 1991. *Importance of purine and pyramidine content of local nucleotide sequences (six bases long) for evolution of human immunodeficiency virus type 1*, Evolution **88**, no. 3, 9282–9286.

Dudoit, S. and T. P. Speed. 1999. *A score test for linkage using identity by descent data from sibships*, Ann. Statist. **27**, no. 3, 943–986.

Durbin, R., S. Eddy, A. Krogh, and G. Mitchison. 1998. *Biological sequence analysis*, Cambridge University Press, Cambridge, UK.

Eaton, M. L. 1983. *Multivariate statistics – a vector space approach*, Wiley, New York, NY.

______. 1989. *Group invariance applications in statistics*, IMS-ASA, Hayward, California.

Evans, S. N. and T. P. Speed. 1993. *Invariants of some probability models used in phylogenetic inference*, Ann. Statist. **21**, no. 1, 355–377.

Faris, W. 1996. *Review of Roland Omnès, the interpretation of quantum mechanics*, Notices of the American Mathematical Society **43**, no. 11, 1328–39.

Farrell, R. H. (ed.) 1985. *Multivariate calculation*, Springer-Verlag, New York, NY.

Ferguson, T. S. 1967. *Mathematical statistics – a decision theoretic approach*, Academic Press, New York, NY.

Ferris III, F. L., V. Freidlin, A. Kassoff, S. B. Green, and R. C. Milton. 1993. *Relative letter and position difficulty on visual acuity charts from the early treatment diabetic retinopathy study*, American Journal of Ophthalmology **116**, no. 6, 735–40.

Fisher, R. A. 1930. *Genetical theory of natural selection*, Oxford University Press, Oxford, UK.

______. 1942. *The theory of confounding in factorial experiments in relation to the theory of groups*, Ann. Eugenics **11**, 341–353.

______. 1947. *The theory of linkage in polysomic inheritance*, Philosophical Transactions of the Royal Society of London, Series B, Biological Sciences **233**, 55–87.

Gao, Y. and J. Marden. 2001. *Some rank-based hypothesis test for covariance structure and conditional independence* (M. Viana and D. Richards, eds.), Vol. 287, American Mathematical Society, Providence, RI.

Gupta, S. and R. Mekerjee. 1989. *A calculus for factorial arrangements*, Springer-Verlag, New York. Lecture Notes in Statistics.

Hannan, E. J. 1965. *Group representations and applied probability*, Journal of Applied Probability **2**, 1–68.

Harris, A. B., Randall D. Kamien, and T. C. Lubensky. 1999. *Molecular chirality and chiral parameters*, Rev. Mod. Phys. **71**, no. 5, 1745–1757.

Harris, D. C. and M. D. Bertolucci. 1978. *Symmetry and spectroscopy – an introduction to vibrational and electronic spectroscopy*, Oxford University Press, New York, NY.

Helland, I. 2004. *Statistical inference under symmetry*, Int. Stat. Rev. **72**, 409–422.

Herzel, H., W. Ebeling, and A. O. Schmitt. 1994. *Entropies of biosequences: the role of repeats*, Physical Review E **50**, no. 6, 5061–5071.

Hoffman, W. 1966. *The Lie algebra of visual perception*, Journal of Mathematical Psychology **3**, 65–98.

Huang, K. 1987. *Statistical mechanics*, second, Wiley, New York, NY.

Humphrey, W. E. 1976. *A remote subjective refractor employing continuously variable sphere-cylinder corrections*, Opt. Engineering **15**, 286–291.

James, A. T. 1954. *Normal multivariate analysis and the orthogonal group*, Annals of Mathematical Statistics **25**, 40–75.

______. 1957. *The relationship algebra of an experimental design*, Annals of Mathematical Statistics **28**, 993–1002.

James, G. D. 1978. *The representation theory of the symmetric groups* (A. Dold and B. Eckmann, eds.), Springer-Verlag, New York.

Jaeger, F. M. 1919. *Lectures on the principle of symmetry and its applications in all natural sciences*, Cambridge University Press, New York.

Khinchin, A. I. 1957. *Information theory*, Dover, New York, NY.

Kullback, S. 1968. *Information theory and statistics*, Dover, New York, NY.

Lakhsminarayanan, V. and S. Varadharajan. 1997. *Expressions for aberration coefficients using nonlinear transforms*, Optom Vis Sci **74**, no. 8, 676–86.

Lakshminarayanan, V., R. Sridhar, and R. Jagannathan. 1998. *Lie algebraic treatment of dioptric power and optical aberrations*, J. Optical Society of America, A **15**, 2497–2503.

Lakshminarayanan, V. and M. Viana. 2005. *Dihedral representations and statistical geometric optics I: Spherocylindrical lenses*, J. Optical Society of America A **22**, no. 11, 2483–89.

Lam, T. Y. 1998a. *Representations of finite groups: A hundred years, Part I*, Notices of the American Math Soc **45**, no. 3, 361–372.

______. 1998b. *Representations of finite groups: A hundred years, Part II*, Notices of the American Math Soc **45**, no. 4, 465–474.

Ledermann, W. 1968. *Representation theory and statistics*, Secrétariat mathématique, Paris.

Lee, H. and M. Viana. 1999. *The joint covariance structure of ordered symmetrically dependent observations and their concomitants of order statistics*, Statistics and Probability Letters **43**, 411–414.

Leggett, A. J. 1987. *The problems of physics*, Oxford, New York, NY.

Mann, H. B. 1960. *The algebra of a linear hypothesis*, The Annals of Mathematical Statistics **31**, no. 1, 1–15.

Marshall, A. W. and I. Olkin. 1979. *Theory of majorization and its applications*, Academic Press, New York, NY.

von Mises, R. 1957. *Probability, statitistics and truth*, Dover, New York, NY.

Muirhead, R. J. 1982. *Aspects of multivariate statistical theory*, John Wiley & Sons Inc., New York. Wiley Series in Probability and Mathematical Statistics.

Nachbin, L. 1965. *The Haar integral*, Van Nostrand, Princeton, N.J.

Naimark, M. A. and A. I. Štern. 1982. *Theory of group representations*, Springer-Verlag, New York, NY.

Olkin, I. and S. J. Press. 1969. *Testing and estimation for a circular stationary model*, The Annals of Mathematical Statistics **40**, 1358–1373.

Olkin, I. and M. Viana. 1995. *Correlation analysis of extreme observations from a multivariate normal distribution*, Journal of the American Statistical Association **90**, 1373–1379.

Olkin, I. 1973. *Testing and estimation for structures which are circularly symmetric in blocks* (D. G. Kabe and R. P. Gupta, eds.), North-Holland, Amsterdam.

Omnès, R. 1994. *The interpretation of quantum mechanics*, Princeton Press, Princeton, NJ.

O'Neill, E. 1963. *Introduction to statistical optics*, Dover, New York, NY.

Pachter, L. and B. Sturmfels (eds.) 2005. *Algebraic statistics for computational biology*, Cambridge University Press, New York.

Papakostas, A., A. Potts, D. M. Bagnall, S. L. Prosvirnin, H. J. Coles, and N. I. Zheludev. 2003. *Optical manifestations of planar chirality*, Phys. Rev. Lett. **90**, no. 10, 107404.

Peng, C. K., S. V. Buldyrev, A. L. Goldberger, S. Havlin, F. Sciotino, M. Simons, and H. E. Stanley. 1992. *Long-range correlations in nucleotide sequences*, Letters to Nature **356**, 168–310.

Perlman, M. D. 1987. *Group symmetry covariance models*, Statistical Science **2**, 421–425.

Petitjean, M. 2003. *Chirality and symmetry measures: A transdisciplinary review*, Entropy **5**, 271–312.

Pistone, G., E. Riccomagno, and H. Wynn. 2001. *Computational commutative algebra in discrete statistics*, Algebraic Methods in statistics and probability (M. Viana and D. Richards eds.) Vol. 287, Contemporary Mathematics Series, American Mathematical Society Providence, RI.

Pólya, G. 1954. *Induction and Analogy in Mathematics*, Princeton University Press, Princeton, NJ.

Prosvirnin, S. L. and N. I. Zheludev. 2005. *Polarization effects in the diffraction of light by a planar chiral structure*, Physical Review E (Statistical, Nonlinear, and Soft Matter Physics) **71**, no. 3, 037603.

Rao, C. R. 1973. *Linear statistical inference and its applications*, Second, John Wiley & Sons, New York-London-Sydney. Wiley Series in Probability and Mathematical Statistics.

Reif, F. 1965. *Fundamentals of statistical and thermal physics*, McGraw-Hill, New York, NY.

Riley, K. F., M. P. Hobson, and S. J. Bence. 2002. *Mathematical methods for physics and engineering*, 2nd ed., Cambridge University Press, New York, NY.

Rosen, J. 1975. *Symmetry discovered*, Dover, Mineola, NY.

______. 1995. *Symmetry in science, an introduction to the general theory*, Springer-Verlag, New York.

Rotman, J. J. 1995. *An introduction to the theory of groups*, Springer-Verlag, New York.

Ruhla, C. 1989. *The physics of chance*, Oxford Press, New York, NY.

Salamon, P. and A. K. Konopka. 1992. *A maximum entropy principle for the distribution of local complexity in naturally occuring nucleotide sequences*, Computers Chem. **12**, no. 2, 117–124.

Sarton, G. 1921. *The Principle of Symmetry and Its Applications to Science and to Art*, Isis **4**, no. 1, 32–38.

Savage, L. J. 1954. *The foundations of statistics*, John Wiley and Sons, New York.

Searle, S. R. 1971. *Linear models*, John Wiley & Sons Inc., New York.

Serre, J. P. 1977. *Linear representations of finite groups*, Springer-Verlag, New York.

Silverman, M. P. 1995. *More than one mistery – explorations in quantum interference*, Springer-Verlag, New York.

Simon, B. 1996. *Representations of finite and compact groups*, American Mathematics Society, Providence, RI.

Sloan, L. L. 1959. *New test charts for the measurement of visual acuity at far and near distances*, Am J ophthalmol. **48**, 807–813.

Sternberg, S. 1994. *Group theory and physics*, Cambridge, New York, NY.

Streater, R. F. and A. S. Wightman. 1964. *PCT, spin and statistics, and all that*, W. A. Benjamin, New York, NY.

Thibos, L. N., W. Wheeler, and D. Horner. 1994. *A vector method for the analysis of astigmatic refractive error*, Vision Science and its Applications **2**, 14–17.

Van de Ven, P and A. Di Bucchianico. 2006. *Factorial designs and harmonic analysis on finite Abelian groups*, EURANDOM, Eindhoven University of Technology, Eindhoven, The Netherlands. Electronic version http://www.eurandom.tue.nl.

Van de Ven, P. 2007. *Equivalences in designs of experiments*, Eindhoven University Press, Eindhoven, The Netherlands.

Vencio, R., L. Varuzza, C. de B Pereira, H. Brentani, and I. Shmulevich. 2007. *Simcluster: clustering enumeration gene expression data on the simplex space*, BMC Bioinformatics **8**, 246.

Viana, M. and V. Lakshminarayanan. 2005. *Data-analytic aspects of chirality*, Symmetry, Culture and Science **16**, no. 4, 401–421.

Viana, M. and D. Richards. eds. 2001. *Algebraic methods in statistics and probability*, Vol. 287, Contemporary Mathematics Series, American Mathematical Society, Providence, RI.

Viana, M. 1998. *Linear combinations of ordered symmetric observations with applications to visual acuity* (N. Balakrishnan and C. R. Rao, eds.), Vol. 17, Elsevier, Amsterdam.

______. 2003. *Invariance conditions for random curvature models*, Methodology and Computing in Applied Probability **5**, 439–453.

______. 2006. *Symmetry studies and decompositions of entropy*, Entropy **8**, no. 2, 88–109.

______. 2007a. *Symmetry studies for data analysis*, Methodology and Computing in Applied Probability **9**, 325–341.

______. 2007b. *Canonical invariants for three-candidate preference rankings*, Canadian Applied Mathematics Quarterly. Vol. 15, no. 2, 203–222.

______. 2008. *Canonical decompositions and invariants for data analysis* (M. Hazenwinkel, ed.), Vol. 6, Elsevier, Amsterdam.

Viana, M. and V. Lakshminarayanan. 2006. *Dihedral representations and statistical geometric optics II: Elementary instruments*, J. Modern Optics **54**, no. 4, 473–485.

Viana, M. and I. Olkin. 1997. *Correlation analysis of ordered observations from a block-equicorrelated multivariate normal distribution* (S. Panchapakesan and N. Balakrishnan, eds.), Birkhauser, Boston.

Votaw, D. F. 1948. *Testing compound symmetry in a normal multivariate distribution*, The Annals of Mathematical Statistics **19**, 447–473.

Weyl, H. 1950. *The theory of groups and quantum mechanics*, Dover, New York, NY.

______. 1952. *Symmetry*, Princeton U. Press, Princeton, NJ.

______. 1953. *The classical groups, their invariants and representations*, Princeton University Press, Princeton, NJ.

Wijsman, R. A. 1990. *Invariant measures on groups and their use in statistics*, Vol. 14, Institute of Mathematical Statistics, Hayward, California.

Wilks, S. S. 1946. *Sample criteria for testing equality of means, equality of variances, and equality of covariances in a normal multivariate distribution*, Annals of Mathematical Statistics **17**, 257–281.

Wit, E. and P. McCullagh. 2001. *The extendibility of statistical models*, Algebraic methods in statistics and probability (M. Viana and D. Richards, eds.) Vol. 287, Contemporary Mathematics Series, American Mathematical Society, Providence, RI.

Youden, W. J. 1951. *Statistical methods for chemists*, Wiley, New York, NY.

Index

Printed in the United States
by Baker & Taylor Publisher Services